THE DEVELOPMENT OF LIABILITY IN RELATION TO TECHNOLOGICAL CHANGE

Technological developments posed a challenge to the established law, especially tort law, at approximately the same time across Europe. This book focuses on the similarity and diversity of responses to such developments in different jurisdictions. Three examples have been studied in depth: the escape of sparks from steam engines in the middle of the nineteenth century; exploding boilers in the latter part of the nineteenth century; and asbestos-related industrial disease in the middle and late twentieth century. The book shows how the rules of tort law were used and adapted and demonstrates how other systems of regulation and compensation were introduced to prevent injuries or to provide compensation to victims outside tort law. The relatively marginal role of tort law in these areas reveals much about legal development in general.

MIQUEL MARTÍN-CASALS is Professor of Civil Law at the University of Girona.

COMPARATIVE STUDIES IN THE DEVELOPMENT OF THE LAW OF TORTS IN EUROPE

Series editors
John Bell and David Ibbetson

THE DEVELOPMENT OF LIABILITY IN RELATION TO TECHNOLOGICAL CHANGE

Volume 4

Edited by

MIQUEL MARTÍN-CASALS

CAMBRIDGE
UNIVERSITY PRESS

CAMBRIDGE
UNIVERSITY PRESS

University Printing House, Cambridge CB2 8BS, United Kingdom

Cambridge University Press is part of the University of Cambridge.

It furthers the University's mission by disseminating knowledge in the pursuit of education, learning and research at the highest international levels of excellence.

www.cambridge.org
Information on this title: www.cambridge.org/9780521199537

First published 2010
First paperback edition 2014

A catalogue record for this publication is available from the British Library

Library of Congress Cataloguing in Publication data
The development of liability in relation to technological change / edited by Miquel Martín-Casals.
p. cm. – (European legal development series)
Includes index.
ISBN 978-0-521-11679-4 (hardback) 1. Liability (Law)–Europe.
2. Technological innovations–Law and legislation–Europe.
3. Technological innovations. 4. Technology and law.
I. Martín-Casals, M. (Miquel) II. Title. III. Series.
KJC1663.D48 2010
346.403–dc22
2009026028

ISBN 978-0-521-11679-4 Hardback
ISBN 978-1-107-47580-9 Paperback

Hardback only available as:
ISBN 978-0-521-19953-7 6-volume set

CONTENTS

SERIES EDITORS' PREFACE

The European Legal Development series has arisen from a project funded by the AHRC from January 2005 until February 2008.

The aim of the project as a whole is to examine the nature of legal development in Western Europe since 1850, focusing sharply on liability for fault. Behind this there is a more abstract purpose, to attempt to cast some light on the factors which have influenced the way in which the law has changed over this period. Legal historians have looked at the general question, usually focusing on the rather facile distinction between the English common law and continental European legal systems. Though rooted in the sources, these works have been marred by a somewhat unsophisticated methodology and an inevitably selective use of evidence. Comparative lawyers have developed far more sophisticated methodologies, but their theoretical perspectives have too often borne little relation to empirical data. Over the last twenty years, tort lawyers have looked at the same types of question; but their analysis has invariably been at a high level of generality and has rarely looked at the historical components. By bringing together experts with different disciplinary backgrounds – comparative lawyers and legal historians, all with an understanding of modern tort law in their own systems – and getting them to work collaboratively, we have aimed to produce a more nuanced comparative legal history, and one which is theoretically better informed.

The topic of legal development is broad and, to make it manageable, we have undertaken a programme of work which has built up from a number of case studies and has moved towards a more general analysis and conclusions. Although we have been concerned with the development of the law, and although many of those involved in the project have been lawyers, we have also been concerned to include and benefit from the insights of historians and scholars in other disciplines.

Liability for fault between 1850 to 2000 has been our major area of study. Around 1850, there were many similarities in approaches to liability for fault across the legal systems of Western Europe. But since then, there has been significant divergence. Our method has been first to chart the changes and then to seek the explanations for what happened. Although there have been many changes in tort and delict laws over the period, the idea of liability for fault remains central to private law approaches to the compensation of victims of harms caused by the actions of others.

As a first stage, the project worked on six case studies which illustrate the general theme of liability for fault and its development within the period:

- product liability;
- legal doctrine;
- medical liability;
- relations between neighbours;
- technological change;
- traffic and railways.

This research involved scholars from a range of countries, in particular, England and Scotland, Spain, the Netherlands, Germany, France, Sweden, Austria and Italy. Each working group drew on the expertise of both senior and more junior scholars familiar with different European legal systems, and contained a mixture of comparative lawyers and legal historians.

A second stage has involved further groups examining a number of salient factors in legal development. The topics covered in this stage are:

- institutions and professions;
- social and political ideas;
- the economy (including the impact of insurance).

A final strand to the work is an overview book.

This book, edited by Miquel Martín-Casals, provides us with a good example of the benefits of this approach. We gain a clear insight into the way in which established principles of tort engaged with the problems thrown up by new technologies. We see the way in which regulatory law, private insurance and state-run compensation schemes developed to deal with the issues the law now confronted. Regulatory law and inspections by officials and private insurers and associations dealt with many of the issues of preventing accidents. Compensation systems outside tort offered remedies to many of the victims of accidents. In this matrix of legal interventions, we can see that the place of tort law and of fault in particular changes. We become aware of its limitations. The factors for legal development identified in this book are ones which will be developed in later volumes.

Particularly in relation to this book, the editors are grateful to the contribution of Matthew Dyson McGrath, Ph.D student on the AHRC project, who made a significant contribution to the formulation, discussion and execution of this part of the project. His own work on the relationship of criminal and civil fault will be published separately and will complement significantly the understanding of this area presented in this book.

John Bell
David Ibbetson

ABBREVIATIONS

AcP	*Archiv für die civilistische Praxis*
AiB	Arbeitsrecht im Betrieb
AJDA	*Actualité Juridique Droit Administratif*
App.	Corte di appello (Italy)
Arch. Circolaz.	*Archivio giuridico della circolazione e dei sinistri stradali*
ARP	Aranzadi Penal
AS	Aranzadi Social
BGB	Bürgerliches Gesetzbuch
BGBl.	Bundesgesetzblatt
BGHZ	Entscheidungen des Bundesgerichtshofes in Zivilsachen
BOCG	Boletín Oficial de las Cortes Generales
BOE	Boletín Oficial del Estado
Bull. Civ.	Bulletin de la Cour de cassation, civil
BVerfG	Bundesverfassungsgericht
Cass. (Fr.)	Cour de cassation
Cass. (It.)	Corte di cassazione
Cass. civ	Cour de cassation, chambre civile
Cass. com.	Cour de cassation, chambre commerciale
Cass. Pen.	Corte di cassazione, sezione penale
Cass. Unica	Corte di cassazione, sezione unica
CC	Código Civil
c.c.	Codice civile
C civ.	Code civil
C.E. (Ass.)	Conseil d'Etat (Assemblé plénière)
chr.	Chronique
Cons. Stato	Consiglio dello Stato
Corte cost.	Corte costituzionale
CP	Codigo Penal
c.p.	Codice penale
CPA	Comité permanent amiante
C séc. soc.	Code de la sécurité sociale
D.	Receuil Dalloz

Danno e resp.	*Danno e responsabilità*
D.A.	Receuil Dalloz Analytique
D.C.	Receuil Dalloz Critique
D.H.	Receuil Dalloz Hebdomadaire
Dig. It.	*Novissimo Digesto Italiano*
D. Lgs.	Decreto legislativo
D.M.	Decreto Ministeriale
D.P.	Receuil Dalloz Périodique
D.P.R.	Decreto Presidente della Repubblica
EGBGB	Einführungsgesetz zum BGB
Enc. Dir.	*Enciclopedia del Diritto*
Enc. Giur	*Enciclopedia Giuridica Treccani*
ERPL	*European Review of Private Law*
EVBl.	Evidenzblatt der Rechtsmittelentscheidung
Ferr. Ital	*Ferrovie Italiane*
FIVA	Fonds d'indemnisation des victimes de l'amiante
Foro it.	*Foro italiano*
Foro sic.	*Foro siciliano*
FS	Azierda Autonoma delle Ferrovie dello Stato
Gaz. Pal.	*Gazette du Palais*
Gaz. Trib.	*Gazette des Tribunaux*
GewO	*Gewerbeordnung*
Giur. Cost.	*Giurisprudenza costituzionale*
Giur. It.	*Giurisprudenza italiana*
Giur. Sic.	*Giurisprudenza siciliana*
Giust. Civ.	*Giustizia civile*
Harv. LR	*Harvard Law Review*
HVBG	Hauptverband der gewerblichen Berufsgenossenschaften with C. Drechsel-Schlund, M. Butz, G.Drexel, W. Plinske and H.-P. Francks, *Asbestverursachte Berufskrankheiten in Deutschland – Entstehung und Prognose* (Sankt Augustin, 2003)
INAIL	National Institute of Insurance for Occupational Accidents and Disease
INSERM	Institut national de la santé et de la recherche médicale
JCP	Juris-Classeur Périodique, la Semaine Juridique
J Econ. Hist.	*Journal of Economic History*
J.O.	Journal Officiel
J.O. (Doc.)	Journal Officiel (Documents)
JUR	Jurisprudencia
Lav. giuri.	*Lavoro giurisprudenziale*
Leb.	Receuil Lebon

MDR	Monatschrift für Deutsches Recht
Monit. Trib.	*Monitore dei Tribunali*
MSUA	Manchester Steam Users' Association
NJW	Neue Juristische Wochenschrift
NJWE-MietR	Neue Juristische Wochenschrift – Entscheidungen Mietrecht
OAG	oberappellationsgericht
OHofG	Oberhofgericht
O.J.	Official Journal of the European Communities
OLG	Oberlandesgericht
Or. Giur. Lav.	*Orientamenti della giurisprudenza del lavoro*
Parl. Deb.	Hansard, *Parliamentary Debates*
PGE	Preußisches Gesetz über die Eisenbahnunternehmungen
Pret.	Pretore (tribunale del)
Rass. dir. civ.,	*Rassegna di diritto civile*
R.D.	Reale Decreto
Rec. pér. assur.	*Recueil périodique des assurances*
RENFE	Red nacional de ferrocarriles españoles
Req.	Cour de cassation, chambre des requêtes
Resp. civ. et assur.	*Responsabilité civile et assurances*
Resp. Civ. Prev.	*Responsabilità civile e previdenza*
Rev. trim.	*Revue trimestrielle de droit civil*
RGBl.	Reichsgesetzblatt
RG EE	Reichsgericht, Eisenbahn- und Verkehrsrechtliche Entscheidungen und Abhandlungen
RGZ	Entscheidungen des Reichsgerichts in Zivilsachen
RHPflG	Reichhaftpflichtgesetz
Riv. Crit. Dir. lav.	*Rivistà critica*
Riv. Dir. Lav.	*Rivistà del diritto del lavoro*
Riv. Dir. Pubb.	*Rivistà del diritto pubblico*
Riv. Inf. Mal. Prof.	*Rivistà degli infortuni e delle malattie professionali*
Riv. It. Dir. Proc. Pen	*Rivistà italiana del diritto e della procedura penale*
RJ	Repertorio de Jurisprudencia
ROGE	Reichs-Oberhandelsgericht, Entscheidungen des Reichs-Oberhandelsgerichts
S.	Recueil Sirey
SAP	Sentencias Audiencias Provinciales
sez.	Sezione
somm.	sommaire
STS	Sentencias Tribunal Supremo
STSJ	Sentencias Tribunales Superiores de Justicia
SU	Sezioni uniti (of the Corte di cassazione)

TA	Tribunal administratif
Temi rom.	*Temi romani*
Trib.	Tribunale
Trib. civ.	Tribunal civil
Trib. com.	Tribunal commercial
UVG	Unfallversicherungsgesetz
VersR	Versicherungsrecht
WuM	Wohnungswirtschaft und Mietrecht
ZMR	Zeitschrift Mietrecht

1

Technological Change and the Development of Liability for Fault: A General Introduction

MIQUEL MARTÍN-CASALS

1. Introduction

The group researching this part of the European Legal Development Project has been examining the problems arising from several specific technological developments occurring at approximately the same time across Europe. Our main interests have been the responsiveness of law to technological change, the similarity and diversity of the legal solutions provided in different jurisdictions and how social change has contributed to consolidate them, make them evolve or substitute them by others.

Jonathan Morgan (University of Cambridge) has written the English report and Yvonne Salmon (EUI Florence), the French one. Jens Scherpe (University of Cambridge) has contributed the German report and Chiara Favilli (University of Pisa) has drafted the Italian report. Jordi Ribot and I (University of Girona) have been in charge of the Spanish report. Over a time-span of two years we met several times in Cambridge and in Girona to discuss our papers as they were progressing. To all of them I owe my gratitude, since working with them and learning from them has been a challenging experience.

In our part of the Project we have studied three examples of changes brought about by technological change in certain detail:

(1) The first refers to the escape of sparks from steam engines from the middle of the nineteenth century (the historical starting point of this research project) until steam engines were replaced by more efficient diesel or electric locomotive engines.
(2) The second example refers to the response of the legal system and, in particular of tort law, to damage caused by exploding boilers and, more specifically, to personal injury caused to factory workers until

the time when Workers' Compensation Acts came to the rescue of the victims at the end of the nineteenth century.

(3) Finally, our third example is asbestos-related industrial disease, from the middle of the twentieth century to our present time.

2. Railway sparks burning crops

In 1846, Robert Ritchie, an English civil engineer, expressed in his book on railways the spirit of the time as regards the prospects of the technological development:

> That railways should have effected such vast and important changes in this country in so short a period as thirty years, through the medium of steam as a motive power, must stamp the era in which we live as one of the most enterprising in British history ... and this spirit is fast extending itself over our Transatlantic Colonies, both in British America and The West Indies, and will soon be developed in Australia.

These important changes, he added, were not confined to Britain and its Colonies, since 'In the Spanish settlement of Cuba, railways have for some years been in existence' and 'on the continent of Europe great progress in railways has already been made, chiefly in France, Belgium and Germany'. In his enthusiasm he concluded that:

> in a short time every country in Europe will be intersected with railways; and perhaps at no very distant period continuous railways will be constructed across Europe into the centre of Asia, and even perhaps to the capital of China itself.[1]

As is well known, Great Britain was the pioneer of steam railway transportation, with a 26-mile (40 km) line from Stockton to Darlington (1825), followed by a 35-mile (54 km) line from Liverpool to Manchester (1830) which was the first inter-city passenger railway in the world. In Germany, the track length expanded rapidly from only 61 km in 1835/36 (and none in Prussia) to almost 43,000 km in 1879 (almost 27,000 km of which were in Prussia). In France, by contrast, although trains both for freight and for passengers were put to use as early as 1837, the development was much slower, with only 3,546 km of track in use in 1852 and 17,440 in 1870. In Italy and Spain, the developments were both much more modest and much slower. When the

[1] Robert Ritchie, *Railways: their Rise, Progress and Construction, with Remarks on Railway Accidents and Proposals for their Prevention* (London: Logman, Brown, Green and Longmans, 1846), pp. 1–3.

Kingdom of Italy was proclaimed in 1861, only 2,521 km of rail-track were in operation, reaching the figure of 7,780 km by 1876. Spain, after a first train operated in Cuba in 1837, had to wait until 1848 to see a train operating on Spanish soil, and the roughly 220 km of track that were in operation in 1850 expanded to only around 9,000 km by the middle of the 1880s.

The increase in steam locomotive lines and traffic also gave rise to an increase in accidents. Ritchie, with the insight gained through living in a pioneering country, asserted that the increase in the number of accidents had 'created well-grounded alarm as to the efficiency of the system itself and led, as we have seen, to the consideration of whether a safer method of traction could not substitute it'.[2] The most alarming consequences of these accidents were, without any doubt, death and personal injury caused to railway workers and passengers. Great concern to land-owners was also caused by sparks escaping from locomotives and burning crops and fields.

Escaping sparks is the paradigmatic example illustrating the legal conflict between activities related to the rising industrialisation and traditional land use. A simplistic approach might entertain the idea that the conflict was solved either in favour of railway companies or in favour of land-owners. In those nations where land property was more important due to the predominantly agrarian character of the country, the conflict was bound to be resolved in favour of land-owners. Conversely, where land property was not so important and the controlling interest was the expansion of the new technology of railway transportation, the conflict would have been resolved in favour of railway companies. Such oversimplification could be used to explain why Prussia, with an 1838 Railways Act establishing strict liability of railway operators, which modern scholarship has qualified as 'a law designed to benefit the landowning *Junkers*',[3] seemed to follow a pro-land-owner approach. By contrast, the English approach, clearly leaning towards a negligence rule, seemed to favour the interests of the rising railway industry. However, a more nuanced multi-factor approach is required. Sometimes, in the same historical period, different courts of the same country reached conclusions which seem incompatible with each other and even nonsensical. The variety of solutions was even greater when we bear in mind that at the beginning of the period, Italy and Germany were not unified, and

[2] *Ibid.* p. 361.

[3] John M. Kleeberg, 'From Strict Liability to Workers' Compensation: The Prussian Railroad Law, the German Liability Act and the Introduction of Bismarck's Accident Insurance in Germany, 1838–1884' (2003) 36 *New York University Journal of International Law and Policy* 53 at 72.

the courts and legislatures of the different kingdoms followed different approaches. Industry and doctrinal writing lamented a lack of certainty. In fact, none of the solutions seems to have had the purpose of subsidising the nascent industry through legal rules, or of moulding private law to accommodate the needs of the entrepreneurial classes. However, all the solutions adopted, with nuances that varied from country to country and over time, enabled the development and expansion of the technological and social changes that were brought about by the railways.

(a) *Legal devices used to solve conflicting interests*

In most countries studied in this volume, from the start of the operation of railways it was discussed whether, in the case of a crop being burned by escaping sparks, land-owners could avail themselves of an injunction in order to stop railway activity, and in the case that they could not, to which other remedies they could have access. In particular, the main point of discussion was whether a claim for damages could be brought without the need to prove fault or whether land-owners could obtain compensation only after fault of the railway company had been established.

(b) *Application of property law rules vs. liability rules*

The possibility of bringing an injunction was generally rejected. The railway operators had obtained an authorisation from the public authorities. Therefore, it was considered that a private individual could not stop an activity that had been expressly warranted by the public authority.

Then a second question arose: did this authorisation also entail a permit to cause damage to others, a sort of *carte blanche* for the damage resulting from its operation?

In Germany, cases dealing with fires caused by escaping sparks were mostly argued on the basis of property law rules rather than tort law rules.[4] After some initial hesitation, it was held in 1859 that neither the authorisation of the railway operation nor the fact that it used state-of-the-art technological equipment to prevent the escape of sparks gave rise to a defence. The court compared the position of the claimant to a person who would be compensated had his or her land been expropriated, and awarded damages.[5] Similarly, in a landmark decision in 1886, the Reichsgericht awarded damages under the *actio negatoria*, despite

[4] Germany, below p. 136 *et seq.* [5] Germany, below p. 145.

explicitly finding that the railway company was not at fault. It argued that since the claimant was barred from obtaining an injunction due to the fact that the railway operated under authorisation, not allowing him to claim damages would amount to an unacceptable infringement of his property rights, i.e. to a sort of latent or indirect expropriation.[6] The same interpretation was maintained after the Bürgerliches Gesetzbuch (BGB) was passed. In two 1904 decisions, the Reichsgericht held that, although the BGB did not provide for a damages award in these cases, compensation had to be awarded regardless of whether the defendant was at fault or not. The reason was that when there was no authorisation by the public authority, the claimant could obtain an injunction (§1004 BGB). Such authorisation, however, deprived him of this possibility and if no damages were awarded, the claimant would be left without any protection, and this situation would amount to an expropriation.[7]

Some decisions followed a similar line of thought in Italy and France. Starting from the idea of the sanctity of property, damage to property was considered a sort of indirect expropriation which nobody could be compelled to endure without receiving just compensation in exchange. Since the defendant was acting in the general interest, the claimant was not entitled to an injunction that prevented the operation of the railway, but only to compensation for the harm suffered.

Thus, a decision on 12 March 1877 of the Court of Appeal of Florence held that the railway operator was obliged to compensate the land-owner because:

> property is sacred and inviolable, both for private individuals and for the State; and nobody is obliged to transfer it or suffer any sort of material diminution in it, even for works and enterprises of public interest (this would be in its effects a sort of indirect expropriation) without appropriate and just compensation (art. 438 c.c.). Because the licence obtained did not certainly include the right to set neighbouring property on fire ... but current and permanent damage only and solely for this damage established a preventive compensation.[8]

Almost forty years later in a decision of 18 May 1914, the Court of Cassation of Florence insisted on this idea and added that the harm had been imposed on the claimant in the name of public interest and that 'the *immissio in alienum* was therefore the infringement of the right of another person, a *damnum non iure datum*, and, therefore a "colpa aquiliana"

[6] Germany, below p. 171. [7] Germany, below pp. 173–4.

[8] App. Firenze, 12 March 1877, *Foro it.*, 1877, I, c. 36, and Italy, below pp. 192 *et seq.*, 198.

[*sic*]. An exception to this rule would amount to a limitation of property. Such limitation can only by established by law.'[9] Similar decisions were handed down by courts in Naples (1911) and Bologna (1914), usually mixing justifications from property rules with fault liability rules, until 1915, when case law underwent a U-turn.[10] As Rodotà has pointed out, on occasions like these the courts did not take the opportunity offered by liability for fault to turn the balance in favour of the interests more favourable to industry, but quite the opposite: their approach aimed at reducing, as much as possible, the importance of the yardstick that fault offered.[11]

The intertwining of property rules and liability rules and a similar departure from a stringent interpretation of fault as an infringement of a duty can also be found in France, where courts considered that establishing an activity which, by its nature, inconvenienced neighbours to a greater extent than traditional activities would constituted fault and had to be compensated for. In this sense, courts held, for instance that 'it would be unjust if a proprietor found himself suddenly harmed and inconvenienced by the use to which his neighbour put his property' (Metz, 16 August 1820) or that 'the exercise of the right to property ceases to be legitimate and becomes a fault from the moment that it seriously harms the right of the neighbour' (Cass., 3 December 1860).[12] Thereby, the courts considered that, although establishing an industry that abnormally inconvenienced neighbours was not fault in itself, the effects of the operation of such industries was 'fault' and, accordingly, neighbouring land-owners could claim damages for the burned crops even if the railway company had taken all possible measures to prevent the harm from taking place.[13]

It is within this context that decisions, which would seem nonsensical when analysed with our present understanding of fault liability, regain a coherent meaning. This is, for instance, the case of the well-known 1861 decision of the Court of Appeal of Munich, where the court held that the condition of the existence of a culpable act (*culpose Handlung*) was always met since 'the operation of a locomotive as such is necessarily a culpable act'.[14]

Along the same lines, other decisions paid lip-service to the fault requirement and compensated for damage suffered by land-owners by applying a

[9] Cass. Firenze, 18 May 1914, *Giur. It.*, I, 1, 1914, p. 999, and Italy, below p. 194 *et seq*.

[10] See below Italy, p. 199 *et seq*.

[11] See S. Rodotà, 'Proprietà e industria. Variazione intorno alla responsabilità civile' *Pol. dir.* 1978, 421–6.

[12] France, below p. 97. [13] France, *ibid*. [14] Germany, below p. 145.

very loose and vague notion of fault. This was done, for instance, in Italy, in cases decided in the 1860s, either by considering that the emission of sparks showed a defective adoption of the precautionary measures (a sort of *culpa in re ipsa*) or by introducing a sort of 'presumption of fault' of the railway companies which could be rebutted only by proving force majeure or contributory negligence of the victim, but not by proving that the defendant had acted with due care.[15] Something similar happened in Germany where, for instance, the Court of Appeal of Munich in 1853 held a railway company liable for the ensuing damage because escaping sparks 'could only be explained by the employee's negligence or insufficient measures to prevent the sparks'.[16]

Although these solutions did not formally disregard the principle of fault liability, in practice they converted it into strict liability.

(c) *Establishing a strict liability rule*

Another possible device to protect land-owners was to establish strict liability in order to force railway companies to internalise all the costs created by their activity. This was done by the legislature in Prussia and by the courts in France in the wake of the *Teffaine* case.

The paradigmatic (and quite a minority example) of a strict liability rule established by the legislature is to be found in §25 of the Prussian Railway Act of 1838, which provided that:

> the company is liable for all damage occasioned during train transport to goods transported and persons as well as to other persons and goods, and the company can only exonerate itself from this liability by proving that the damage was caused through fault of the claimant or by an inevitable event.

and added that 'the dangerous nature of the undertaking as such is not such an event as to exonerate from liability'. This provision was considered to cover personal injury, damage to transported goods and damage to 'any other goods', the latter including damage to neighbouring land.[17] The Act, passed in 1838, was enacted only four days after the first Prussian railway line was opened and the driving force behind it was not an obscure and petty civil servant, but Friedrich Karl von Savigny himself, the same great man who had looked with indifference on delictual liability without fault in his academic works.[18]

[15] Italy, below p. 192 *et seq.* [16] Germany, below p. 145.
[17] See Germany, below p. 140 *et seq.* [18] See Germany, below p. 141.

Savigny poured new wine into old wineskins to meet the needs of regulation brought about by technological change. He had in mind two classes of injured plaintiffs, i.e. passengers and land-owners, and drew inspiration from different sources to attain the same result. According to his proposal, passengers should be able to bring their claims under strict liability because the railway was like an innkeeper and, traditionally, innkeepers were strictly liable for injuries to travellers. In the case of damage caused to land by escaping sparks, he contended that often nobody is at fault and since the damage they cause cannot be prevented, the company should be liable for it unless the claimant had contributed to the damage himself or the damage had been caused by force majeure.[19]

It seemed difficult, however, to justify why Savigny favoured strict liability for damage caused by railway operation and not in the case of other industrial activities, such as the operation of factories or steamboats, which also involved the use of machines posing similar dangers. Savigny contended that the grounds for this exception to liability based on fault were that, by contrast to other industrial activities, the operation of railways by its own nature could not be subject to preventive measures, since tracks cover vast distances. This presented a risk to land-owners and people living close to the railway, and making them bear the risk and the damage would have amounted to an uncompensated expropriation for the benefit of railway operators.[20] The Prussian state had already granted railway companies the land they required through expropriation and Savigny saw strict liability as a fair measure in exchange for it.[21]

It must be stressed, however, that strict liability did not mean that landowners suffering fire caused by escaping sparks obtained compensation in any case. First of all, other rules within the tort law system then existing prevented that from happening. This was the case, in particular, for contributory negligence, a defence which proved quite successful since it was understood as an 'all or nothing' rule and the slightest contributorily negligent conduct of the victim was sufficient to rule out the claim.[22] Furthermore, assumption of risk or limitation of liability through contract were possible (in this latter case, at least until the defence was declared void in 1869).[23] Moreover, the courts were quite restrictive, excluding

[19] See Germany, below p. 141 *et seq*. See also Kleeberg, above note 3, at 68–9.

[20] See Germany, below p. 142. [21] See, in this sense, Kleeberg, above note 3, at 69.

[22] See Germany, 143. See also Kleeberg, above note 3, at 73–4.

[23] See Germany, below pp. 146–7. See also Kleeberg, above note 3, at 75 *et seq*.

consequential damage and, by using one device or another, rejecting compensation in the majority of cases.[24]

In France, the 1896 *Teffaine* case gave a new meaning to article 1384, paragraph 1 of the Code civil (C civ.) by establishing a new system of liability based on the custody (*garde*) of any injury-producing thing. After this, the courts also extended strict liability (or at least, a 'stricter' form of liability) to damage caused by escaping sparks. So, for instance, and among many other decisions, the Court of Appeal of Toulouse, on 6 May 1902, held a railway company liable under article 1384, paragraph 1 C civ., in spite of the fact that the locomotive had been furnished with equipment to prevent sparks from escaping and that no recklessness or negligence of the railway company had been shown.[25]

(d) *Establishing a stringent fault liability rule*

A stringent notion of fault liability of railway operators for damage caused by sparks was, from the outset, a well entrenched position in England and in Spain, and it became the prevailing approach of the Italian courts after 1915.

As early as in 1832 (*R* v. *Pease*) it was decided in England that when an Act of Parliament had authorised a railway, an action for nuisance could not lie for running trains along the railway. It would not have been acceptable that an activity which Parliament had expressly sanctioned and which inevitably interfered with passage through the land was effectively rendered unlawful by the law of nuisance and thus could presumably be prohibited by an injunction. Therefore, from the very early cases, claims for damage caused by sparks escaping from locomotives had to be framed as an action for negligence against the defendant railway. Moreover, fault was not presumed and had to be established by the defendant.[26]

It has been pointed out that during the railway boom, the legislated expropriation of property by railway companies became national policy in England, and that by the 1840s, the power to expropriate private land was routinely bestowed on any group which put forward a plan which they declared to be for public benefit.[27] Since railway entrepreneurs had been granted this extraordinary power of expropriation, the single absolute

[24] See in more detail Germany, below p. 144. See also Kleeberg, above note 3, at 75 *et seq.*

[25] See in more detail France, below p. 98 *et seq.*

[26] See England, below p. 41 *et seq.*

[27] See R.W. Kostal, *Law and English Railway Capitalism 1825–1875* (Oxford: Clarendon Press, 1994), p. 177.

right of a land-owner was considered to be his right of compensation and, therefore, railway entrepreneurs had no right to complain of being obliged to purchase land at very high prices if they wanted to establish new lines.[28]

From the start, railway companies had to pay very high prices for the land they bought, and as the railway industry became more competitive, large payments for small amounts of land became more commonplace:[29] the equation of economic development with public interest provided the central ideological justification for the invasion of the land by the railway. As Kostal has pointed out, it seemed that land-owners and railway operators had secured what they each wanted most: whereas the former became rich through the high monetary compensation and could also enjoy the world's most advanced domestic transportation system, the latter had promptly obtained the material foundation for their 'vast operational enterprises'.[30] Within this context, it does not seem surprising that, unless the railway company was at fault, the courts were reluctant to further compensate land-owners when the operation of railways burnt crops. Moreover, from a doctrinal point of view, such compensation would not have fitted very well within an action for nuisance which, in the face of parliamentary authorisation, would have had to be brought without the possibility of claiming an injunction.

There were, however, attempts to change the prevailing approach of the courts. Notably, Lord Bramwell argued, for instance, that the authorisation clause which gave railway companies the power to run trains had nothing to do with the authorisation of nuisance; or, that the lack of compensatory provisions would amount to derogation of property rights without compensation; or, that in order to produce an efficient resource-allocation, an activity must bear all the costs that it creates for society.[31]

In practice, however, fault liability was impossible to enforce, inter alia, because bringing an action was very expensive and proving fault of the defendant very difficult. In England, this gave rise, at the beginning of the twentieth century, to the felt need to provide some sort of compensation for farmers whose crops had been burnt by sparks without requiring them to prove fault. After several setbacks, the Railway Fires Act 1905 was finally passed, establishing that farmers could recover damage caused to crops without the need to prove fault up to the amount of £100; when damage exceeded the statutory limit, farmers might still bring an action in negligence. It appears that in 99 per cent of the cases, this amount

[28] *Ibid.* p. 180. [29] *Ibid.* pp. 150–1. [30] *Ibid.* p. 180. [31] See England, below p. 43 *et seq.*

(which was increased to £200 a few years later) was sufficient to fully compensate the damage caused.[32]

(e) *Effects of regulations upon the required standard of care*

In Spain, in Italy after 1915 and, for some time, in France, compliance with safety regulations played an important role, both in establishing the required standard of care of railway operators and in determining the land-owner's lack of self-protection, which could give rise to contributory negligence.

As regards land-owners, safety regulations made those who owned land neighbouring railway tracks subject to various restrictions, such as not being able to build within a certain distance from the rails of a track or place inflammable materials, such as piles of wood, crops, straw or bales of hay, less than a certain distance from them. As regards railway operators, safety regulations required them, inter alia, to keep the tracks and the surrounding land, up to a certain distance, cleared of undergrowth, and to equip locomotives with technical devices which prevented sparks from escaping, or at least reduced their occurrence as much as possible.[33]

In Spain, the property law approach seems to have been totally absent and fault liability, with a very stringent proof of fault, was required from the outset. The first cases decided by the Spanish Supreme Court (some of them dealing with fires caused by sparks in the *ingenios* or Cuban plantations) held that even when it had been proved that sparks escaping from the defendant's locomotive caused the fire, the clamant was not entitled to compensation if he had failed to prove that the fires had resulted from fault of the railway company. This stringent proof of fault was somewhat alleviated by regulations which distributed (so to say) the safety burden between the railway operators and land-owners. Accordingly, courts started to consider that a breach of the safety regulations imposed on railway operators automatically triggered fault and gave rise to liability, if the other conditions for liability were met. By the same token, compliance with regulations excluded any possible proof of fault of the defendant. However, in the early 1930s, courts (probably aware that, more often than not, such regulations were only formally complied with) departed from this position and declared that fault of the defendant could be established even when the railway operator had complied with the measures provided by the regulations.

[32] See England, below p. 50.

[33] See France, below pp. 93–4 and Spain, below pp. 233–5.

A similar solution was adopted in France, until the emergence of strict liability for the 'acts of things' established itself at the end of the nineteenth century as an alternative ground for liability; and, after 1915, in Italy when, during a second period, courts changed their approach and held that the defendant railway companies could be held liable only when they were at fault. It was for the claimant to prove fault, but fault could easily be established by proving that the defendant had not complied with protective measures provided by regulations. Conversely, and until well into the twentieth century, defendants could escape liability by showing that they had complied with safety regulations. Moreover, liability would not exist if the damage could be attributed to the fact that the claimant had not complied with the duties imposed by regulations for his or her self-protection (i.e. to contributory negligence).[34]

3. Exploding boilers

(a) *Legal and extra-legal factors*

The spread of steam engines in the countries studied was very uneven. According to some data it has been estimated that the steam engines in operation in 1900 were able to provide 13.7 million horsepower in the United Kingdom, 8.08 in Germany, 5.92 in France, only 1.52 in Italy and 1.2 million horsepower in Spain. One of the reasons for the rather modest expansion of steam power in Italy and in Spain was the lack of abundant coal at affordable prices.[35]

In spite of this uneven expansion, in all these countries accidents caused by exploding boilers were already common from the 1850s onwards. In the United Kingdom, for instance, where steam power was more developed, it is estimated that between 1865 and 1882 some 1,051 persons were killed in boiler explosions, whereas 1,519 were injured. A common thread in all the countries under survey is, however, that litigation ranged between very low and non-existent.

To a certain extent, legal factors can explain this situation. The need for proof of fault of the defendant, plus the existence of other legal rules such as the doctrine of 'common employment', assumption of risk or contributory negligence, or even the possibility of excluding tort liability by contract, made a favourable outcome for the claimant quite unlikely. But

[34] See France, below p. 93, quoting a judgment of the Cour de cassation of 28 October 1903, and Italy, below p. 200 *et seq*.

[35] See Italy, below p. 204 and Spain, below p. 239.

the main reasons seemed to have been extra-legal. The typical defendant in these cases was the factory owner and the typical claimant the employee. Social, economic and psychological factors seem to explain the absence of actions brought by employees against their employers: fear of instant dismissal, the enormous costs of bringing an action, and a social attitude which considered employees to be self-protecting individuals who were assuming risks and who were bargaining with their employers for wages that corresponded to the danger of their tasks.[36]

(b) *Importance of regulations establishing preventive measures*

In order to try to reduce the number of accidents and to organise the activity of producing and operating steam engines, most countries introduced step by step piecemeal regulations.

In France, for instance, an Ordonnance royale of 22 May 1843 unified the many different regulations to which steam boilers were subject and placed a stringent control on the materials and methods for boiler construction. It also set the conditions for the placement and operation of steam boilers according to their capacity and the level of steam pressure, provided for professional qualifications required for boiler operators and established maintenance procedures and a regime of inspection. However, this intensive regulation was rendered ineffective by technological development and an increased familiarity with the new technology. In 1865, an Imperial decree deregulated the installation of steam boilers, and although it relaxed certain safety regulations it kept control over the observance of these prescriptions in the hands of governmental authorities (*préfet*) and engineers of mines, bridges and roads. Infringement of the regulations gave rise to criminal sanctions.[37] In Italy, several provisions were also adopted in response to a high number of accidents brought about by the poor state of maintenance and construction of boilers and the lack of inspection by expert engineers until 1888, when a general regulation, modelled on the double level of surveillance and supervision (by local governmental authorities and by technicians) which characterised French regulation, was enacted. Requirements for steam boilers were introduced in the legislation on public order, which forbade putting a boiler into operation without a declaration attesting to its safety, or

[36] See France, below p. 109 *et seq*, England, below p. 51 *et seq*., and Germany, below p. 160 *et seq*.

[37] See France, below pp. 103–8.

operating it without the continuous assistance of specifically appointed personnel. These controls were prescribed by the local representative of the central government (*prefetto* or *sotto-prefetto*) and performed by an expert selected among engineers or machinists. As in France, the infringement of these requirements gave rise to criminal sanctions.[38]

Germany introduced a general duty to obtain a licence in order to operate steam boilers as early as 1831. This legislation provided technical specifications for the construction and placement of boilers. However, regular inspections of steam boilers were not established until 1856 and in the beginning these were carried out by civil servants.[39] In Spain, by contrast, the first national regulation was not passed until the second quarter of the twentieth century and supervision was left in the hands of incompetent local authorities. In the absence of civil and criminal litigation against steam boiler owners, in practice they enjoyed total immunity. Engineers concerned about the great number of explosions that they considered avoidable, and pointing out the measures taken in other European countries, repeatedly demanded similar safety legislation to be passed in Spain, with no success.[40]

England preferred the route of self-regulation, already adopted in 1854 with the establishment of the Manchester Steam Users' Association, which provided inspection with the aim of preventing boiler explosions. Attempts to introduce compulsory government inspection of steam boilers were rejected in England as early as 1869 on the grounds of excessive interference with industry. In 1881, the Boiler Explosions Act was passed in order to improve investigation into cases of accidents, but it did not aim at establishing compulsory boiler inspection.[41]

Associations of steam boiler users flourished not only in England, but also in France and Italy. In the wake of the French Imperial decree of 1865, owners of steam boilers formed an association which allowed them to intervene in the selection of engineers, and in 1919 another decree allowed associations of steam boiler owners to have authority over the control and inspection of their machines.[42] They obtained the privilege that only their own engineers, who had to meet the technical conditions established by the law, were to perform inspections of their steam boilers. Insurers also contributed to safety, since they inspected boilers before insuring them. Italy also followed the French model in this respect[43] and technical supervision associations also played an important role in

[38] Italy, below p. 207. [39] Germany, below pp. 158–9. [40] Spain, below pp. 240–1.
[41] See England, below p. 61 *et seq.* [42] France, below p. 108. [43] Italy, below p. 208.

Germany, after an initial period where inspections were carried out by civil servants only.[44] However, this was not the case in Spain, where owners of steam boilers never agreed to set up associations so as to share the costs of their maintenance and periodical safety inspections.[45]

In general terms, regulations establishing preventive measures did not include any mention of compensation for personal injury or other harm suffered by victims of exploding boilers and at most, as in the case of French regulations, referred to the general rules.[46] In England, indirect means were tried to encourage the awarding of compensation to the victims but did not establish a legal duty to do so. This was the case, first, with the so-called 'deodands', i.e. monetary payments that had to be paid by the owners of boilers that had caused death;[47] and, later on, with the awards referring to the amounts corresponding to investigation costs that under the Boiler Explosions Act 1881 were to be paid by the negligent boiler owner and which could be reduced when he had already made provision for a dead worker's family.[48]

(c) *Limited role of specific developments in tort law*

Increasing concern as to the devastating consequences of personal injury suffered in different sorts of accidents seems to have swept all countries under survey from the 1850s onwards. Traditional tort law rules could not meet the need to compensate victims of industrial accidents, not only because liability was based on fault and claimants had to furnish proof of it, but also because of the defences that defendants could avail themselves of. French statistics from 1888, for instance, suggest that 68 per cent of industrial accident cases were considered to be instances of *cas fortuit* or force majeure, 20 per cent due to contributory negligence of the worker himself, and only 12 per cent were found to be due to fault of the employer.[49]

The three traditional defences of assumption of risk, common employment and contributory negligence seem to have played, in most countries under survey, an important role as difficult legal hurdles to overcome when litigating and which contributed decisively to outcomes in favour of defendants. Thus, according to the notion of assumption of risk or *volenti not fit iniuria*, it was a widespread opinion that employees freely accepted the risks that their dangerous jobs involved and that payments already included a premium for that danger, so that if danger materialised, they

[44] Germany, below p. 169. [45] Spain, below p. 241. [46] Italy, below p. 208, note 86.
[47] England, below pp. 57–8. [48] England, below p. 64. [49] France, below p. 109.

had no reason to complain.[50] The doctrine of common employment, which established that where a servant was injured at work by another servant of the same master the principle of vicarious liability did not apply, was explicitly invoked in England[51] and similar considerations led to the rejection of claims in other countries.[52] So did contributory negligence, understood, as in its Roman origins, as an 'all or nothing rule'. Finally, at least for some time, liability in tort could also be limited by contract.[53]

In Germany, concern for victims of industrial accidents led in 1871 to the enactment of the Liability Act of the Reich (Reichshaftplichtgesetz). In France, it was the Cour de cassation which, from 1896 on, attempted to provide a solution by giving the Code civil a different interpretation in the *Teffaine* case and its successors. Tort law developments, however, do not seem to have been effective in coping with the social problems created by the need to compensate the victims of steam boiler explosions, among other reasons because they operated within the context of well-established tort law rules, including the above-mentioned defences.

Thus, for instance, the Liability Act of the Reich of 1871 provided for new liability rules for certain undertakings, such as the operation of railways, mines and factories, and also introduced changes in the procedural rules. The Act established strict liability for personal injury occurring 'during the operation of trains' but only vicarious liability for the operators of 'mines, quarries, pits and factories', which obviously required fault of the employee. However, courts interpreted the Act quite narrowly, particularly as regards the meaning of 'factory' and 'scope of employment' and it appears that only 20 per cent of the actions brought were successful.[54]

In France, the Cour de cassation developed the so-called 'liability for the act of things' (*responsabilité du fait des choses*) on the basis of article 1384, paragraph 1 C civ., a section that hitherto had been regarded as serving a merely introductory purpose and which was considered for the first time as an autonomous provision in the *Teffaine* case. In a development that crystallised three decades later in the *Jand'heur* case, this doctrine (which

[50] France, below p. 109, Germany, below p. 166, England, below p. 53 *et seq.*

[51] England, *ibid.*

[52] For the discussion of liability of employers for the acts of their employees in Germany and in France in situations corresponding to the English doctrine of common employment, see Kleeberg, above note 3, at 89, where he reports that the Association of German Lawyers recommended the rule to be made general throughout Germany; and that whereas the Cour de cassation in Paris held that employers were also liable for accidents caused by fellow workers, the courts in Toulon and Lyons disagreed. See also Spain, below p. 243 *et seq.* and references in notes 56 and 57.

[53] See, for instance, Germany, below p. 146 *et seq.*

[54] Germany, below p. 168 *et seq.*

currently pervades French tort law and makes it so distinctive) allowed the replacement of liability for fault in many areas of French law, including, at the time, liability for exploding boilers.[55] However, in spite of the new reading of article 1384, paragraph 1 C civ., court proceedings were still lengthy and costly, damages awards arbitrary and unpredictable, and employees could still be found contributorily negligent, something which happened in a large number of cases.[56]

Boilers were expressly mentioned in both the Italian Civil Code 1865 and the Spanish Civil Code 1889. In the first case, the rules did not refer directly to liability for damage caused by boilers, but to preventive measures that should be adopted, due to the risks they entailed, in order to prevent harm to neighbours (cf. article 574 c.c.); since the specific preventive measures were to be laid down by regulations and such regulations were issued many decades later, the rule, which might have had some bearing when establishing fault of boiler owners and operators, remained a dead letter for a long time.[57] In the Spanish case, however, article 1908 CC did establish a tort law rule by providing for a reversal of the burden of proof of fault in some cases of industrial accidents, including those caused by the 'explosion of machines' (*explosion de máquinas*). At the time, 'machines' were mainly steam driven and, accordingly, the provision should have included exploding boilers. However, article 1908 CC went unnoticed by the courts. For decades, they stubbornly continued to require the claimant to establish proof of fault and dismissed claims when he had not furnished sufficient proof of it, until the rule was no longer necessary due to the general reversal of the burden of proof in situations of risk operated by case law pursuant to article 1902 CC, i.e. the general provision establishing fault liability along the same lines as the French article 1382 C civ.[58]

The time seemed to be ripe for much more effective measures, i.e. for the transfer of this area of accidents from tort law to social insurance law by introducing an insurance system for industrial injuries which, albeit still far from full compensation, could provide an acceptable level of compensation to victims and could shield employers from tort actions brought by their employees. This time Germany took the lead with its Accident Insurance Act 1884 (Unfallversicherungsgesetz);[59] England followed suit

[55] See France, below p. 112 *et seq*. In more detail, see Edward A. Tomlinson, 'Tort Liability in France for the Act of Things: A Study of Judicial Lawmaking' (1988) 48 *Louisiana Law Review* 1299.

[56] France, below p. 114. [57] See Italy, below p. 205. [58] Spain, below p. 244 *et seq*.

[59] Germany, below p. 168 *et seq*. See also Kleeberg, above note 3, at 108 *et seq*.

with the Worker's Compensation Act 1897, and in 1898 so did both Italy,[60] and France, with the Loi du 9 avril 1898 concernant les responsabilités des accidents dont les ouvriers sont victimes dans leur travail.[61] Finally, Spain passed a similar provision in 1900.[62]

4. Personal injury caused by asbestos

(a) A multifaceted problem

In just one century, asbestos has transmuted from a 'magic mineral' into a 'killer dust',[63] and while it became one of the leading causes of occupationally related deaths in the twentieth century, the spectre of hundreds of thousands of deaths caused by this mineral still hovers over the first decades of the twenty-first century.

What makes the problems related to asbestos difficult to master is that asbestos is a multifaceted phenomenon: asbestos is a mineral that does not exist as such but is found in different varieties that also have different qualities. Asbestos has not been used to produce one single product or just a few, but thousands of different products through many different industrial processes. Accordingly, victims affected by asbestos are neither of a single class nor located in a single branch of industry. Although asbestos-related diseases are caused by the inhalation of its fibres, it gives rise to dramatically different types of diseases which range from alterations that can hardly be classified as a disease at all to fatal diseases which, despite the fact that they may have long periods of latency, can cause death only a few months after they have been diagnosed.[64]

[60] Italy, below p. 209. [61] France, below pp. 114 and 115 *et seq.*

[62] Ley de 30 de enero de 1900, de accidentes de trabajo en los establecimientos industriales (*Gaceta de Madrid*, 31 January 1900). See Spain, below pp. 243–4.

[63] See the well-known book by Geoffrey Tweedale, *Magic Mineral to Killer Dust: Turner and Newall and the Asbestos Hazard* (Oxford: Oxford University Press, 2000).

[64] The literature on asbestos and its past and foreseeable future consequences has become too vast to list in full. Suffice here to mention, among others: in English, P. W. J. Bartrip, *The Way from Dusty Death: Turner and Newall and the Regulation of Occupational Health in the British Asbestos Industry, 1890s–1970* (London: The Athlone Press, 2001) and Rachel Maines, *Asbestos and Fire: Technological Trade-offs and the Body at Risk* (New Brunswick: Rutgers University Press, 2005); in French, Sophie Musso-Gabai, *Amiante: obligations et responsabilités* (Grenoble/Paris: Société Alpine de publications, 2000), François Malye, *Amiante: 100.000 morts à venir* (Paris: le Cherche midi, 2004) and Odette Hardy-Hémery, *Eternit et l'amiante, 1922–2000: aux sources du profit, une industrie du risque* (with l'UMR 8529 CERSATES) (Villeneuve-d'Ascq: Presses Universitaires du Septentrion, 2005); in Italian, Luigi Montuschi and Gaetano Insolera (a cura di), *Il Rischio da amianto: questioni sulla responsabilità civile e penale* (Bologna: Bononia University Press, 2006); in Spanish,

(i) Asbestos: a generic term

As is well known, asbestos is just a generic term used to describe six naturally occurring fibrous minerals which can be grouped on the basis of their mineralogical properties into two families: serpentines (which include chrysotile (white asbestos)), and amphiboles, of which there are five types. Of these, crocidolite (blue asbestos) and amosite (brown asbestos) are the most widely used in industry.[65] The name derives from the Greek adjective 'asbestos' (meaning 'unchangeable', 'non-perishable', 'non-combustible'). Asbestos has been known since antiquity but its commercial exploitation only became widespread in the late nineteenth century. In the twentieth century, it was used extensively for purposes as varied as protecting buildings and ships against fire, reinforcing road surfaces, strengthening plastics, constructing sheeting and pipes from asbestos-cement, and making fireproof cord, gaskets, brake linings and heat-protective clothing, to name just a few among many other possible industrial applications.[66]

Regardless of type, scientists are of the unanimous opinion that asbestos is carcinogenic to humans. The health risk posed by asbestos stems from the inhalation of fibres released into the air and exists at all stages of the process: mining, preparation, transport, processing, installation, normal use, machining, removal, demolition and waste management.

(ii) Types of asbestos-related diseases

Three main distinct diseases are associated with asbestos. Two are quite specifically associated with this exposure (asbestosis and mesothelioma of the pleura and peritoneum) and one is also associated with other exposures (lung cancer).

Although the dangers of asbestos inhalation seem to have been known since antiquity, the links between asbestos and severe lung disease were first discussed in the early twentieth century by French and American experts. However, although the annual report of the French Factory Inspectorate for 1906 already pointed out that asbestos textile workers suffered pathological changes in the lungs and that the rates of death among

Albert Azagra Malo, *La tragedia del amianto y el derecho español* (Barcelona: Atelier, 2007); and in German, Maria Roselli, *Die Asbestlüge: Geschichte und Gegenwart einer Industriekatastrophe* (Zürich: Rotpunktverlag, 2007).

[65] The other varieties of asbestos are tremolite, actinolite and anthrophyllite.

[66] See, e.g., among a vast literature, International Social Security Association, *Asbestos: Towards a Worldwide Ban* (Geneva: ISSA, 2006), p. 11.

these workers were very high,[67] a clear link between asbestos inhalation and asbestosis was not conclusively established until 1930 by Merewether. Asbestosis is defined as lung fibrosis caused by the inhalation of asbestos fibres. Diagnosis is made on the basis of clinical features, X-ray appearances and a history of heavy asbestos exposure. As is widely known, the occurrence of asbestosis is associated with high levels of prolonged exposure to asbestos dust (with amfibole fibres considered more potent than chrysotile). Its period of latency is from twenty-five to forty years.

Suspicion of asbestos inhalation as a factor in the occurrence of lung cancer begun to arise between the 1930s and 1940s, but it is widely acknowledged today that Richard Doll's celebrated epidemiological study on 'Mortality from Lung Cancer in Asbestos Workers' in 1955[68] confirmed these suspicions. However, even when death statistics are readily available, the number of deaths attributable to lung cancer caused by asbestos cannot be determined directly, since there are a number of agents that can cause the disease (most importantly, tobacco smoke) and lung cancers resulting from asbestos exposure are clinically indistinguishable from those caused by these other agents. In the early 1960s, Chris Wagner, together with other South African doctors who treated miners in asbestos fields, linked the contact with asbestos with mesothelioma, a rare and especially deadly cancer.[69] Mesothelioma may be defined as a cancer of the pleura (the smooth lubricated double membrane sacs containing the lungs) or of the peritoneum (the smooth double membrane lining the inside of the abdominal cavity). This form of cancer is peculiar because the only known cause is from asbestos exposure. Since 1965, it was known that low doses and transient exposure to asbestos suffice to give rise to mesothelioma. Although the period of latency can last from fifteen to thirty years, the disease is inevitably fatal, with death occurring within a year or two of diagnosis.[70]

Finally, exposure to asbestos can lead to pleural plaques. Pleural plaques are areas of fibrosis, with pleural thickening and sometimes calcification. Unlike asbestosis, these benign pleural plaques do not in general cause impaired lung function or any other problems. They are usually identified

[67] Morris Greenberg, 'Knowledge of the Health Hazard of Asbestos Prior to the Merewether and Price Report of 1930' (1994) 7 *Social History of Medicine* 493.

[68] (1955) 12 *British Journal of Industrial Medicine* 81.

[69] Jock McCulloch, 'The Discovery of Mesothelioma on South Africa's Asbestos Fields' (2003) 16 *Social History of Medicine* 419 at 427.

[70] For a detailed account of the research process that led to the establishing and publication of knowledge of this disease, see Bartrip, above note 64, at p. 165.

on a chest X-ray. Regarded as a 'marker' of asbestos exposure, they are not a predictor of any other asbestos-related diseases.[71]

(iii) Types of victims ('waves of asbestos victims')

It is recognised nowadays that the threat to life from breathing asbestos extends in some degree even to individuals who have been only slightly exposed to the dust, and that the levels of exposure insufficient to produce asbestosis can nonetheless cause cancer. For simplification's sake, we can identify different types of asbestos victims.[72] Since claims for asbestos diseases have also appeared subsequently according to the type of victim, we could also speak about different 'waves' of asbestos victims. This classification can be useful as regards the genesis of the legislation, its terms, the speed with which it has been revised, updated and extended, the decisions of the courts awarding or rejecting compensation for damage and, in general, the current state of affairs in the evolution of asbestos litigation in the different countries.

The first group or wave refers to victims who were asbestos handling workers ('direct' occupational exposure), i.e. workers in the asbestos manufacturing industry. The second group ('indirect' occupational exposure) refers to victims outside the asbestos manufacturing industry itself, such as shipbuilding or construction, who have not worked with asbestos themselves but with products containing it such as insulation, cement, brake linings, etc. It includes those persons who have worked near workers directly or indirectly exposed, but who have not worked with asbestos or asbestos products themselves (such as clerks and accountants working in a factory where asbestos materials are produced or are used for further production). In both cases we are talking about occupational diseases and prolonged exposure to asbestos. Although mesothelioma can also be the result of asbestos exposure in these cases, a typical result is asbestosis. Asbestosis and lung cancer in non-production factory workers (such as machine adjusters, plant or departmental managers) were already being reported in the 1940s by Canadian and British writers.[73]

The third group refers to workers' family members, i.e. to victims who do not suffer an occupational disease and who suffer so-called 'domestic' or 'household contact exposure' (typical case: mesothelioma suffered by a wife who washed the clothes of her husband, a worker directly exposed

[71] International Social Security Association, above note 66, at p. 16.

[72] See, inter alia, Barry I. Castelman, *Asbestos: Medical and Legal Aspects* (5th edn, New York: Aspen, 2005), p. 429.

[73] *Ibid.* p. 432.

to asbestos) and to persons who were neighbours of sources of asbestos air pollution and solid wastes such as mines or industries, or who worked or lived in asbestos contaminated premises ('environmental exposure').[74]

(iv) Seriousness of the problem

In 1999, a leading asbestos researcher, Julian Peto, and his colleagues predicted that, in Western Europe alone, past asbestos exposure would cause a quarter of a million deaths from mesothelioma over the following thirty-five years. The number of lung cancer deaths caused by asbestos would be at least equal to the number of mesotheliomas, suggesting that there would be more than half a million asbestos cancer deaths in Western Europe over the following thirty-five years.[75] However, the reality seems to have been worse than these predictions.[76]

In the United Kingdom, some 1,969 persons died from mesothelioma in 2004 and, taking into account that the peak of importation of asbestos was in the early 1970s, the annual total number of mesothelioma deaths is estimated to peak around 1,950 to 2,450 deaths per annum, sometime between 2011 and 2015.[77] In France, 50,000 to 100,000 deaths are expected over the coming twenty years. Experts say this number will increase the 25,000 cases of lung cancer declared each year by approximately 10 per cent.[78]

In Germany, just over half of the 1,800 deaths in the occupational context occurred in 1999, i.e. 903 cases or 50.2 per cent were attributed to asbestos-related diseases and, at the beginning of 2002, the German occupational health services (Berufsgenossenschaften) estimated that 460,000 workers had been exposed to asbestos.[79] The number of deaths will slightly increase in the future, with a peak between 2005 and 2015.[80]

In Italy, the national register of mesothelioma created in 2004 certified 3,670 cases, but this was considered partial since it excluded the data of some regions and of people who died outside hospitals. The peak of

[74] *Ibid.*

[75] J. Peto, A. Decarli, C. La Vecchia, F. Levi and E. Negri, 'The European Mesothelioma Epidemic' (1999) 79 *British Journal of Cancer* 566.

[76] With more recent data, Laura Salvatori, Alessandro Santoni and Darren Michaels, 'Asbestos: The Current Situation in Europe', available at www.actuaries.org/ASTIN/Colloquia/Berlin/SalvatorI_SantonI_Michaels.pdf (visited 1 May 2009).

[77] See England, below p. 64.

[78] See France, below p. 120.

[79] Germany, below p. 182. See further Salvatori, Santoni and Michaels, above note 76.

[80] Germany, below p. 182.

mortality for asbestos-related diseases is expected between 2015 and 2025 and it is estimated that between 4,000 to 5,000 persons will die every year.[81] Finally, in Spain some sources claim that more than 60,000 workers were exposed to asbestos dust up until 1985, that by 2002 more than 1,000 would have died as a result of an asbestos-related disease, and it is estimated that around 50,000 deaths will occur over the next thirty years.[82]

(b) Regulation

Although one of the first references to asbestos in industrial relations legislation is to be found in 1909 in a Italian Royal decree banning unsafe work for women and children or admitting it with special protective measures,[83] the first legislation in the world controlling asbestos was the British Asbestos Industry Regulation 1931, requiring manufacturers to provide exhaust ventilation in areas of factories concerned with the various manufacturing processes listed in the legislation.[84]

The dangers of asbestos in the industrial context were admitted in most countries by including certain illnesses caused by asbestos in tables of industrial diseases, something which created the presumption that the disease, whenever it occurred, resulted from the exercise of the affected worker's profession. It appears that Germany was the first country to recognise asbestosis as an industrial disease in 1936.[85] In Italy, together with silicosis, this was accepted in 1943[86] and Spain followed suit in 1947.[87] In France, although pulmonary fibrosis resulting from the inhalation of silicon or asbestos dust was included in the tables of industrial diseases in 1945, asbestosis as such was not included in the tables until 1950.[88]

Lung cancer and cancer of the throat caused in connection with asbestosis as a separate occupational disease was recognised in Germany in 1943.[89] In Italy, lung cancer was included in the tables of industrial diseases in 1973, in France in 1976 and in Spain in 1978.[90] Mesothelioma was recognised in Germany and in France in 1976 and in Spain in 1978.[91] In Italy, however, although mesothelioma, if associated with asbestosis, had

[81] Italy, below p. 214. [82] Spain, below p. 253. [83] Italy, below p. 213.
[84] England, below p. 67 *et seq.* [85] Germany, below p. 180. [86] Cf. Italy, below p. 220.
[87] Spain, below p. 254. [88] France, below pp. 120–1.
[89] Germany, below p. 180.
[90] Italy, below p. 221, France, below p. 121, Spain, below p. 254.
[91] France, below p. 121, Spain, below p. 254, Germany, below p. 181.

been recognised as an occupational disease as early as 1973, it was not independently recognised until 1994.[92]

As regards safety standards, all countries under survey followed a similar path in order, first, to diminish the health risks related to asbestos exposure and, later on, to try to avoid the risk by banning the import, manufacture, sale or use of asbestos.

The main characteristic of the regulations regarding safety standards in the case of asbestos exposure was that they had been based on research that had been conducted mainly in the asbestos manufacturing industry, neglecting the exposure of workers in other branches of industry using asbestos products; measures relating to asbestos dust and asbestos fibres were continuously readapted to new evidence of the dangerousness of asbestos and had to be re-examined in detail once the link between asbestos and cancer was made in the 1950s.[93] Moreover, little pressure was exerted to ensure the enforcement of regulations protecting workers from asbestos exposure. Thus, for instance, in the United Kingdom there were only two prosecutions for breach of such regulations between 1964 and 1970, and apparently none before. At the same time, trade unions and workers were not very active in highlighting the dangers of asbestos exposure until the 1980s.[94] Governments also showed an ambivalent attitude, in an apparent trade-off between the social costs (in terms of diseases and deaths) and benefits (of the jobs in all branches related to the asbestos industry and the benefits of asbestos manufacturing to society at large). A 2005 Report of the French Senate even refers to the 'state being "anaesthetised" by the asbestos lobby'.[95]

The tolerant attitude towards asbestos changed when it was clear that asbestos gave rise to fatal diseases like mesothelioma and could also be detrimental to 'third wave' victims, i.e. victims who were not workers of asbestos manufacturing industries or other industries employing asbestos. It was clear then that the social costs were not confined to workers but also affected users, who until then had been believed only to be reaping the benefits of the wonder product at a very low cost. These facts, and pressure from the mass media, led finally to the total asbestos ban.

[92] Italy, below p. 221.

[93] See, England, below p. 68 *et seq.*, France, below p. 121 *et seq.*, Germany, below p. 179 *et seq.*, Italy, below p. 219 *et seq.* and Spain, below pp. 254–5.

[94] England, below pp. 69–70 *et seq.*

[95] See G. Dériot and J.P. Godefroy, *Rapport d'information fait au nom de la mission commune d'information sur le bilan et les conséquences de la contamination par l'amiante* (No. 37 Sénat, 2005), p. 71. See in detail, France, below p. 122 *et seq.*

Italy was one of the first countries to ban the production, use and distribution of asbestos or material containing asbestos. A 1992 Act forbade the extraction, production and sale of products containing asbestos and ordered the destruction of all products manufactured with asbestos and its removal from buildings.[96] Germany and Spain followed suit in 1993.[97] France banned the use of all amphibole type of asbestos in 1994, but the prohibition of the manufacture, importation and sale of asbestos and asbestos containing materials did not come into effect until 1 January 1997.[98] Finally, the United Kingdom, which had banned blue and brown asbestos with effect from 1 January 1986, did not establish a complete ban on the import and supply of asbestos until late 1999 with the Asbestos (Prohibitions) (Amendment) Regulations 1999 (SI 1999/2373).[99]

(c) Asbestos claims

Compared to the United States, the level of asbestos litigation is rather low in Europe. Moreover, in the countries studied here, litigation ranks from almost non-existent in Germany to increasingly important in the United Kingdom. The reasons for these differences cannot be found in dramatically different factual data, but in radically different legal approaches. In some countries, such as Germany and Spain, the social security approach prevails, either excluding recourse to tort law almost completely or restricting it to a greater or lesser extent. In Italy and in the United Kingdom, tort law plays an important role, while in France a no-fault compensation scheme (which, however, does not exclude tort liability altogether) has been established. Accordingly, in most of the countries under survey, tort law plays rather a subsidiary role.[100]

(d) Social security solutions vs. tort law claims

Neither in Germany nor in Spain has asbestos litigation so far had a social, legal and political impact comparable to other jurisdictions. The overwhelming majority of the cases litigated have been confined to occupational diseases, and personal injury claims outside this context are

[96] Italy, below p. 213.

[97] Germany, below p. 179 and Spain, below p. 253 and notes 93 and note 102.

[98] France, below p. 120. [99] England, below p. 72 and note 114.

[100] For the relationship between tort law and insurance law, see, in general terms, Ulrich Magnus (ed.), *The Impact of Social Security Law on Tort Law* (Vienna/New York: Springer, 2003) with reports referring to all countries under survey.

still unheard of in Germany, and are just starting to reach the civil courts in Spain.[101]

The German Accident Insurance Act 1884 liberated employers from tort liability for accidents and, in 1925, insurance cover was extended to occupational diseases. Occupational diseases are described as those that employees suffer as a result of their working activity and that are either listed in the Occupational Diseases Ordinance or have been caused by professional activity, according to the latest medical knowledge. They are covered by the so-called Berufsgenossenschaften, or institutions for statutory accident insurance and prevention, which also assume liability for the consequences of occupational accidents and commuting accidents. Insurance is financed by employers only and individual contributions are (basically) computed by taking into account the probability of accidents occurring in certain enterprises and the total income of the insured employees in question.[102] Compensation is only granted if the damage was caused by an industrial accident or an occupational disease (principle of causality). The benefits awarded, however, do not compensate for the actual damage (i.e. the precisely calculated loss in earnings), but are calculated on a flat rate basis. Resort to tort law is not possible in Germany in cases of occupational victims, since the Social Security Code bars tort law claims unless the injury has been caused with intent.[103]

In Spain, asbestosis, lung cancer and mesothelioma are classified as professional diseases and, until the mid-1990s, problems caused by asbestos-related diseases were tackled through incapacity allowances or surviving dependants' benefits awarded within the framework of the social security system. Since in Spain, social security law does not completely replace tort liability and social security benefits for personal injuries are fully compatible with tort liability,[104] the number of claims against employers and their liability insurers seeking additional or supplemen-

[101] See Germany, below p. 182 and Spain, below p. 255 *et seq.*

[102] Cf. www.hvbg.de/e/pages/versich/index.html and Joerg Fedtke and Ulrich Magnus, 'The Impact of Social Security Law on Tort Law in Germany' in Magnus, above note 100, at p. 99 and nn. 12 and 14. See also, for an interesting comparison between the German and the English approaches to asbestos-related occupational diseases, the notes to *Barker* v. *Corus UK Ltd* [2006] UKHL 20, [2006] 2 AC 572 by Gerhard Wagner, 'Asbestsschäden – Bismarck was right' *ZEuP* 4/2007, 1122–36.

[103] Germany, below p. 182.

[104] See M. Paz Garcia Rubio, Javier Lete, F. Gómez Abelleira and C. Regueiro Ferreiro, 'The Impact of Social Security Law on Tort Law in Spain' in Magnus, above note 100, at p. 170 n. 15. A very recent trend, however, has begun to favour deductibility, see STS 24 July 2008 (RJ 2008, 4626).

tary compensation has increased dramatically. Most of these claims have been dealt with by labour courts and a recent account reported no fewer than sixty judgments on appeal on the issue.[105]

In France, the workers' compensation gives the employer immunity from another civil action being brought against him, except where there is *faute inexcusable* of the employer or of the person who had the power to supervise the work of the victim. In these cases, the injured worker can bring an action against the person who committed the *faute inexcusable* in order to obtain damages compensating the part of his loss that is not covered by social benefits.[106] For many decades, the Cour de cassation defined *faute inexcusable* very narrowly, so that the number of cases decided upon this principle were insignificant. Since 2002, the concept has been redefined by broadening its scope, giving rise to a substantial increase in such cases (from 300 in 2002 to over 1,500 in 2004) and, what is more important, to a success rate of more than 98 per cent of all claims.[107]

In Italy, the practice of the courts as regards civil liability in the cases of occupational diseases was quite restrictive until recently. As early as 1949, it was considered that employers were exempted from civil liability for occupational diseases covered by insurance except when the disease was caused or increased by the infringement of safety measures. However, since safety measures remained unimplemented for many years, courts had to reduce the instances of exemptions for employers, most notably through criminal law and through vicarious liability.[108] Compensation is currently pursued mainly through civil law actions which are ancillary to criminal proceedings for the employers' infringement of duties that result in the crimes of manslaughter or personal injuries. The consequences of this development are that while labour judgments are few, criminal proceedings have become common.[109]

In England, although occupational diseases are covered by the National Health Service, disability benefits for industrial injuries are quite modest, currently reaching a maximum weekly amount of £143.60.[110] Since social security claims and tort claims are fully compatible, with the only consequences that the value of the tort claim is affected by the claimant's receipt of a social security benefit, employees

[105] Spain, below p. 255 *et seq.*

[106] See Suzanne Carval, 'The Impact of Social Security Law on Tort Law in France' in Magnus, above note 100, at p. 77 n. 10.

[107] France, below p. 132. [108] Italy, below p. 225. [109] Italy, below pp. 225–6.

[110] See www.jobcentreplus.gov.uk/JCP/Customers/WorkingAgeBenefits/008433.xml.html (visited 30 April 2009).

suffering an asbestos-related disease have a clear incentive to sue their employers in tort.[111] Until the 1980s, asbestos manufacturers seem to have escaped tort law litigation thanks to *ex gratia* payments.[112] Once the first tort law actions were brought against asbestos manufacturers they tried to settle the claims rapidly to avoid publicity or, otherwise, if possible, to wear claimants down by utilising all the hurdles that the tort law system offers to their benefit.[113]

(e) No-fault compensation schemes

Of all the countries under survey, France seems to have been the only one to adopt a no-fault compensation scheme. First, it allowed litigation against the state for failure in its obligations to protect individuals. The state had failed to obtain precise information of the risks involved and had been too tardy to react. Later, after allowing coverage of asbestos-related occupational diseases under a compensation scheme for terrorism and other crimes for a short period, in 2000 the Fonds d'indemnisation des victimes de l'amiante (FIVA) was set up, which aims at full compensation of the harm suffered by asbestos victims, whether occupational or not, according to a non-compulsory tariffication scheme (*bàreme indicatif*), based on the seriousness of the injury and the age of the victim.[114]

The scheme has been instituted as an independent body, presided over by a senior judge, and establishes a non-adversarial system, which enables victims to receive compensation sums quite quickly. Acceptance of the compensation offered by the fund entails the victim giving up any other claims for compensation. However, if he has not received a satisfactory offer from the fund, the victim may bring his case before a court of appeal (with some appeal courts awarding significantly higher sums), and may bring an action for *faute inexcusable* against the employer, a criminal action or even an action against the state for failure to exercise its powers of regulation and control. *Faute inexcusable* will be established if the employer was aware (or should have been aware) of the danger to which he exposed the employee. Although 86 per cent of asbestos-related cases were resolved by the FIVA in 2004, the continued use of other

[111] See Richard Lewis, 'The Impact of Social Security Law on Private Tort Law in England and Wales' in Magnus, above note 100, at p. 61 n. 18.

[112] England, below p. 77.

[113] England, below p. 78 *et seq.*

[114] France, below pp. 131–3. See also A. Guégan-Lecuyer, *Dommages de masse et responsabilité civile* (Paris: LGDJ, 2006), p. 159.

means of recourse creates some confusion and shows that victims may be dissatisfied with the levels of compensation offered.[115]

(f) *Tort law solutions: an array of legal hurdles*

As is well known, neither non-fault compensation schemes nor social security payments aim at full compensation and, whenever it is possible, victims tend to resort to tort law in order to obtain additional compensation. Moreover, except in the case of the French scheme, recourse to tort law is the only possible action in the case of non-occupational or 'third wave' victims, a stage which has been reached in England, to a lesser extent in Italy and in Germany, and which is only about to be reached in Spain.

However, when claims are brought in tort many difficulties arise. The most notable are the following.

(i) Fault

Social security claims do not require fault on part of the employer or of a co-employee as a precondition for obtaining a benefit. By contrast, tort law claims generally require, in these cases, proof of fault,[116] except in France where tort law seems to have done away with fault liability also in this area by imposing *gardien*'s liability, as in the case of the wife of an employee who contracted asbestosis by washing and leaving her husband's work clothes in their bedroom to dry. A French Court of Appeal held that the husband's employer had failed to exercise its power of control over the dangerous fibre and was liable as the *gardien* of asbestos pursuant to article 1384, paragraph 1 C civ.[117]

Particularly striking as regards fault is the Spanish case. In Spanish tort law, when damage has been caused by an activity entailing a risk, fault is presumed and, in practice, there has been until quite recently a strong tendency to presume fault even beyond cases of specific or qualified risks. Thus, the burden of proof is regularly displaced to the defendant, who has to prove that he acted with due care if he is to escape liability. However, this is not the case with asbestos-related diseases. When the defendant is an asbestos manufacturer, courts usually require full proof of fault from the claimant.[118]

[115] France, below p. 133.

[116] Ulrich Magnus, 'Comparative Report' in Magnus, above note 100, at p. 284, n. 43.

[117] See France, below p. 129.

[118] See Spain, below p. 257.

However, not complying with safety regulations can also play an important role in ascertaining the fault of the employer. In Italy, it is contended that not complying with regulations amounts to fault, even if these regulations could not foresee the type of harm ensuing. Thus, even if regulations (made for asbestosis) were not meant to prevent the particular harm that has ensued (mesothelioma), the defendant who breached the regulations is liable. It is sufficient that the employer could 'foresee that by adopting the safety measures a serious damage to health or to life could be avoided' (App. Turin, 29 November 1995).[119] The same reasoning can be found in Spain, according to one line of argument which is applied in some cases.[120]

In England, not complying with safety regulations gives rise to a breach of statutory duty, a specific tort which, by contrast to the tort of negligence, does not require fault for liability to be established.[121] Additionally, a similar reasoning regarding the foreseeability of the risk, linked now to the breach of the duty of care in the tort of negligence, can be found in *Margerson* (1996), where the court held that it was sufficient that the evidence demonstrated that the defendants should have reasonably foreseen a risk of some pulmonary injury, not necessarily mesothelioma.[122] However, in a 2005 case (*Maguire* v. *Harland & Wolff plc*),[123] the English Court of Appeal adopted a quite restrictive approach and dismissed the claim of a woman who had contracted mesothelioma through washing the asbestos-saturated clothes of her husband, who had been employed by the defendant in the early 1960s. The reason was that in absence of contemporaneous evidence of the danger of such secondary victims' exposure, the defendant could not be negligent in having failed to foresee the risk to the health of the worker's wife. A similar narrow approach has been adopted in Germany in several decisions dealing with claims brought against manufacturers of off-peak storage heaters containing asbestos, arguing that a manufacturer cannot be considered at fault if his product is found to be harmful at a later point in time. To hold him liable according to §823 BGB, the risk must be known at the time of production.[124]

If not complying with safety regulations ordinarily gives rise to establishing fault, the converse is not true: complying with regulations does not exclude fault. A manufacturer or employer is not required to follow

[119] Italy, below p. 224. See also Naddia Coggiola, 'Nesso di causalità e colpa nel danno da amianto. Le esperienze italiana e inglese' *Rivista di Diritto Civile* 2008, 381–424, at 396.

[120] Spain, below p. 258 *et seq.*

[121] England, below p. 78 *et seq.*

[122] England, below p. 82 *et seq.*

[123] England, below pp. 84–5 *et seq.*

[124] Germany, below pp. 183–4.

the most advanced scientific knowledge at the time, but he is expected to comply with safety levels which are higher than those established by regulations if they are common in the sector at the relevant time.[125] Obviously, the safety levels must be at least those established by legal regulations and enactments, but the standard required may well exceed them.

(ii) Causation

The role that causation plays also differs in social security law compared with tort law. In the case of occupational diseases, the link between work carried out by the worker and the disease included in the list establishes a presumption of causation and, additionally, workers are also allowed to furnish proof of the disease's occupational origin if it is not included on the list. Under tort law, however, the victim must establish a causal link between the conduct of the tortfeasor and the damage suffered.[126] The different levels of exposure to asbestos dust required to contract the different asbestos-related diseases, and the long latency period of these diseases, usually pose difficult problems in establishing causation.

Establishing causation in the case of asbestosis is relatively easy, since medical science contends that there is a clear causal link between exposure to asbestos dust and this disease, and it should therefore suffice to prove that the employer has negligently exposed his employee to asbestos dust. However, it may become difficult to establish in which employment he has contracted the disease, if the employee has worked for several employers during his life, or where he has been exposed to asbestos for some time as self-employed. The problem is even more difficult to solve in the case of mesothelioma, since while asbestosis requires long exposure to asbestos dust, it is now generally accepted that mesothelioma can be caused by very slight and transient exposure. When, as in this latter case, the inhalation of a single fibre suffices, it is clear that the harm cannot have been caused by all employers, so that imposing joint and several liability upon them may give rise to the unjust situation of holding an employer liable for a harm that he may not have caused. This result can be even more unjust when one takes into account that, due to the long latency period between exposure to asbestos dust and contracting mesothelioma, the other employers may no longer exist or may no longer be covered by insurance when the claim is filed. If this is the case, the remaining employer will not be able to obtain contribution from the others and will have to bear the total amount of the damages award.

[125] See Italy, below p. 223. [126] Magnus, above note 116, at p. 284, nn. 44–7.

Causation does not seem to have posed serious problems so far in France and in the more pro-claimant case law in Spain. Italian courts have not elaborated much on causation problems; whereas they have had no problem establishing causation in the case of asbestosis, they have applied different criteria to facilitate its proof in the cases of mesothelioma, such as that exposure has increased the risk of suffering the disease to a high level of probability, or that the employer has 'omitted the reduction of the risk' when he has not taken the required safety measures. According to this 'high level of probability', courts have awarded compensation to a woman who died of mesethelioma after washing her husband's contaminated clothes for many years.[127] In the case of lung cancer, courts have also resorted to a rebuttable presumption of causation when exposure has been proven.[128]

In England, causation issues have given rise to several important decisions of the House of Lords over the last few years and to a weighty debate amongst legal writers.[129] In *Fairchild* v. *Glenhaven Funeral Services*,[130] the most significant asbestos-related decision, the claimants had contracted mesothelioma through exposure to asbestos over a lifetime of work for different employers. Although all employers were in breach of the relevant duties of care and had, therefore, increased the risk of mesothelioma, none could be said to have caused the disease. Medical evidence showed that mesothelioma did not necessarily build up gradually but could have been caused by the inhalation of one or several asbestos fibres, and the claimants had not been able to prove which employer or employers were responsible for the fatal fibre or fibres causing the disease. For these reasons, the Court of Appeal dismissed the claim on the grounds that the claimants had not been able to establish causation, but the House of Lords allowed the appeal and found for the employees on the grounds that in this exceptional case, it was sufficient for the claimants to show that the defendant's negligence had increased the risk of the disease. In another important asbestos-related decision, *Barker* v. *Corus plc*,[131] the House of Lords dealt with an issue that *Fairchild* had left open, namely whether, in the case of multiple defendants who had negligently exposed the claimant to asbestos dust, liability was joint and several, and therefore the claimant could obtain compensation in full from any

[127] Italy, below p. 219.
[128] Italy, below p. 217 *et seq*. See also Coggiola, above note 119, at p. 398.
[129] England, below pp. 85–8. [130] [2002] UKHL 22, [2003] 1 AC 32.
[131] [2006] UKHL 20, [2006] 3 All ER 785.

of the defendants, or was liability proportionate to the exposure for which each defendant was responsible and, accordingly, the claimant could recover from each defendant his share of the damage only. In this case, the House of Lords found for the employers and held that they were liable for a proportionate share, because they had all contributed to the risk of mesothelioma.[132] The practical effect of this decision was that the risk of any of the defendants being insolvent and unable to pay the appropriate share would fall on the claimant, and that before liability could be apportioned and full compensation paid, the claimant would have to trace all relevant defendants or file multiple claims to recover from each of them. This outcome, which would have made full compensation for the victims more difficult and time-consuming, was not acceptable to public opinion. After the decision in *Barker*, there was a swift and fierce political backlash, with large numbers of workers, families, trade unions, and Members of Parliament calling for the reversal of the ruling. Soon enough, the Compensation Act 2006 was introduced, specifically to reverse the ruling. Section 3 reversed the House of Lords' decision by making all defendants jointly and severally liable for the damage, so that a claimant could recover the totality of his damage. However, the Act applies to mesothelioma only and not to all asbestos-related diseases.[133]

(iii) Damage

Asbestos-related diseases have also called into question the very same concept of damage as a basic condition of liability. What is an actionable injury? A physiological change in the body of the victim or only a change that has an adverse effect on his bodily function or on his appearance? Is it possible to distinguish between an 'injury or a disease', 'in some sense', which would not deserve redress, and 'damage' which would be compensable on the grounds that the 'claimant is worse off'.[134] Additionally, does anxiety or fear of suffering an asbestos-related disease in the future qualify as compensable damage?

The problem has arisen with regard to victims who have been exposed to asbestos and have been diagnosed with pleural plaques. Although these plaques constitute a physiological change in the body, they do not normally cause any symptoms, nor do they threaten health nor are they

[132] See England, below pp. 87–8. [133] England, below p. 88.

[134] Posing this question, Lord Hoffman in *Johnston* v. *NEI International Combustion Ltd* [2007] UKHL 39, [2007] 4 All ER 1047, at para. 19.

a precondition of other asbestos-related diseases, nor do they lead to any of them.

French law has no doubt that pleural plaques are recoverable damage. Since 1985, they have been included in the list of occupational diseases caused by asbestos exposure and they constitute about 75 per cent of the cases resolved by the FIVA. This scheme assigns to them a rate of impairment of 5 per cent and provides for a total damages award of €22,000 in its *barème indicatif.* Acceptance of this award will bar any action for compensation; however, if the petitioner belongs to the 7 per cent of those who refuse the sum offered by the scheme because he considers that it is too low, he can still bring an action before the social security courts. In such a case he would seek an award assessed in the discretion of the judge (and not limited by the parameters of the scheme) and he could obtain a higher award if it could be established that the damage was the result of *faute inexcusable* of his employer, i.e. the breach of a contractual '*obligation de sécurité de résultat*'.[135] It is true, however, that the legal devices used in French law to deal with the problem are rather different from the English ones: social security law, a non-fault scheme and contract remedies related to industrial relations law. However, it can be contended that a similar outcome of recoverability would be reached if the traditional route of *responsabilité civile* had to be followed and that, as in the cases of HIV or hepatitis C, courts might recognise, in the case of pleural plaques, the existence of a '*préjudice de contamination*'.[136]

In England, High Court decisions in the 1980s had also awarded damages for negligent exposure to asbestos which had given rise to pleural plaques. Thus, in *Church* v. *Ministry of Defence*[137] and *Sykes* v. *Ministry of Defence*,[138] Peter Pain J and Otton J, respectively, held that symptomless pleural plaques, on their own, were sufficient to constitute compensable damage. In *Patterson* v. *Ministry of Defence*,[139] Simon Brown J formulated what was to be known as the 'aggregation theory' by holding that, although a 'symptom-free physiological change' such as pleural plaques was not in itself an actionable injury, when combined with the risk of future disease and anxiety, it could create a cause of action. A successful claim regularly

[135] See France, below p. 132 and, in more detail, Sebastien Beaugendre, 'Regard d'outre-manche sur un arrêt de la Chambre des Lords relatif à l'exposition de salariés à l'amiante' (2009) 2 *European Review of Private Law* 177 at 189.

[136] *Ibid.* 194. [137] (1984) 134 *NLJ* 623. [138] *The Times*, 23 March 1984.

[139] [1987] CLY 1194.

received an award for provisional damages of between £5,000 and £7,000 (leaving open the possibility of a further claim if the claimant subsequently developed mesothelioma or another asbestos-related disease), or a full and final award of between £12,500 and £20,000.[140] However, in *Johnston* v. *NEI International Combustion Ltd*,[141] the House of Lords, upholding unanimously the previous decision of the Court of Appeal,[142] held that the presence of pleural plaques, whether or not combined with a risk of developing an asbestos-related disease and anxiety about that risk, did not constitute actionable or compensable damage and, therefore, could not form the basis of a claim in negligence.[143]

Johnston has been applauded by lawyers close to the interests of the insurance industry[144] and even by some authors who, although they concede that the decision may be considered unsympathetic to claimants, believe it is to be welcomed for the clarity it brings to the tort of negligence by defining the concept of damage more precisely.[145] The decision, however, has also given rise to enough concern for the government to undertake a consultation process, in which the government will consider several options, such as 'increasing support, help and information for people with pleural plaques', 'changing the law of negligence' or establishing a no-fault scheme, either for those exposed to asbestos in the workplace and diagnosed with pleural plaques within a fixed period prior to the House of Lords' judgment and who had not already received compensation, or without this time restriction, also for those exposed and diagnosed after the House of Lords' judgment.[146] Although the English government is reluctant to change the law of negligence, the Scottish Parliament has already amended the law of delict in the

[140] See Ministry of Justice, *Pleural Plaques* (Consultation Paper CP 14/08, 9 July 2008), paras. 4, 7, available at www.justice.gov.uk/docs/cp1408.pdf (visited 1 May 2009).

[141] [2007] UKHL 39, [2007] 4 All ER 1047. See England, below, pp. 81–2 and note 207.

[142] [2006] EWCA Civ 27, [2006] 4 All ER 1161.

[143] Among many other English comments on this decision see Ken Oliphant, 'England and Wales' in Helmut Koziol and Barbara C. Steininger (eds.), *European Tort Law 2007* (Vienna/New York: Springer, 2008), p. 239, para. 5–14 and Gemma Turton, 'Defining Damage in the House of Lords' (2006) 71 *Modern Law Review* 987. See also, from a comparative perspective, Miquel Martín-Casals, 'Introduction to the Annotations to *Johnston* v. *NEI International Combustion Ltd*' (2009) 2 *European Review of Private Law* 177.

[144] David Pugh and Michael Rawlinson, 'Legal Analysis: Pleural Plaques – No Harm Done', *Post Magazine*, 25 October 2007, p. 14.

[145] See, in this sense, Turton, above note 143.

[146] See Ministry of Justice, Consultation Paper CP 14/08, above note 140, at p. 15.

Damages (Asbestos-related Conditions) (Scotland) Act 2009.[147] Pursuant to section 1 of the Act:

> (1) Asbestos-related pleural plaques are a personal injury which is not negligible.
> (2) Accordingly, they constitute actionable harm for the purposes of an action of damages for personal injuries.
> (3) Any rule of law the effect of which is that asbestos-related pleural plaques do not constitute actionable harm ceases to apply to the extent it has that effect.
> (4) But nothing in this section otherwise affects any enactment or rule of law which determines whether and in what circumstances a person may be liable in damages in respect of personal injuries.

In Italy, as a general rule, courts compensate the claimants for bodily and psychiatric injuries (*danno biologico*) only when they are clinically ascertained. Moreover, if the injury has negative consequences for the claimant's daily life they also compensate for the diminished ability to carry on daily activities (*danno esistenziale*). Since pleural plaques are neither a bodily harm nor do they cause any detriment to the daily activities of the claimant, it is contended that they would not be recoverable for.[148]

Finally, in Spain there are so far no decisions of the Supreme Court, but only a few decisions of the High Court of Galicia where the claimant suffered pleural plaques but claimed for compensation for asbestosis. All such decisions considered that pleural plaques were not an occupational disease, did not affect the daily activities of the claimant and, since the claimant did not allege any other damage, dismissed the claim.[149]

Compensating for mere anxiety which does not amount to a medically assessed disease entails the moral hazard of compensating for fake and unworthy claims. To ward off this danger, some legal systems link compensation for anxiety to the existence of a personal injury, as in the English case, or to other compensable damage, such as impairment to the victim's quality of life. In this sense, Italian courts have rejected compensation for anxiety resulting from exposure to asbestos when it was not consequential upon a physical or psychiatric illness (*danno biologico*) or upon a detriment to the daily activities of the victim (*danno esistenziale*).[150]

[147] Asp 4, available at <www.opsi.gov.uk/legislation/scotland/acts2009/pdf/asp_20090004_en.pdf (visited 1 May 2009).

[148] See Nadia Coggiola, 'The Johnston Case, the Compensation of Pleural Plaques and Psychiatric Harm in Italian Tort Law' (2009) 2 *European Review of Private Law* 239 at 247.

[149] See Spain, below p. 260 and note 128, and, in more detail, Albert Azagra, 'Spanish Case Note' (2009) 2 *European Review of Private Law* 219 at 222.

[150] Coggiola, above note 148, at p. 243.

Other legal systems, such as the French or Spanish, seem to be more open when compensating for non-pecuniary loss. Accordingly, anxiety need not necessarily be consequential upon a personal injury or other recoverable damage. If anxiety is real, effective and certain, it should be recoverable per se. In France, anxiety is covered under the FIVA asbestos scheme, while civil courts, under the general private law rules, have already compensated for mere anxiety in other cases.[151] In the absence of decisions on this point referring to pleural plaques, there are now enough decisions of the Spanish Supreme Court in other areas that show that according to the Spanish legal system, anxiety as such, as long as it is real, effective and certain, can be compensated for and, in fact, it has been so compensated many times without being submitted to a *de minimis* test.[152]

(iv) Limitation periods

A damages claim can only be enforced when all the conditions of liability are met, the occurrence of damage being one of them. This is why in the case of prescription nearly all legal systems, including those under analysis in this book, focus on the occurrence of the damage rather than on the damaging event. This is the case of France, where the new article 2770-1 C civ. provides that prescription begins to run from the time of the manifestation of damage or its aggravation. The same standard applies in Spain (article 1968.2 CC in connection with article 1969 CC), and also in Italy, where Italian courts, in spite of the express wording of article 2947.1 of the Italian Civil Code to the contrary ('si prescrive in cinque anni dal giorno in cui il fatto si è verificato'), also consider it to be the applicable rule. In Germany, although §199(1), no. 1 BGB refers to the moment when the claim comes into being (*der Anspruch entstanden ist*), it is generally understood that it refers to the moment when the claim becomes enforceable, something which in the instant case requires, inter alia, the existence of damage. Additionally, all these legal systems provide in one form or another for a 'subjective' prescription regime, i.e. a regime where prescription does not run unless the victim knows, or could reasonably have known, of the

[151] Beaugendre, above note 135, at p. 193.

[152] See Joan Carles Seuba Torreblanca, Sonia Ramos González and Álvaro Luna Yerga, *Falsos positivos. La responsabilidad civil derivada del diagnóstico erróneo de enfermedades* (InDret 3/2002), available at www.indret.com, stressing this trend in favour of compensation for mere anxiety (at p. 7). Recently, the Spanish Supreme Court has compensated for the anxiety resulting from a fear of developing AIDS in the case of a patient who had been erroneously diagnosed (STS 19 July 2007 (RJ 2007\5143)). See the decision discussed by Jordi Ribot and Albert Ruda, 'Spain' in Koziol and Steininger, above note 143, at p. 557, paras. 37–43.

existence of damage. This is also the case in England, where the Limitation Act 1980 provides currently for a six-year period running from the time of damage or three years from the date on which the claimant became aware, or ought to have been aware, of the damage, whichever is later.[153]

As has already been pointed out, asbestos-related diseases have long latency periods and usually manifest themselves only a considerable period after exposure has occurred. This gives rise to a serious problem when the relevant legal system does not follow a subjective prescription regime. That was the case in England under the Limitation Act 1939, which had laid down a six-year period both for contract and tort running from the date the cause of action accrued, i.e. the occurrence of the damage. Both the Court of Appeal and the House of Lords held in *Cartledge* v. *E. Jopling and Sons*[154] (a case where the claimants had contracted pneumoconiosis by the inhalation of metal particles over many years), that the damage had occurred (and the cause of action accrued) at the time when the claimants were breathing the dust and expressly rejected the argument that the limitation period began when the claimants knew of the existence of the damage. Since the symptoms had manifested themselves more than six years after the employment had ceased, and in spite of the fact that the disease had been unknown to anyone and not even reasonably discoverable, the claim was time-barred and dismissed.[155] Since both courts had expressed their dissatisfaction with this outcome, Parliament intervened and introduced a subjective standard to establish the beginning of the limitation period for personal injury claims, with a six-year period running from the time of damage, supplemented by a three-year period running from the claimant's date of knowledge.[156]

The introduction of a subjective standard has been counter-balanced in some legal systems with the establishment of a 'long-stop' period, i.e. a maximum period tied to an objective standard at the expiry of which the claim is time-barred regardless of the claimants' knowledge. In the legal systems under analysis, this is the case in the German Civil Code (§199 BGB) and, recently, in the French Civil Code (article 2232 C civ.); by

[153] See especially for a comparative perspective, Reinhard Zimmermann and Jens Kleinschmid, 'Prescription: General Framework and Special Problems concerning Damages Claims' in Koziol and Steininger, above note 143, at pp. 26–78.

[154] [1963] AC 758.

[155] See England, below pp. 80–1 and Zimmermann and Kleinschmid, above note 153, at p. 48. See also Richard James, 'The Limitation Period in Medical Negligence' (1998) 6 *Medical Law Review* 62–98.

[156] See Limitation Act 1980, ss. 11 and 14, and England, below p. 80 *et seq*.

contrast, neither English law nor the Italian or the Spanish Civil Codes provide for a 'long-stop' period.[157] However, the English Law Commission has also proposed introducing long-stop periods.[158]

Long-stop periods tend to be between ten and thirty years[159] and they may give rise to problems in the case of asbestos-related diseases. A clear case in which this problem would arise is in the application of any national law implementing the European Directive on Product Liability, since its article 11 provides for a short long-stop period of ten years running from the date on which the producer put into circulation the actual product which caused the damage.[160] In this case, the consequences might not be too serious, since the victim could still avail himself of the general tort law rules of the corresponding legal system. However, the problem may persist even when the long-stop period adopted by the relevant legal system tends to the maximum in the scale between ten and thirty years, since it can ocasionally have lapsed before the disease has become reasonably discoverable.

Comparative research shows that the long-stop period is considered as particulary problematic with regard to personal injury claims and, therefore, legislatures tend currently either to enact a long-stop for these claims which is considerably longer than for the other damages claims, or to do away altogether with long-stops for these cases. The first solution was adopted by the 2001 reform of the German Civil Code, which provides now for a thirty-year period for damages claims for the infrigmement of life, bodily integrity, health or liberty (§199(2) BGB). The second solution, which seems preferable,[161] has been adopted by the 2008 amendment of the French Civil Code, which now provides that the general long-stop period of twenty years does not apply, among other specific cases, to claims for personal injuries (article 2232 C civ. in connection with article 2226 C civ.).

[157] In Catalonia, by contrast, arts. 121–4 of the Catalan Civil Code provide for a long-stop of thirty years. See Albert Lamarca Marqués, in Antoni Vaquer Aloy and Albert Lamarca Marqués (eds.), *Comentari a la nova regulació de la prescripció i caducitat en el Dret civil de Catalunya* (Barcelona: Atelier, 2005), pp. 281–5.

[158] *Limitation of Actions* (Consultation Paper no. 151, 1998), p. 241. See James, above note 155, at 94 and Zimmermann and Kleinschmid, above note 153, at p. 51.

[159] Zimmermann and Kleinschmid, above note 153, at p. 30.

[160] Council Directive 85/374/EEC of 25 July 1985 on the approximation of the laws, regulations and administrative provisions of the Member States concerning liability for defective products (OJ L210, 7 August 1985). Member States must provide in their legislation that the rights conferred upon the injured person pursuant to the Directive will be extinguished upon the expiry of a period of ten years from the date on which the producer put into circulation the actual product which caused the damage, unless the injured person has in the meantime instituted proceedings against the producer.

[161] See, in this sense, Zimmermann and Kleinschmid, above note 153, at p. 57.

2

Technological Change and the Development of Liability for Fault in England and Wales

JONATHAN MORGAN

1. Fires caused by sparks from railway engines

Great Britain was the pioneer of steam railway transportation.[1] The first line was the Stockton to Darlington railway, opened in 1825. Then in 1830, the hugely important Liverpool–Manchester line opened; the ceremony is memorable for the first recorded passenger fatality, none other than William Huskisson, the Member of Parliament for Liverpool. The network continued to expand, especially in 1845, the 'Railway Mania', and the years following. The railways were constructed and operated purely as a matter of private enterprise.[2] Nevertheless, such privately financed great projects invariably required parliamentary sanction, in the form of a Private Act of Parliament, primarily because the projected railway would need powers to compel the sale of land by reluctant land-owners, which only a special Act could confer. Secondly, prior to the General Incorporation Act of 1844, there was no general power under English law to incorporate, and a Private Act was the most straightforward way to bring a company into being.

No official figures seem to have been kept for the incidence of fires started by sparks from railway locomotives, and the value of property destroyed. However, judging by the numerous reported decisions in the higher courts, it seems to have been a common occurrence. By the time Parliament intervened to provide a remedy in the reign of King

[1] In this chapter, however, legal attention, especially regarding the common law, will be confined to England unless stated otherwise. Parliamentary action (legislation, select committees, etc.) generally applied to the whole of Great Britain.

[2] Although from 1840 the Railway Regulation Act required the notification of various data to the Board of Trade, and empowered it to appoint inspectors of railways.

Edward VII (1901–1910), the number of such fires seems to have been a major source of grievance for agriculture. In a case of 1946, the (large, regional) London and North Eastern Railway gave evidence that in 1943 it paid out some £50,000 for damage to growing crops.[3]

The question of liability for sparks can be set within wider areas of socio-political conflict: industrialisation versus agriculture; the frequently alleged subsidisation of nascent industry at the expense of those damaged by it.[4] Furthermore, the sparks situation has proved to be an archetypical problem for the analysis of private law from the standpoint of economic efficiency.[5] In legal-doctrinal terms, it reflects the fundamental question of strict versus fault liability in tort, which was debated from the mid-nineteenth century onwards. A particular feature of this debate is the protection of private property, for it was damage to land (growing crops) which formed the basis for all of the reported cases. The prevalence of statutory authority is a complicating factor in tracing the legal development, and it will be seen that the presence of a statute made it harder to establish liability than when the 'pure' common law applied. However, the controversy seen throughout this development shows that there was little consensus over how to solve the problem; nor was there concerted parliamentary intervention (it was, in fact, parliamentary silence which arguably created most of the legal problems). There is certainly little support here for the 'industrial subsidy' thesis of Horwitz and others.

(a) *Liability of statutory undertakings: negligence*

It was decided very early on, in *R* v. *Pease* (1832), that where an Act of Parliament had authorized a railway, an action in nuisance could not lie for running of trains along the railway, at least where all practicable means had been used to prevent damage.[6] Ordinarily, taking reasonable care – or even all possible care – provided (and indeed still continues to provide) no answer to an action in nuisance. If the result of the defendant's activity is to interfere with his neighbour's land use beyond the degree to which the

[3] *Parker* v. *LNER* 147 EG 269, [1946] WN 63, first instance. Most of the individual payments were under the (then) £200 limit in the Railway Fires Act 1905.

[4] Seminally, M.J. Horwitz, *The Transformation of American Law, 1780–1850* (Cambridge, Mass.: Harvard University Press, 1977).

[5] Seminally, R.H. Coase, 'The Problem of Social Cost' (1960) 3 *J Law and Economics* 1. See now E. Claeys, 'Jefferson Meets Coase: Train Sparks, the Harm–Benefit Distinction, and Natural Property Rights' (2006, working paper available via Social Science Research Network).

[6] 4 B & Ad 30.

neighbour can be expected to put up with, the defendant will be liable irrespective of the care which he takes in carrying out the activity. The only way to avoid liability in such a situation, therefore, is to cease the activity altogether. This last point has strongly influenced the statutory authority defence in actions against railways. As the court said in *Pease*, the legislature must be presumed to have known that users of the adjacent highway 'would be in all probability incommoded' by the noise from the railway locomotives (which tended to startle horses, as in that case). It would be odd if an activity which Parliament had expressly sanctioned, and which inevitably interfered with passage along the highway, was effectively rendered unlawful by the law of nuisance. The result would be criminal liability in a case of public nuisance like *Pease*; in private nuisance cases, moreover, the activity would presumptively be prohibited by an injunction.

An early case specifically about sparks from locomotives was *Piggot* v. *Eastern Counties Railway*.[7] The case was overtly framed as a negligence action against the defendant railway. *Pease* was not cited by the court, but Tindall CJ held that:

> The defendants are a company intrusted by the legislature with an agent of an extremely dangerous and unruly character for their own private and particular advantage: and the law requires of them that they shall, in the exercise of the rights and powers so conferred upon them, adopt such precautions as may reasonably prevent damage to the property of third person through or near which their railway passes.

On the facts of the case, negligence was made out. Some of the judges' comments suggest a doctrine of prima facie negligence, or '*res ipsa loquitur*': the fact that the fire had been started by sparks from the defendants' locomotive, while not conclusive, 'at least affords a strong presumption of negligence, in the absence of evidence to shew that something had been done by the company to lessen the chances of danger', said Maule J. However, this must be read in the light of evidence provided by the plaintiff at trial that sparks could be prevented by installing guards or grills in the locomotive chimney. The evidence had simply been left to the lay jury, who were not directed by the trial judge in terms of a presumption of negligence. The jury found for the plaintiff, and the full court sitting *in banc* saw no reason to interfere with this verdict.

R v. *Pease* was definitively affirmed in the context of sparks by the Exchequer Chamber in *Vaughan* v. *Taff Vale Railway* (1860).[8] At the trial, Bramwell B had directed the jury that the defendant railway was liable

[7] (1846) 3 CB 229. [8] (1860) 5 H & N 679.

irrespective of fault: 'if, to serve his own purposes, a man does a dangerous thing, whether he takes precautions or not, and mischief ensues, he must bear the consequences'.[9] Bramwell B indeed offered to direct the jury to find for the plaintiff, but plaintiff's counsel preferred that the question be left for their determination, possibly showing concern that Bramwell's uncompromising approach might not survive higher scrutiny. The jury, in any event, found for the plaintiff, and the court *in banc*, speaking through Bramwell B, unsurprisingly upheld his Lordship's earlier decision at trial.[10] On appeal to the Exchequer Chamber, the defendants predictably relied upon the precedents of *Pease* and *Piggot*. The plaintiff, it seemed, tried to draw a distinction between public nuisance (as in *Pease*: rights over the public highway) and private nuisance, arguing that while statutory authority protected the company from indictment by the public, nevertheless if damage to an individual was caused, it should be compensated. The appeal court was unimpressed, and found for the defendant railway. Cockburn CJ simply affirmed *Pease*, stating that the need to prove negligence to establish liability for an undertaking authorised by statute was 'consistent with policy and justice'. Blackburn J admitted that the case might have involved a difficult question if a matter of first impression, but held that *Pease* had definitively settled the matter.

Vaughan was affirmed in turn by the House of Lords, in *Hammersmith and City Railway* v. *Brand*.[11] This case raised two questions: first, whether the standard authorization clause in section 86 of the Railway Clauses Consolidation Act 1845, which gave railway companies the power to run trains, charge tolls, etc., provided a defence to a common law action for nuisance (this time, by vibrations); secondly, whether the plaintiffs had a right to claim under the detailed compensation scheme provided in the same statute. As was common practice at that time, the law lords invited the common law judges to advise them as to the relevant law. This provided Bramwell B with an opportunity to attack the decision of the Exchequer Chamber in *Vaughan*. He declared that that case proceeded on a 'misapprehension' of the purpose of section 86. The point of that standard clause, he argued, had nothing to do with the authorization of nuisance. Instead, it was to establish as one of the objects of the railway company in question the business of *operating* a railway, so that it could not be objected that the continued operation was ultra vires the terms of incorporation in its constitutive Private Act (i.e. once the railway had been *constructed*). Bramwell argued that further support for this interpretation came from

[9] See (1858) 3 H & N 743. [10] *Ibid.* [11] (1869) LR 4 HL 171.

the absence of any express provisions for compensation of those affected by nuisance. The presumption was that Parliament would not abrogate property rights without compensation. Indeed, elaborate machinery was provided by the Railway Clauses Consolidation Act for compensating outright takings of land, or physical intrusion onto land, so it would be strange if the same statute gave *carte blanche* to commit nuisance, which is simply indirect interference with land. (Bramwell B did, however, go on to advise that the statutory compensation scheme could be interpreted to include the plaintiff's case; these may be seen as alternative arguments.) The only argument presented in favour of the nuisance anomaly, he said, was that of the public benefit accruing from the operation of the railway. But, retorted Bramwell, if the public benefits, the public should compensate those individuals damaged.

All of the judges (except Blackburn J) agreed with Bramwell B that the plaintiffs should succeed. Even Blackburn J offered but weak support for *Pease* and *Vaughan*, admitting that there was 'much to say' for Bramwell's construction of section 86, but that *Pease* had for so long been taken to represent the law that even if it had been wrongly decided, *communis error facit jus*. Notwithstanding this advice, the House of Lords held that the plaintiffs had no right to compensation under the scheme contained in the statute (Lord Cairns dissenting on this point). Moreover, their Lordships were unanimous in upholding *Vaughan* on the authorization of nuisance (although for Lord Cairns, this provided added force to his argument that the statute should be interpreted to give a right to compensation). Lord Chelmsford argued that one could not expect Parliament to authorize a nuisance expressly. To conclude that the mere running of trains, however careful, was a nuisance would render the company's statutory right to run trains 'nugatory'. Lord Cairns elaborated further, although describing his argument as a *reductio ad absurdam*, stating that if simply to run trains was a nuisance, the Court of Chancery could issue an injunction to prevent what Parliament had authorized. Bramwell B had sought to pre-empt this formidable objection, stating that he 'apprehended that injunctions would not be granted in such cases', although not giving any reason for this view.

With this unanimous affirmation by the highest English court, the 'statutory negligence' doctrine of *Vaughan* v. *Taff Railway* was entrenched in the law. It is fair to say that the issue at the heart of the case has never really gone away, however.[12] In essence, how should interference with the

[12] See, e.g., *Wildtree Hotels Ltd* v. *Harrow London Borough Council* [2001] 2 AC 1. Lord Hoffmann commented (at para. 3): 'The nineteenth century authorities on this subject

property rights of another in the public interest be analysed by the law? An absolute property analysis would prevent this altogether; at the other extreme, individual rights may be freely sacrificed to the public good, which is the effect of the decision in *Hammersmith Railway* v. *Brand*. The middle course is that interference should be permitted, but only when the invasion of rights is compensated. In recent years, these issues have continued to be canvassed in the courts, with some sharp differences of opinion in at least two contexts: first, whether to grant an injunction to restrain a nuisance arising from some activity which is in the public interest, or to refuse an injunction and award damages in lieu of future infringement;[13] secondly, even the implicit statutory authorization of a nuisance has continued to be controversial, dividing the House of Lords in *Allen* v. *Gulf Oil Refining* (1981).[14] In theoretical terms, the question is whether it is optimal to protect rights in this concept by a property rule, or a liability rule.[15] The continuing theoretical and doctrinal debate reveals a timeless quality to the question.

(b) Common law: strict liability

In *Jones* v. *Festiniog Railway Co.* (1868), the Court of Queen's Bench confirmed that in the absence of statutory authorization, a railway was liable for fire caused by sparks from its engines, even when all possible care had been taken.[16] As stated above, it would normally be impossible for a railway company to be formed, or a railway constructed, without such statutory authorization. In *Jones* itself, the situation was rather unusual. The Festiniog Railway Co. had been formed by a Private Act in 1832, and enabled to lay tracks for the transportation of minerals, etc. by rail. Crucially, however, at this date (at the dawn of the railway age), the Act had not authorized the use of steam locomotives. Thus, when the railway later developed a steam passenger business along the same tracks, such activity was outside the authority of its governing Act of Parliament. *Vaughan* v. *Taff Railway* therefore did not apply.

do not display a steady development of the law but the shifting ascendancies of one view or the other.'

13 See *Miller* v. *Jackson* [1977] QB 966; *Kennaway* v. *Thompson* [1981] QB 88; *Dennis* v. *Ministry of Defence* [2003] EWHC 793 (QB).

14 [1981] AC 1001. Cf. Lord Keith's Bramwellian dissent.

15 Seminally, G. Calabresi and A.D. Melamed, 'Property Rules, Liability Rules and Inalienability: One View of the Cathedral' (1972) 85 *Harv. LR* 1089.

16 (1868) LR 3 QB 733.

The difference drawn between the statute in *Jones* and that typically encountered in railway sparks cases was undoubtedly a subtle one.[17] However, given the unusual facts, we might doubt the contemporary criticism of *Jones* as 'jeopardizing the position of the railways'.[18] However, it is possible that *Jones* was seen as an indication of a general move towards strict liability, and a repudiation of *Vaughan*. It is significant in this respect that both Blackburn J and Lush J expressly based their decisions on the recent decision of the Exchequer Chamber in *Rylands* v. *Fletcher*.[19] Whether that famous case meant to create a general principle of strict liability for dangerous things has been controversial since at least the first edition of *Beven on Negligence* in 1889.[20] Beven argued for a narrow treatment of *Rylands* v. *Fletcher*, said to create a special rule about escapes from land onto the land of one's neighbour.[21] Sir Frederick Pollock, on the other hand, commented that the case 'was certainly intended to enunciate something much wider',[22] although he also had some doubts on the historical provenance of the decision. Ultimately, Pollock rejected *Rylands* as too harsh, in favour of a rule requiring 'all reasonable practical care and caution' in the custody of dangerous things, in his draft Code on Civil Wrongs for British India.[23] D.J. Ibbetson, in our own time, suggests that the reason for the strictness in *Rylands* v. *Fletcher* was that it involved the infringement of property rights: the case would have been argued and decided differently had it involved personal injury.[24] Burning crops through the escape of sparks would anyway fall within the narrowest interpretation of the rule, of course, so *Jones* v. *Festiniog* was quite consistent even with Beven's thesis.

Jones v. *Festiniog* was followed in *Powell* v. *Fall*, where sparks from a steam-powered traction engine driven along the highway caused fire,

[17] J.R. Spencer, 'Motor-cars and the Rule in *Rylands* v. *Fletcher*' [1983] *CLJ* 65 at 68 n.16.

[18] 'Fire from Locomotives' (1870) 5 *Law Journal* 685.

[19] (1866) LR 1 Ex 265. Affirmed (1868) LR 3 HL 330.

[20] T. Beven, *Principles of the Law of Negligence* (London: Stevens and Haynes, 1889), pp. 1133–6.

[21] This view eventually prevailed: *Reid* v. *Lyons* [1947] AC 156.

[22] F. Pollock, *The Law of Torts: A Treatise on the Principles of Obligations Arising from Civil Wrongs in the Common Law* (1st edn, London: Stevens and Sons, 1887), p. 399.

[23] See notes to clause 68 of the Draft Bill in Appendix, *Pollock on Torts* (5th edn, 1897).

[24] D.J. Ibbetson, *A Historical Introduction to the Law of Obligations* (Oxford: Oxford University Press, 1999), p. 186 n.112. More generally on proprietary reasoning in tort law, see *ibid.* pp. 111–12, 186. Both nuisance and *Rylands* v. *Fletcher* have controversially been re-affirmed as property torts: *Hunter* v. *Canary Wharf* [1997] AC 655; *Transco* v. *Stockport MBC* [2004] 2 AC 1.

without any negligence on the part of the operators.[25] This was if anything a stronger case than *Jones*, for the Locomotive Acts 1861 and 1865 expressly preserved the right of action/indictment for nuisance (whether public or private) caused by a road locomotive. Bramwell LJ used the opportunity to record his continued – indeed, 'hardened' – opinion that *Vaughan* v. *Taff Vale* had been wrongly decided (ignoring its express approval by the House of Lords in the *Brand* case!). Again, Bramwell LJ showed a keen grasp of economic theory, sketching the idea that in order to produce efficient resource-allocation, an activity must bear all of the costs which it creates for society. If, he said, a dangerous machine could be profitably operated while compensating all of the damage caused by it, then there could be no objection to strict liability. If, however, the machine could not be run with profit under such a regime (i.e. internalizing all of its social costs), then it should not be operated at all.[26] Bramwell LJ here anticipated arguments popular a full century later: economists argue that strict liability is optimally efficient, as maximizing cost internalization (so that the cash cost of a product/service represents the true social cost of its manufacture or provision). But then George William Wilshere, Lord Bramwell, as he finally became, was a rather exceptional judge.[27]

An afterword could be made concerning the petrol-driven motor-car. J.R. Spencer has recounted how, after an early flirtation with a *Rylands* v. *Fletcher* strict approach, the English courts in 1909 affirmed the necessity of fault for an action by one injured by the new machines, on the highway.[28] Essentially, they chose to follow the old approach used for horses and carts (negligence), rather than the strict approach of *Rylands*, *Jones* v. *Festiniog* and *Powell* v. *Fall*. Spencer plausibly suggests a number of factors at work, including the general reversion towards fault as the proper moral basis for liability in Anglo-American legal thought. Also, motor-cars were (upper-class) pleasurable amenities, whereas traction engines were obviously connected with trade. Even if it was appealing for businesses

25 (1880) 5 QBD 597.

26 See also Bramwell B's judgment in *Bamford* v. *Turnley* (1862) 3 B & S 62.

27 See E. Heward, *Lives of the Judges: Jessel, Cairns, Bowen and Bramwell* (Chichester: Barry Rose, 2004). Recently in *Transco* [2004] 2 AC 1, Lord Hobhouse of Woodborough evinced in his speech enthusiasm for *Rylands* v. *Fletcher*, and the loss-internalization which its strict liability approach embodies. But cf. Lord Hoffmann, *ibid.* at para. 29: 'With hindsight, *Rylands* v. *Fletcher* can be seen as an isolated victory for the internalisers' – naming Bramwell B in particular. Lord Hoffmann in common with a majority of their Lordships emphasized the many restrictions which have come to hedge in *Rylands* liability.

28 See Spencer in [1983] *CLJ* 65. *Wing* v. *London General Omnibus Co.* [1909] 2 KB 652; *Parker* v. *LGOC* (1909) 73 JP 283.

to internalize their costs in the way suggested by Bramwell LJ, this was patently not so for private motorists, who would normally at this date pay the damages awarded against them personally. Spencer notes that *Powell* v. *Fall* was affirmed for fires caused by traction engines even after the courts had opted for fault liability for other motoring accidents: *Mansel* v. *Webb*.[29] Spencer comments that the latter case displays, in full glory, 'that manly disregard for vulgar logic which makes the common law so much superior to other systems'![30] The later developments of motor negligence are, of course, inextricably tied up with the question of liability insurance and its effect on tort law; such insurance became compulsory in the United Kingdom in 1930.

It will be noted that in this small corner at least, the legal development seems to be the exact opposite of what the 'industrialization subsidy' thesis would predict. Strict liability applies to the industrial steam locomotive, but fault to the motor-car designed for personal use. Global consistency has never been a feature of the common law process, nor social reform. As S.F.C. Milsom says about the classical common law, 'The life of the common law has been in the abuse of its elementary ideas … In the twentieth century, with varying success, we use legislation; and our familiarity with deliberate change makes it easy for us to misread history … Certainly if we view the common law on the eve of reform as a piece of social engineering, we see the spirit of Heath Robinson at his most extravagant.'[31] What, then, was the role of statutory reform on the problem of sparks?

(c) *Parliamentary intervention: the Railway Fires Act 1905*

A Bill was first introduced in Parliament in 1900 to provide for compensation without the need to prove fault, when crops were destroyed by sparks from railway engines.[32] Interestingly, the recitals expressed the desire to place railway engines on the same footing as steam locomotives used on the highway, where strict liability was the rule, suggesting (contrary to what common law romantics would have us believe) that it is actually Parliament, and not the courts, which desires consistency across the legal system, or 'integrity' in law. Of course, the consistency argument

[29] (1919) 88 LJKB 323. [30] [1983] *CLJ* 65 at 70.

[31] S.F.C. Milsom, *Historical Foundations of the Common Law* (2nd edn, London: Butterworths, 1981), pp. 213–19.

[32] 63 Victoria, Bill 51. NB indexed as 'compensation for damage to crops' and not 'railway fires'.

proved little, for why should we not 'harmonize downwards' by applying the fault standard of the railway cases to the traction engines?

A very similar Bill was introduced in 1901, and progressed to a Second Reading. The debate revealed a great deal of concern from farmers throughout the country, as to the current legal position. While, in theory, they could recover compensation on the proof of fault, pursuant to *Vaughan* v. *Taff Railway*,[33] this right was impossible to enforce in practice. As ever, the costs of bringing an action were deplored, particularly, in the present context, the difficulty and expense of proving negligence on the part of the railway. This confirms that any presumption of negligence (the *res ipsa loquitur* approach) had been abandoned by the courts in these cases, to the detriment of plaintiff farmers, by the end of the nineteenth century. Indeed, some Members of Parliament explicitly suggested a reversed burden of proof as the answer to the farmers' complaints. Opponents of the Bill argued that farmers could insure themselves very cheaply against such damage, in the rare cases that it was caused without the fault of the railway. Moreover, they attacked the further anomalies that the proposed legislation would bring, asking why agriculture should be in a privileged position. The farmer would recover compensation without fault: why not the householder or mill-owner? Although the House of Commons voted to give the Bill a Second Reading by 307 votes to 80, it progressed no further.

A similar Bill was again introduced in 1905. Its promoter, Mr Mount, stated on Second Reading that the situation had if anything worsened since 1901, citing more powerful steam engines as the culprit.[34] He enlarged upon the practical barriers faced by the current law: the need to identify the particular locomotive which emitted the fatal spark – nearly impossible on busy lines – and the expense of providing expert evidence about engine design, etc. The cost was prohibitive for all but the most valuable claims. Finally, Mount cited the positive experience with compensation regimes in Canada, the United States and 'many continental countries'. The President of the Board of Agriculture cited the calls for legislative action by 'almost every farmers' club of importance'. Opponents argued that land-owners had been compensated for damage to crops by the compensation paid on the initial construction of the railway. In response, supporters of the Bill doubted whether the relevant statutory payments had really contained an element for the perpetual licence to burn down woods or fields of crops. In any event, farmers were usually tenants of much larger

[33] (1860) 5 H & N 679. [34] Parl. Deb. (4th Series), vol. 142, cols. 348–74.

land-owners, and so had never been the recipients of such compensation, although their crops were being burned. 'Fairness' was appealed to, in a rather abstract way, on both sides of the debate: it was fair that railways paid for all crops which they destroyed through their profitable activities; it was not fair to hold railways liable for damage which they could not have prevented. Another 'consistency' argument was urged in favour of farmers (it seemed that the highway locomotives point had been quietly dropped): if farmers were strictly liable to their workmen for accidents in the course of employment,[35] then railways should be strictly liable to farmers.

The Bill again received a Second Reading. In Standing Committee, it was watered down in a number of ways, not least limiting railways' liabilities for burned crops to £100. On return to the Commons, some members attempted to raise or abolish this limitation, but were warned that if they demanded too much the Bill would fail altogether, and farmers would get nothing.[36] Mr Mount defended the compromise as 'reasonable', covering 99 per cent of cases. The Attorney-General pointed out that the Bill would protect small farmers, in particular, from hardship. In the House of Lords, it was noted that the railway interest had made a 'good bargain' in amending the initial proposals, but that half a loaf, for the farmers, was better than no bread at all. The Bill passed into law as the Railway Fires Act 1905. It seems to have generated very little case law. Where the damage exceeds the statutory limit (raised later to £200 and now £5,000), the farmer might still bring a negligence action.[37] The Act remains on the statute book, although such fires are presumably now rare.

The wider principles of the liability and compensation debate, however, are timeless. During the period examined above, Parliament seemed reluctant to lay down general principles to settle the legal debate.[38] The question of whether nuisances were implicitly authorized by the railway statutes, and whether they fell within the land-owners' compensation schemes in those Acts, divided the judiciary, coming to a head in *Hammersmith Railway* v. *Brand*.[39] The legislative silence effectively delegated the dispute to the judiciary (notwithstanding the elusive judicial search for 'the intention of Parliament'). And when reform finally came in

35 The Workmen's Compensation Act 1897 had been extended to agriculture in 1900, and was extended to all forms of employment in 1906.

36 Parl. Deb. (4th Series), vol. 148, cols. 1478–92.

37 *Parker* v. *LNER* (1946) 147 EG 269, [1946] WN 63.

38 Cf. now Land Compensation Act 1973.

39 (1869) LR 4 HL 171.

1905, it was a narrow measure designed to address one particularly pressing issue (crop fires caused by railway sparks), and not to lay down general principles. Maitland's remark on considering the jumble of eighteenth-century statutes, that 'the British parliament seems rarely to rise to the dignity of a general proposition', might apply equally here.[40]

2. Exploding steam boilers

Steam power has been harnessed by man since ancient times. Thomas Newcomen's steam-powered 'atmospheric engine' was first commercially installed in 1711. It was much improved in efficiency by James Watt in the late eighteenth century (patented 1769). Watt's partnership with manufacturer Matthew Boulton (1775) led to rapid sales and proliferation of the steam engine across Great Britain. Steam thus replaced water, and animal and human muscle, as the energy source for the world's first Industrial Revolution.

The early boilers were not always reliable. Later, more powerful versions were apt to cause greater damage when they failed. Between 1840 and 1870, the death rate from boiler explosions increased more rapidly than the total horsepower produced.[41] A Select Committee of 1871 estimated that there were about 50 explosions per annum, that is, around one for every 2,000 boilers in the country. Between 1865 and 1882, some 1,051 persons were killed in Britain in boiler explosions, and another 1,519 injured (we may assume, seriously).[42] Societies were set up to promote boiler safety,[43] Bills were introduced in Parliament, and committees inquired into the safety of steam boilers. A real puzzle, then, is the apparent lack of litigation over boiler explosions.

(a) Actions in tort

While negligence was yet to be unified as a single nominate tort in English law,[44] it seems plausible to suppose that one who injured the person or property of another through the negligent operation of a steam

[40] F.W. Maitland, 'English Law (History)' in *Encyclopaedia Britannica* (11th edn, 1911).

[41] P.W.J. Bartrip, 'The State and the Steam-Boiler in 19th Century Britain' (1980) 25 *International Review of Social History* 77.

[42] Earl of Derby, Parl. Deb. (3rd Series), vol. 269, col. 1244.

[43] Especially the Manchester Steam Users' Association in 1855.

[44] The honour goes either to Lord Esher MR in *Heaven* v. *Pender* (1883) 11 QBD 503 or Lord Atkin in *Donoghue* v. *Stevenson* [1932] AC 562.

boiler in the mid-nineteenth century would be held liable in damages. It is therefore surprising to find no mention of exploding boilers in any of the nineteenth-century tort textbooks, the indices of which are highly factual and casuistic.[45] Similarly, electronic trawls of the consolidated nominate English Reports, and the later Law Reports, reveal no discussion of the issues in the higher courts. Given the frequency and severity of boiler explosions, this is truly surprising. Yet when we consider that most victims of such explosions would have been workmen, the omission becomes explicable.

Faults in boilers, leading to explosions, may be divided between faults in manufacture, failure to maintain, and faults in operation.[46] Accordingly, the potential defendants would be, respectively, the boiler manufacturer; the owner of the boiler, or his engineers; the actual boiler operative. The third group would never have means to be worth suing; the doctrine of vicarious liability, however, meant the employer would be liable for the torts of his servants (although not of independent contractors, provided he took reasonable care to engage a competent person).[47] Manufacturers, on the other hand, would effectively be immune from suit at the instance of one injured by an exploding boiler. It was determined in *Winterbottom* v. *Wright* that a manufacturer was not liable *in tort* for injuries caused by defects in the product.[48] Although a manufacturer would be liable to the purchaser under the contract of sale, to extend that liability more widely via the law of torts would 'unpick the bargain' between the manufacturer and customer. This stringent application of privity of contract, now recognized as fallacious, persisted as a major brake on the expansion of the law of torts until *Donoghue* v. *Stevenson*.[49] Therefore, with the other possible defendants excluded either practically (workmen) or legally (manufacturers), the sole realistic defendant in a boiler explosion case would be the owner, typically an industrialist of some sort, a steamship company, or a railway.

Potential plaintiffs may be divided into two main categories: employees of the boiler user, and others. The law applicable to these two groups developed in sharply divergent directions in the nineteenth century: favourable to the general public, but strictly unfavourable to employees,

[45] C.G. Addison, *Treatise on the Law of Torts* (London: Stevens, 1860); Beven, above note 20, although the index does include 'bastard', 'bull' and 'butty-man'; Pollock (1st edn, 1887), above note 22 ('balloon – trespass by').

[46] Accidents were referrable 36 per cent to initial construction, 39 per cent to wear and deterioration, 25 per cent to negligence in operation: Appendix 3, Select Committee on Steam Boiler Explosions, 1870–71.

[47] *Reedie* v. *LNWR* (1849) 4 Ex. 244. [48] (1842) 10 M & W 109. [49] [1932] AC 562.

who effectively had no right to sue.[50] Doctrinally, the reason for the employee's inability to sue his employer rested on the 'three ugly sister' defences: common employment; assumption of risk (*volenti non fit injuria*); and contributory negligence. Any one of these acted as a complete defence to an action in tort. The doctrine of common employment was probably the salient reason for the employee's legal disability. Where a servant was injured at work by another servant of the same master, the principle of vicarious liability did not apply. A servant was taken to assume the risk of being injured by others employed in common with him. Traced back to *Priestley* v. *Fowler*[51] (amazingly, the first recorded case in which a servant had sued his master in England, for accidental injury) the doctrine assumed real importance following *Wigmore* v. *Jay*.[52] In that case, it was held that labourers were in common employment with their foremen/managers. Given the division of labour within large companies such as railways, and the fact that a corporation can only act through its servants and agents, this effectively precluded all actions by a workman against his corporate employer.[53] The affirmation of the 'assumption of ordinary risks' (*volenti*) principle by the House of Lords was merely the icing on a rather unpalatable cake: *Bartonshill Coal Co.* v. *Reid*.[54]

Moreover, powerful extra-legal factors militated against employees bringing suit against their employers. R.W. Kostal refers to 'social, economic and psychological barriers' to explain the absence of another action by an employee, in the eleven years following *Priestley* v. *Fowler*.[55] Instant

50 Opponents of Lord Ashley's Factories Bill in 1833 forced him to withdraw it, arguing that the common law was sufficient to deal with negligent manufacturers. Even at this early stage in the doctrinal development, such an appeal to the common law met with bitter criticism. P.W.J. Bartrip and S.B. Burman, *The Wounded Soldiers of Industry* (Oxford: Oxford University Press, 1983), p. 18.

51 (1837) 3 M & W 1; cf. A. W. B. Simpson, 'A Case of First Impression: *Priestley* v. *Fowler* (1837)', in *Leading Cases in the Common Law* (Oxford: Oxford University Press, 1995), ch. 5.

52 (1850) 5 Ex. 354.

53 Evidence to the Royal Commission on Railway Accidents (1877) stated that of the several thousand employees killed on the railways 1872–1875, railway companies had paid compensation in just four cases, in which common employment was inapplicable because the railwaymen involved worked for different companies: Bartrip and Burman, above note 50, at p. 136.

54 (1858) 3 Macq. 266 (Scotland). This was seen at the time as a test case, with the miners' union sparing no expense in their (unavailing) support for the claim: Bartrip and Burman, above note 50, at p. 110.

55 *Law and English Railway Capitalism 1825–1875* (Oxford: Clarendon Press, 1994), p. 266 n. 65

dismissal may be assumed to have been the result of such temerity.[56] Moreover, of course, the cost of bringing an action would have been prohibitive to all but the comfortably well-off. The theoretical right of suing *in forma pauperis* seems to have been extremely restricted in practice.[57] A philanthropic patron might take up the injured labourer's cause, but he will have had to have been a veritable plutocrat – one such, Lord Ashley, estimating that the costs of an action could reach £600, then an enormous amount.[58] The 'Poor Man's Lawyers' giving charitable advice began in London only in 1890.[59] The cost factor in litigation will have deterred many in Victorian society from bringing actions, of course, and not only the employees of boiler users. Employees, however, will have formed the majority of those injured by boiler explosions. Kostal notes that railway workers were killed five times as often as passengers.[60] All of the evidence suggests that given the formidable legal and social obstacles in the path of an action, labourers killed or injured in a boiler explosion simply could not sue. It is significant that while the Steam Boiler Assurance Co. offered to insure workmen (at a fixed amount per head) as well as boilers and surrounding property, on its inception in 1859, there was little interest in this: 'scarcely worth calling a branch of our business'.[61] Policies were later written to insure liability to labourers on the Forth Bridge and Manchester Ship Canal, under the Employers' Liability Act 1880. However, only after the enactment of no-fault liability of masters to servants in the Workmen's Compensation Act 1897 did third party liability insurance really take off, for this specialist boiler insurance firm.[62]

Why, then, the dearth of cases by upper and middle class victims of boiler explosions? One answer may be that there simply weren't any. It may well be the case that the people killed in boiler explosions (only one or two, in the typical incident) were invariably the ones in close physical proximity: factory workers, and, especially, the hapless boiler attendant.[63] But given their sometimes catastrophic effect, and the presence of

[56] Bartrip and Burman, above note 50, at p. 28.

[57] *Ibid.* pp. 26–7. There were on average just five such applications per year, between 1858 and 1882.

[58] *Ibid.* p. 27. [59] *Ibid.*

[60] Kostal, above note 55, at p. 225 n. 7. In 1865 it was estimated to be fifty times more dangerous to work than to travel on the railways: *ibid.* p. 259, n. 20.

[61] W.H. Chaloner, *Vulcan: The History of One Hundred Years of Engineering and Insurance, 1859–1959* (Manchester: Vulcan General and Boiler Insurance Co., 1959).

[62] *Ibid.*

[63] In C.H. Hewitson, *Locomotive Boiler Explosions* (Newton Abbot: David and Charles, 1982) there is a useful collection of the Railway Inspectorate reports into all such explosions since 1840. The fatalities recorded are invariably to the driver and fireman,

all social classes on railways and steamships, it seems hard to believe that there were no wealthy victims of exploding boilers at all.[64] R.W. Kostal's study of railway accident litigation, 1840–1875, is in point. He notes, with some puzzlement, that although several hundred passengers, of all social classes, had been killed on the railways by the mid-1840s, only two minor cases appear in the law reports before 1847. Why should this be? Before Lord Campbell's Act (1846) introduced an action for the dependants of those killed in accidents, the rule in English law was *actio personalis moritur cum persona*. As for non-fatal injuries, Kostal finds some evidence of a willingness by companies to settle such claims out of court, and also notes the opinion of *The Times* in 1845, railing that the difficulties of proof made passenger actions pointless.[65] Another possibility, perhaps, may be that those who were wealthy enough to bring an action for personal injury in this period did not indulge in anything so vulgar as earning wages, but rather were persons of property and fortune. Of course, a non-fatal accident would in no way reduce landed income (although on death, family members would undoubtedly suffer an altered situation). However, we might still expect claims appearing for medical expenses and perhaps pain and suffering – perhaps such relatively small claims were, indeed, routinely settled.

There was, however, an explosion of actions arising out of railway accidents following Lord Campbell's Act of 1846, brought by the families of injured passengers.[66] This provoked anger from the railway industry. It is not clear, however, that the courts did much more than to apply the standard law of negligence, and give full reign to damages awards by juries under Lord Campbell's Act; on occasion, enormous sums were awarded to dependants, when a very wealthy person was killed.[67] Indeed, following an early flirtation with a presumption of fault, or *res ipsa loquitur*, the courts came to require positive proof of negligence from an accident plaintiff.[68] One final attempt to convince the Exchequer Chamber to hold railways strictly liable for latent defects was made in *Readhead* v. *Midland*

typically through scalding when the boiler collapsed into the engine's furnace, releasing steam at high pressure. As he comments (at p. 7), the accidents 'rarely killed or injured passengers'.

64 Chaloner, above note 61, states that many members of the general public were killed by boiler explosions in the mid-nineteenth century.

65 Kostal, above note 55, at p. 283.

66 Contrast labourers: in the 1860s, the Mines Inspectors stated that the Act was ineffective, and resorted to in very few cases of fatal accidents underground: Bartrip and Burman, above note 50, at p. 94.

67 E.g. *Pym* v. *Great Northern Railway* (1862) 2 B & S 760.

68 *Stokes* v. *Eastern Counties Railway* (1860) 5 QB 692.

Railway; the claim failed.[69] Liability to passengers was straightforwardly based on negligence.[70]

What is remarkable about the passenger cases, then, is the contrast with the restrictive approach to actions by employees.[71] For example, passengers were at least as likely to be aware of the risks of railway travel as the companies' uneducated workforce. Yet *volenti non fit injuria* was not raised against passengers, while it provided a blanket defence against employees. Again, this rather undermines the 'industrialization subsidy' argument.[72] It might be possible to argue that the passenger litigation only really came 'on stream' once the initial pioneering period of railway construction had passed. Yet this would be equally true for the timeframe of the developing employment jurisprudence. It seems unlikely that the courts would simultaneously protect and mulct the railway industry, if subsidy was indeed the driving force behind Victorian legal development, as M.J. Horwitz has suggested.

R.W. Kostal more convincingly argues that there was a class-based double standard operating. Employees were to be considered self-protecting individuals, voluntarily assuming risks, and bargaining with their employers for wages commensurate with the danger of their tasks. Yet this contractualism was simply cast aside when it came to passenger liability. The developments went beyond a restricted view of *volenti non fit injuria*, and a continued adherence to the fiction of '*respondeat superior*' on which vicarious liability was based.[73] In an intriguing passage (given the importance of the issue in twentieth-century English contract law), it seems that early experiments with excluding liability to passengers by the contract of carriage were quickly abandoned by the railway companies, in

[69] (1869) LR 2 QB 412. Kostal suggests that, although it was not cited, counsel 'might have been encouraged' by the recent decision in *Rylands* v. *Fletcher* (1866) LR 1 Ex. 265: above note 55, at p. 302 n. 313.

[70] It may be noted that the leading US case, in which *Rylands* v. *Fletcher* was carefully considered but expressly rejected in favour of fault liability, concerned an exploding steam boiler: *Losee* v. *Buchanan* (1873) 51 N.Y. 476.

[71] Between 1867 and 1871, railway companies paid out some £2.3 million in personal injuries compensation to members of the public (including £1.6 million to passengers), but *none* to their (very numerous) injured servants 'since they were not liable': Bartrip and Burman, above note 50, at p. 76.

[72] M.J. Horwitz, *The Transformation of American Law, 1780–1860* (Cambridge, Mass.: Harvard University Press, 1977).

[73] A very consistent devotee of freedom of contract, Bramwell B deplored the whole doctrine of vicarious liability, i.e. making people liable for the wrongs of others, in his evidence to the Workmen's Compensation Select Committee in 1877.

the face of extreme hostility from the judiciary.[74] So much for freedom of contract!

The judges had a clear public policy objective in mind: at a time of limited governmental regulation of railway safety, the threat of common law damages was the most important deterrent in encouraging railways to ensure public safety. Yet while the common law was deeply concerned with passenger/public safety, employees were left to bargain their way out of danger, or stoutly protect themselves against it, as best they could. The high casualty rate of Victorian railways and Victorian industry suggests that the answer was 'not very well'.[75]

(b) *Fatal accidents inquests*

An ancient English officer, the coroner, has a duty (then as now) to hold an inquest, with a jury, into sudden or suspicious deaths occurring in his locale. E.A. Cawthon has stressed the importance of coroner's juries in holding employers to account for fatal accidents, in the 1830s and 1840s.[76] While indictments (leading to criminal proceedings for manslaughter) were rare, increasing use was made of the deodand. An archaic legal doctrine mentioned in the Old Testament, an object which had caused a human death would be offered up as a sacrifice to God, or rather, in England, forfeit to the Crown. This had long since been commuted to a monetary payment, and for centuries had been merely perfunctory. However, from about 1830, coroner's juries began to impose substantial deodands. In some cases of steamship and railway accidents, Cawthon argues that juries were deliberately using the deodand to punish the railway, etc. for callous policy decisions.[77] Sometimes, indeed, a jury may have used the threat of a substantial deodand to ensure that an employer properly compensated an accident victim's family.[78] This indicates that the 'voluntary philanthropy' of employers in such situations was not all that it might have been. Of course, there was no *legal* obligation upon tortfeasors to compensate the dependants of those who were killed until Lord Campbell's Act of 1846.

[74] Kostal, above note 55, at p. 319.

[75] *The Report of the Royal Commission on Railway Accidents* (1877), p. 26, concluded that the common law, on which Parliament had relied for railway safety, was 'a dead letter' as far as railway servants were concerned.

[76] E.A. Cawthon, *Job Accidents and the Law in England's Early Railway Age* (Lewiston, NY: Mellen, 1997).

[77] *Ibid.* p. 136. [78] *Ibid.* pp. 137–8.

The deodand was abolished by legislation linked with that on fatal accidents. Campbell deplored the recent 'vicious' imposition of large ('anachronistic') deodands upon railway engines, etc. Deodands were doubtless an irritant and embarrassment for industry, and their abolition was the *quid pro quo* for the dependants' new right of action. Cawthon suggests that employers, indeed, gained from the reform. The coroner's jury was deprived of its most effective weapon, replaced by a purely theoretical right of suit in the family of a deceased employee.[79] Undoubtedly, employees gained little at first from Lord Campbell's Act. On the other hand, as described above, the wave of litigation from injured middle class passengers meant that swapping the deodand for Lord Campbell's Act proved a very poor financial bargain, all things considered.

Returning specifically to boiler explosions, a theme of evidence to the Select Committee of 1870–1871 was the unscientific nature of coroner's jury inquests into fatal explosions. 'Boiler explosions are never attributable to inevitable accident' writes the author of a work bringing together the results of thirty years of investigations under the Boiler Explosions Acts.[80] Yet in coroner's inquests into boiler explosions in 1875 and 1876, the jury recorded accidental death in thirty-two inquests out of forty-three. As one of the Select Committee witnesses also said, explosions are often the result of a combination of causes, and therefore any investigation is 'one of great complexity and difficulty'.[81] It is not surprising, perhaps, that lay jurors simply recorded accidental death in the majority of cases.

(c) *Parliamentary investigation and self-inspection*

Various Bills were presented to Parliament concerning boiler explosions, and committees set up to investigate these, in the course of the nineteenth century. Yet only one statute specifically concerning boilers was enacted, and a decidedly modest one at that: the Boiler Explosions Act 1881. Much of the reason for this, however, is that reforms proposed for reducing or compensating boiler accidents were overtaken by more general reforms, especially in the field of workmen's compensation. Boiler explosions were not really seen as a special category requiring special treatment by parliamentary law reformers, any more than by the courts of common law.

[79] *Ibid.* p. 153.
[80] E.J. Rimmer, *Boiler Explosions, Collapses and Mishaps* (London: Constable, 1912), p. 15.
[81] Joseph Nasmyth, Select Committee 1870–1871, above note 46, Appendix 4.

An early Select Committee was constituted to 'Prevent the Mischief of Explosion from Happening on Board Steam Boats to the Danger or Destruction of His Majesty's Subjects'. It reported in 1817, recommending, among other things, that boilers be made of copper not cast iron, that they be inspected regularly by a qualified engineer, and that they be fitted with two safety valves. While professing a reluctance to interfere with private property, the Committee recommended these as 'easy means' to protect public safety. A Bill to implement these proposals was opposed in the Commons, an MP arguing that legislation was unnecessary when parties could look after their own interests, and it was unwise to remove primary responsibility from those whose own interests, anyway, encouraged accident prevention, i.e. the owner.[82]

A further Bill in 1864 sought to compensate the families of persons killed in boiler explosions.[83] At first sight this is rather puzzling, since an action would presumably lie under Lord Campbell's 1846 Act in the 'Case of Death resulting from Boiler Explosions … against any Person causing such Death through Neglect', as the marginal note to clause 1 of the Bill puts it. Perhaps this was merely declaratory; the real meat of the Bill seems to have been clause 2, giving a right of action to the Guardians of the Poor of the relevant district, in addition to the executor or administrator of the deceased's estate. Working-class families would rarely seek probate or the appointment of an administrator, for the cost was prohibitive, and accordingly Lord Campbell's Act (which vested the right of action, although on behalf of the dependants, in the personal representative of the deceased) was in such cases something of a dead letter.[84] The Boiler Explosions Bill was withdrawn, but the same promoters shortly afterwards introduced a Bill specifically to amend Lord Campbell's Act. Dependants were empowered to bring the action themselves given inability to appoint personal representatives, or the refusal of the latter to bring a claim. This passed all of its stages in both Houses without debate, and was duly enacted. Boiler explosions might have provoked particular concern, but ultimately the reformers succeeded in enacting a wider reform on behalf of labourers' families.[85]

A further Bill in 1869 recommended compulsory government inspection of steam boilers. This was rejected on the grounds of excessive

[82] 5 May 1818, Parl. Deb., vol. 38, col. 511. [83] Bill 89, 4 May 1864.

[84] Burman and Bartrip suggest this provision was a deliberate ploy to prevent working-class use of the Act: above note 50, at p. 101.

[85] Even after this reform, the financial and social obstacles to suing an employer remained: Bartrip and Burman, above note 50, at pp. 111 and 116–17.

interference, but a House of Commons Select Committee was set up to inquire into boiler explosions. Ultimately, the Committee in its Report of 1871 did not recommend compulsory inspection, arguing that inspection would not prevent all explosions (especially those caused by negligent tending/operation). There was also concern about the scale of the undertaking (there were upwards of 100,000 boilers in the country), given the relatively small number of explosions (around 50 annually). Various state inspectorates had existed for some decades by this time,[86] so we may not assume remorseless hostility to regulation by Victorian legislators, on simplistic *laissez-faire* grounds. Indeed, boilers on board steamships,[87] and railway locomotives,[88] were already subject to inspection by the relevant government agency, although the extension of the factory inspectorate's regime to the (more numerous) stationary boilers had to wait until 1901.[89]

Moreover, in 1854 the Manchester Steam Users' Association (MSUA) had been set up by leading Lancashire mill-owners, to help prevent boiler explosions. The basic means for fostering safety was through periodic inspection of boilers, using inspectors employed by the Association. The MSUA was hostile to providing insurance for its members against explosions, arguing that this would increase carelessness (the 'moral hazard' problem).[90] Hence the formation of a rival organisation in 1859, the Steam Boiler Assurance Co.,[91] which provided insurance in addition to inspection. The Company claimed that it had a much better safety record than the rival MSUA, which offered inspection without insurance. At any rate, by the time of the Select Committee's deliberation, there was a well-established system of industry self-regulation (or self-inspection).

[86] Cf. Factories Act 1833 (factory inspectors).

[87] Steam Navigation Act 1851; Merchant Shipping Acts 1865 and 1862.

[88] Railway Regulation Act 1840.

[89] Factory and Workshop Act 1901. In the 1860s, the Board of Trade had argued that steam boiler explosions were better suited to the proscriptive, safety-focused (factories and mines) inspectors of the Home Office, rather than the Board's own (railway) inspectors. The reason why railways were entrusted to the Board of Trade was their 'political, commercial and financial importance', going beyond mere accident prevention. Bartrip and Burman, above note 50, at p. 78.

[90] The Annual Reports of the Steam Boiler Assurance Co.'s inspectors (from 1860) give many 'hair-raising examples of negligence' on the part of mill-owners and boiler operatives. In 1865, for example, 615 safety valves were found to be out of order or overloaded: see Chaloner, above note 61.

[91] From 1896, the Vulcan Boiler and General Assurance Co. Ltd; see Chaloner, above note 61.

Thus, while employers had been bitterly opposed to the 1869 Bill mentioned above, proposing government inspection of boilers, their actions had shown a preference for self-regulation, as opposed to a complete absence of regulation. Even if part of the motivation for setting up the MSUA had been to forestall 'any pretence for government inspection',[92] the Association was subsequently prepared to work in partnership with the government. Its founder, Sir William Fairbairn, wrote to the Home Office offering that the MSUA inspect all the country's boilers if the government would pay for this, which he argued would improve safety without the need for legislation or enlargement of bureaucratic machinery (and while incidentally increasing MSUA's trade). The offer met with official indifference,[93] as did a suggestion from the National Boiler Insurance Co. that the Board of Trade should advise all businesses in a circular that if they did not have their boilers inspected, they would be held liable for the resulting damage in the event of an explosion.[94] At any rate, the factory inspectors charged with inspecting boilers after 1901 were pleased to find that so many were insured (estimated at 80–95 per cent), because this meant more regular inspections than the law required.[95]

The Boiler Explosions Select Committee of 1870–1871 focused on improving the investigations into the causes of explosions, and upon the liability of owners. The first was to be achieved by requiring coroners to hold inquests into all boiler explosions, whether fatal or not, and for the Board of Trade to direct one of their surveyors 'or some other practical person' to assist the coroner in such an investigation. This aimed to address the drawbacks of lay investigation (especially assignment of cause), noted above. As to liability, the committee recommended 'that it be distinctly laid down by statute that the steam user is responsible for the efficiency of his boilers and machinery, and for employing competent men to work with them'. This is a little difficult to understand, since it is surely repeating or declaring the common law, and it is not obvious what the statutory declaration was intended to change.[96] The

[92] P.W.J. Bartrip, 'The State and the Steam-Boiler in 19th century Britain' (1980) 25 *International Review of Social History* 77, quoting Fairbairn.

[93] *Ibid.*

[94] *Ibid.* The soundness of such advice would have been questionable, given the absence of actions against boiler owners in practice, discussed above.

[95] Chaloner, above note 61. Bartrip, however, estimates that only about 20 per cent of boilers were inspected around the time of the Select Committee's inquiry: above note 92.

[96] Mr Cawley's scoping resolutions proposed to the select committee stated 'it is desirable to define more clearly the responsibilities of boiler owners, and to provide more efficient remedies against those who are negligent in regard to boilers and their use'.

other recommendations of the committee were substantial, however. First, the fact of an explosion should be taken as prima facie evidence of negligence, with the onus upon the user to rebut. Secondly, and more importantly still, 'it shall be no defence in an action by a servant against such owner or user, being his master, that the damage arose from the negligence of his fellow servant'. Such a frontal assault on the doctrine of common employment would undoubtedly have had a major potential impact upon the liability of boiler users.

A Bill followed in 1871, closely modelled on the Select Committee's recommendations.[97] A similar Bill was introduced in the subsequent session,[98] although with the omission of the former clause 2 in the 1871 Bill, headed 'Responsibility of a steam-user'. The deletion of this provision, copied as it had been from the apparently declaratory recommendation of the Select Committee, might suggest a realization by the promoters of the reform that this clause was legally otiose. As suggested above, such an action could anyway be brought at common law, or under Lord Campbell's Act. However, the 1872 Bill, despite its important substance (or perhaps because of it) also failed to progress. It took until the Employers' Liability Act 1880 for Parliament to remove the defence of common employment and so allow an employee to sue as if he had not been in the defendant's service, in a large class of cases.[99] Thus the reform, while delayed, was much wider in scope than that proposed ten years earlier when it finally came.[100] Again, while boiler explosions clearly concerned some Victorian legislators, they would seem to have served as a focus, or seed crystal, for much wider concerns about the legal treatment of industrial accidents.

More successful was the Bill introduced by the President of the MSUA, which became the Boiler Explosions Act 1881. As Joseph Chamberlain approvingly remarked for HM Government, this was not a radical measure, but simply required an investigation by a lawyer and two engineers into every boiler explosion, the idea being that knowledge of the cause of accidents was half the means of their prevention.[101] Some doubted that

[97] 34 & 35 Vict. Bill No. 273. [98] 35 Vict. Bill No. 80.

[99] See generally, Bartrip and Burman, above note 50, chs. 5 and 6.

[100] Even then, it failed to improve safety in the way that campaigners had expected. Bartrip and Burman, above note 50, at p. 188, point out that the other barriers to suit remained (fault, causation, above all expense), and suggest that the deeper reason was a continued unwillingness for a workman to 'declare war' against his employer – moreover, by alleging the negligence of his own fellow workers.

[101] Parl. Deb. (3rd Series), vol. 266, cols. 1348–55.

such a modest measure would have any impact at all, and argued again in favour of compulsory boiler inspection. Further attempts to extend state inspection in the 1880s failed, on opposition from the independent inspection organisations. In 1890, the Boiler Explosions Act was amended to encompass domestic boilers, mines and steamships (with the exception of the Royal Navy).

E.J. Rimmer's 1912 synthesis of the Boiler Explosions Act commissioners' reports is a useful source of information.[102] The commissioners' sole function was to investigate explosions, and to make recommendations as to avoiding these in the future. Thus, they laid down that it was negligent not to inspect a boiler at regular intervals, and grossly negligent not to inspect a second-hand boiler on purchase. It seems that despite the rejection by Parliament of compulsory inspection, it had become de facto obligatory in order to avoid liability, or at least censure by the Boiler Explosions Act commissioners. More importantly still, boiler insurers regularly inspected the boilers which were being insured.[103] The effectiveness of this supervision is shown by the fact that only 8 per cent of the boiler explosions investigated under the Act concerned boilers which had been insured.[104] The Commissioners warned, however, that a user might 'shelter behind' its insurer's certificate of safety only when he had complied in full with the insurer's advice and recommendations (as well as any mandatory requirements).

The Boiler Explosions Act Commissioners enjoyed no real remedial powers, save of course publicity. It may be assumed, also, that their reports enjoyed formidable evidential weight, in any litigation arising out of an investigated explosion. Finally, they had the power to order that a party found at fault should pay the costs of the investigation, such orders ranging from £2 against an attendant to £400 against a manufacturing firm. In awarding costs, the commissioners were guided by moral blame and not formal legal liability, i.e. strict, vicarious liability was not sufficient for a finding of liability under the Act.[105] In fixing the amount of costs, the criteria were the degree of fault and the means at the disposal of the person/company fined, rather than the extent of damage caused by the explosion.[106] This, of course, was an overtly punitive approach, and

[102] *Boiler Explosions, Collapses and Mishaps* (London: Constable, 1912).

[103] H.E. Raynes, *A History of British Insurance* (2nd edn, London: Pitman, 1964), p. 286, states that this inspection and advisory service was more important than the insurance itself, in boiler insurance and the general category of engineering insurance which grew out of it.

[104] Rimmer, above note 102, Preface.

[105] Report No. 1624.

[106] Rimmer, above note 102, at pp. 25–6.

therefore much better suited as a deterrent than damages at common law, based upon the victim's loss. Firms generally found it cheaper to insure their boilers (with the associated inspection service) than to pay the costs of a statutory investigation.[107] Finally, the commissioners might use their costs jurisdiction to encourage munificence from employers: they might greatly mitigate a costs 'fine' when a negligent boiler owner had made provision for a dead worker's family.[108]

Thus, despite the apparent lack of litigation over exploding steam boilers, such accidents were considered and debated in a variety of legal and quasi-legal fora in Victorian and Edwardian England. Ultimately, however, boiler explosions were treated as part of the general problem of accidents in the industrialized society, especially the compensation of workers which gained momentum at the end of the nineteenth century. With the non-thrilling exception of the Boiler Explosions Act 1881, such accidents never qualified for special treatment in English law.

3. Asbestos

Magic Mineral to Killer Dust, the title of an important study of the world's largest asbestos company,[109] sums up well the total transformation of the reputation of asbestos in the course of the twentieth century. In 2004, some 1,969 persons in Great Britain died from mesothelioma, the most lethal asbestos-related illness.[110] The annual total number of mesothelioma deaths is estimated to peak at around 1,950 to 2,450 deaths, some time between 2011 and 2015.[111] The reason for the continuing increase in mesothelioma deaths (surprising, perhaps, in the light of strict controls on asbestos in recent decades) is that the peak for the importation of raw asbestos into the United Kingdom occurred in the early 1970s. The

[107] W.A. Dinsdale, *History of Accident Insurance in Great Britain* (London: Stone & Cox, 1954), p. 140.

[108] Cf. Report No. 431.

[109] G. Tweedale, *Magic Mineral to Killer Dust: Turner & Newall and the Asbestos Hazard* (Oxford: Oxford University Press, 2000).

[110] The prognosis for those diagnosed with mesothelioma is usually a painful death within twelve months. In November 2006, a clinical trial was conducted by Cancer Research UK into the efficacy of radical surgery (removal of the affected lung, pleura and diaphragm) in controlling mesothelioma. A preliminary determination by the National Institute for Clinical Excellence that the only known mesothelioma-specific drug (pemetrexed disodium) was 'not cost effective' for use by the National Health Service caused political and public dissatisfaction.

[111] *Mesothelioma Mortality in Great Britain: Estimating the Future Burden* (London: Health and Safety Executive, December 2003).

grim mortality projections reflect the long latency in the development of mesothelioma, often thirty to forty years, from that peak.[112]

Much of the history of the law's response to asbestos disease in England requires examination of public regulation. Health and safety legislation is a prominent feature of this legal strategy, from pioneering regulations in 1931,[113] to the final prohibition on the importation, use or supply of asbestos in 1999.[114] The earlier legislation has been much debated, given its failure to avert the crisis in public health which is now so visibly with us. This debate encompasses the genesis and terms of the legislation, the speed with which it was revised, updated and extended, and the extent to which it was enforced/complied with. A useful framework for considering these issues is the three 'waves' of asbestos victims: first, workers in the asbestos industry itself; secondly, those injured by the use of asbestos in other industries, such as construction and ship-building; in the third wave, victims outside the workplace, suffering through environmental asbestos contamination. Lack of information has often rendered it difficult to research the impact of such legislation upon the industry regulated, but with asbestos, a full record has been revealed following a US law suit against Turner and Newall Ltd, a multinational group of asbestos companies based in Rochdale, England. Geoffrey Tweedale's book uses the corporate records of this major force in the asbestos industry thus disclosed to great effect. His conclusions are unfavourable in the extreme. To be set against this, however, is the much more favourable thesis of Peter Bartrip, who was given direct access to the corporate archive by Turner and Newall itself.[115]

In a curious parallel with the nineteenth-century findings as to boiler explosions, actions for damages in tort seem barely to figure in the story, until a very late stage. There are no reported decisions of the higher courts before the 1980s, and it is only really since the 1990s that the consequences of asbestos for the law of torts have been considered in the appellate courts. While there were undoubtedly legal obstacles in the path of such claims,

[112] J. Peto *et al.*, 'Continuing Increase in Mesothelioma Mortality in Britain' (1995) 345 *The Lancet* 535.

[113] Asbestos Industry Regulations 1931 (SI 1931/1140).

[114] Asbestos (Prohibitions) (Amendment) Regulations 1999 (SI 1999/2373).

[115] P.W.J. Bartrip, *The Way from Dusty Death: Turner and Newall and the Regulation of Occupational Health in the British Asbestos Industry, 1890s–1970* (London: Athlone Press, 2001). The author records that Turner and Newall also funded his research, although states that the company sought neither to influence his conclusions, nor to vet the text of his book: *ibid.* p. vii.

the explanation for these 'missing claims' again seems overwhelmingly sociological. These issues are considered below.

A major difference between the nineteenth century and the current period was, of course, the development of the welfare state in the course of the twentieth century. The Workmen's Compensation Act 1897 might be seen as the link between the two. The details of this highly complex body of law will not be considered here.[116] Suffice to say that asbestosis had been designated an industrial disease, compensable by employers without proof of fault under the Workmen's Compensation Act, in 1931.[117] The most significant general change in this area occurred in 1948, when the state took over the compensation of industrial injuries through National Insurance, and the possibility of parallel fault-based tort actions against employers revived. It is notorious that state benefits do not provide the full compensation, *restitutio in integrum*, at which tort law theoretically aims.[118] Thus, it is unlikely that the presence of social security in its limited, British form would fully remove the economic incentive to sue – as the recent growth in asbestos tort litigation demonstrates.

Nevertheless, legal liability in England has yet to have the economic impact of the 'asbestos wars' in the United States, where by 2002 some US$70 billion had been paid out by the asbestos industry and its insurers,[119] with future liabilities estimated to lie between US$130 and US$195 billion.[120] The economic dislocation has included some 60,000 job losses, as asbestos-defendant firms have become insolvent; the prognosis for the future in the United States looks gloomy, as insuring against liability becomes ever more difficult.[121] It can be noted that the Federal-Mogul Corporation, which had taken over Turner and Newall in 1997, placed itself in voluntary liquidation under Chapter 11 of the US Bankruptcy Code, in order to restructure its liabilities, in 2001.[122]

[116] See N.J. Wikeley, *Compensation for Industrial Disease* (Aldershot: Dartmouth, 1993).

[117] The Asbestos Industry (Asbestosis) Scheme 1931.

[118] Tweedale, above note 109, at p. 74, comments that the pre-Second World War compensation scheme in Britain was uniquely limited, and was fully intended to make the worker share the financial burden of his injury.

[119] The US$10 billion insurance/reinsurance liability of Lloyd's of London created a financial crisis for that insurance market, causing much satellite litigation, including allegations of negligence by syndicate managers.

[120] See D. de Saulles, 'The Asbestos Wars: The American Experience' [2006] *JPI Law* 215.

[121] *Ibid.*

[122] In August 2005 it was estimated that some 300 former Turner and Newall employees had died without compensation since the beginning of the Chapter 11 process, which froze all compensation pay-outs: www.MesoLink.org. See also *Re T&N Ltd* [2005] EWHC 2870 (Ch), [2006] BPIR 532.

(a) *Legal regulation of asbestos*

The asbestos industry became established in Britain in the early decades of the twentieth century. In many other heavy industries, such as coal-mining and metal-grinding, the danger to the lungs of high concentrations of dust in the atmosphere was well known. Concern about asbestos manufacture, specifically, came to a head with the seminal, government-commissioned report of Merewether and Price in 1930.[123] This study of the asbestos manufacturing workforce provided scientific confirmation of the link between asbestos inhalation and lung fibrosis, designated 'asbestosis'. The Home Office was quick to respond, issuing the Asbestos Industry Regulations 1931, the first legislation controlling asbestos in the world.[124] In essence, the Regulations obliged manufacturers to provide exhaust ventilation in areas of factories concerned with the various manufacturing processes enumerated in the legislation. Despite their world-beating timing, these Regulations have proved highly controversial.

Wikeley has argued that they provided a veritable 'licence to kill'.[125] The dangers of asbestos inhalation had been pointed out in several reports of the public factory inspectors, and elsewhere, as early as 1898, so Merewether and Price's report was not the first opportunity at which asbestosis could have been tackled. This criticism is true so far as it goes, but it requires the perfect vision that only hindsight can bring to accuse the government of culpable slowness, when the earlier comments were widely scattered and gave little prominence to the dangers of asbestos, compared to those in many other industries.[126] The government, as stated, commissioned the Merewether and Price report in 1928, and acted on it very promptly. Another criticism is that the government was influenced by the manufacturers/employers, while failing to consult seriously with trade unions. Again, however, it must be pointed out that consultation with employers was legally required under the empowering statute, and that the Trades Union Congress was certainly involved in the formulation of the 1931 Regulations.[127] Sir Thomas Legge, the TUC's medical advisor (and a former Chief Medical Inspector), praised the proposals. If the union response was weak over all, this might reflect the priority given to resolving the political and economic crises of 1930 (over worker safety).

[123] E.R.A. Merewether and C.W. Price, *Occurrence of Pulmonary Fibrosis and Other Pulmonary Affections in Asbestos Workers* (London: HMSO, 1930).

[124] Tweedale, above note 109, at p. 21.

[125] N.J. Wikeley, 'The Asbestos Regulations 1931: A Licence to Kill?' (1992) *J L and S* 365.

[126] See generally Bartrip, above note 115, ch. 1. [127] *Ibid.*

A more convincing line of attack questions the seminal research of Merewether and Price, on which the Regulations were based. Inevitably, the methods of measuring dust concentration were crude at this point.[128] More fundamentally, however, the authors limited their study to the asbestos industry. Today, it is apparent that much the biggest class of asbestos victims has been those in the 'second wave', workers outside the asbestos manufacturing industry itself and the protection of the Regulations, but working with asbestos products (insulation, cement, brake linings, etc.). Ironically, Turner and Newall invented 'Limpet', the method of spraying asbestos insulation, in 1932. This was undoubtedly a major cause of 'second wave' exposure and deaths. With the admitted benefit of hindsight, eminent cancer epidemiologist Prof. Julian Peto lamented in 1998 that such an exclusive focus on the asbestos factories had been a 'stupid mistake'.[129] Its legal consequences were to be long lasting.

Another crucial feature of the Merewether and Price research was their finding that those spinning asbestos textiles suffered from comparatively much lower levels of asbestosis than workers in other areas of the factories. The level of dust in such processes came to be adopted in the Regulations as a safety standard, and they were not included in the Scheduled areas and processes where the Regulations applied. Tweedale points out that this was 'an entirely rule of thumb measure', criticized by the TUC at the time.[130] Moreover, the Chief Inspector of Factories warned that the 'dust datum' was 'clearly provisional and is subject to alteration in the light of further medical experience'.[131] In line with this warning, the whole basis required re-examination once the link between asbestos and cancer was made in the 1950s; in particular, once the connexion with mesothelioma was published in 1960.[132] However, prior to the links being discovered between transient exposure and malignancy, the consensus of opinion between government, industry and the medical profession was that ventilation in the asbestos manufacturing industry would control if not eliminate the disease of asbestosis. The success of the 1931 Regulations was

[128] Tweedale points out that even much later, research proceeded on the basis that long asbestos fibres were the greater threat, more for the convenience of counting, than on the basis of any evidence as to the relative threat. Optical counting continued to be favoured despite the development of electron microscopy, the only way to measure the smallest asbestos particles (which are readily aspirated into the lungs). Cf. Bartrip, above note 115, at pp. 188–9 on factory inspectors' unavailing attempts to review the 'dust datum'.

[129] Tweedale, above note 109, at p. 284. [130] *Ibid.* 26.

[131] 'Foreword', *Report on Conferences between Employers and Inspectors concerning Methods for Suppressing Dust in Textile Factories* (Home Office, 1931).

[132] Wagner *et al.* (1960) 17 *Brit. J Ind. Med.* 260.

that asbestosis indeed declined, relative to the growing size of the asbestos industry.[133] The limitation to Scheduled areas within asbestos factories was to persist until 1969.

It has also been argued that the enforcement of the 1931 Regulations by the factory inspectorate was ineffectual. There are records of only two prosecutions for their breach in the period between 1964 and 1970, and apparently none earlier.[134] Bartrip points out that the factory inspectors' reports from the 1930s–1960s shows broad satisfaction with industry practice, and certainly little evidence of wilful evasion.[135] Tweedale comments that Turner and Newall's internal records show a poor standard of cleanliness, ventilation and safety: for example, permeable hessian sacks were still being used to carry raw asbestos in the 1960s, on cost grounds, even though their replacement with impermeable material was required by the 1931 Regulations.[136] A report by the Parliamentary Ombudsman in 1976 was very critical of the factory inspectorate,[137] highlighting multiple failures to enforce the Regulations over many decades, concluding that their attention to the problem was 'not as high as it could and should have been'.[138] Tweedale comments that the inspectorate's chronic understaffing, and the modest sanctions forthcoming after long, expensive court enforcement battles, explains (if not excuses) some of these failings.[139] Moreover, the preference for officials to work cooperatively towards a target, in partnership with the industry subject to Regulation, as opposed to viewing legislation as imposing absolute targets to be rigidly enforced, has been well documented by socio-legal scholars.[140]

But many have argued that the asbestos industry, in the present case, was far from cooperative. Tweedale declares that Turner and Newall's safety record demonstrates that it was 'strikingly irresponsible' about the asbestos hazard.[141] Ultimately, the bottom line here was economic.

133 Bartrip, above note 115, at pp. 49, 190, 270.

134 Wikeley, above note 125.

135 Bartrip, above note 115, at pp. 37–9.

136 Tweedale, above note 109, at p. 125. Cf. Bartrip, above note 115, who makes the very sensible point that for reasons of self-interest, a respectable 'complying' firm like Turner and Newall would wish to see the law stringently enforced, so as to impose costs upon small competitors who might otherwise undercut them by ignoring the Regulations: *ibid.* pp. 103–4.

137 Although *The Lancet* thought the criticisms 'comparatively mild': Bartrip, above note 115, at p. 275.

138 Tweedale, above note 109, at p. 240 n. 136

139 Tweedale, above note 109, at p. 163.

140 See, e.g., K.O. Hawkins, *Environment and Enforcement: Regulation and the Social Definition of Pollution* (New York: Oxford University Press, 1983).

141 Tweedale, above note 109, at p. 279.

To the extent that the need to recruit and retain workers imposed costs, Turner and Newall found it cheaper to pay higher wages and to compensate the incapacitated (to a limited extent) than to spend massively on ventilation and other safety improvements.[142] Turner and Newall invested little in developing alternative technologies (mineral wool, fibreglass, etc.), since asbestos was more profitable.[143] Little pressure, then, was placed on asbestos firms to comply with the legislation, whether by public agencies or, crucially, by the workforce itself. Right up until the 1980s, trade unions displayed an ambivalent attitude towards asbestos regulation, correctly recognising the threat it posed to the livelihoods of their members.[144]

As stated above, the warning of the Chief Inspector of Factories in 1930 was not to be heeded. While one of his successors warned of the ever-increasing number of respiratory diseases occurring in ship-building workers as early as 1945, it took fully fifteen years for the government finally to respond.[145] This, surely, was culpable delay.[146] It did not take a great leap of the scientific imagination to foresee the dangers posed in spraying asbestos insulation, or removing dusty asbestos lagging by hand, in the unventilated confined spaces on board ship.[147] The robust fight put up by the asbestos manufacturers against any suggestion that the reach of legislation be extended was a major contributing factor in this long delay.[148]

In the 1950s, concern began to mount over the connexion between asbestos and lung cancer. Ground-breaking research was published by Doll in 1955.[149] This was based on a study of workers in Turner and Newall

[142] Between 1931 and 1948, the firm spent some £73,166 on compensation, just 0.5 per cent of its gross profits of £15 million, over the same period.

[143] Tweedale, above note 109, at p. 140.

[144] Tweedale, above note 109, at pp. 171, 249.

[145] Shipbuilding and Ship-repairing Regulations 1960 (SI 1960/1932).

[146] It is notable that the Admiralty issued very detailed Fleet Orders to safeguard naval personnel at risk from asbestos spraying in 1952: Bartrip, above note 115, at pp. 134–5. He also points out that it is odd to credit the asbestos industry for the delay of a draft revision in 1950, when only 2 out of 75 draft regulations mentioned asbestos.

[147] Even in 1930, the TUC had argued that the proposed Regulations should apply to shipbuilding, where cases of asbestosis were recorded. But the Home Office replied that ventilation in that industry was impossible.

[148] Tweedale, above note 109, at pp. 49–52, 132–7. Bartrip argues that Turner and Newall made a genuine case that those 'in the vicinity' of spraying did not need protection, and were open and honest, with the notable exception of some 'wobbly' figures on asbestosis cases in their sprayers. Bartrip, above note 115, at pp. 143–64.

[149] 'Mortality from Lung Cancer in Asbestos Workers' (1955) 12 *Brit. J Ind. Med.* 81. Doll, later Prof. Sir Richard Doll, was the leading cancer epidemiologist in the world.

asbestos factories; it is telling that the company, alarmed by Doll's findings, had attempted to prevent the publication of the results![150] The storm of publicity which the company had feared over Doll's research never really materialized. But it was different once the connexion was made between asbestos and mesothelioma, a virulent malignant tumour of the pleura or peritoneum, and inevitably fatal. The seminal paper became the most-cited article in occupational medicine.[151] Concern grew as it became obvious that low doses and exposure to asbestos other than in its manufacture could trigger mesothelioma: the paper of Newhouse and Thompson in 1965 identified cases of transient exposure (washing clothes of asbestos workers) and environmental exposure (living near to asbestos factories).[152] The third wave of victims had arrived, and media interest became intense. The cynic might discern self-interest here. The general public was glad to enjoy the benefits of asbestos fire-insulation, and could remain indifferent to the suffering of those involved in its manufacture; but a threat from a material ubiquitous in trains and ships, in shops, offices, schools, hospitals and homes was a different matter entirely.[153]

However, when new legislation was finally forthcoming,[154] not the least of its flaws was a failure to consider the risk of cancer. The research which the government adopted was confined to the risk of asbestosis – ironic, when the government's motivation for action had been the dreaded mesothelioma. The Asbestos Regulations 1969 included a general duty to ensure that asbestos concentration did not exceed two fibres per millilitre of air (except blue asbestos, identified as the most dangerous, subject to a ten times stricter limit). Tweedale states that given the presence of particles invisible under an optical microscope, a worker could inhale as many as a billion asbestos particles in a working day, even at these concentrations. Yet even then, Turner and Newall privately thought that the new limits would be impossible to reach, and that customers and overseas

[150] Although cf. Bartrip, above note 115, at pp. 110–17, who argues that after the initial 'panic' of the T&N Company Secretary, who had sought leading counsel's opinion on legal action to prevent publication, the Board of Directors as a whole took a wiser approach. Doll continued to collaborate with Turner and Newall after the episode.

[151] Wagner *et al.* in (1960) 17 *Brit. J Ind. Med.* 260. Tweedale, above note 109, at p. 154.

[152] M.L. Newhouse and H. Thompson, 'Mesothelioma of Pleura and Peritoneum following Exposure to Asbestos in the London Area' (1965) 22 *Brit. J Ind. Med.* 261.

[153] Tweedale, above note 109, at p. 290. The government has confirmed that asbestos remains 'widespread' in British buildings, and its complete removal is projected to take another fifty years. Parl. Deb., 13 February 2006, col. 1659W.

[154] Bartrip argues that the time taken, once the mesothelioma hazard became known, was not in fact much longer than the response to asbestosis in the 1920s.

competitors would think the legislation 'wildly utopian and even catastrophic'.[155] Furthermore, the guidance provided to factory inspectors was:

- for a reading of 2–12 fibres per millilitre, 'check reading again';
- for a reading over 12 fibres, 'use discretion as to enforcement'![156]

The pattern of industry foot-dragging, combined with a hands-off approach to enforcement of the law, continued.

The catalyst for the final proscription of asbestos seems to have been media pressure.[157] Campaigning journalists produced damning indictments of the asbestos hazard. This, with the formation of a unified Health and Safety Executive in 1974, has been credited with an ultimately more vigorous approach towards enforcement of the 1969 Regulations.[158] Turner and Newall was the first firm to be prosecuted, and indeed suggested that the inspectors had been driven to take court proceedings by the media.[159] A 1979 Report recommending lower limits for asbestos fibres initially had negligible impact in government.[160] It is surely not coincidental that the recommendations were implemented within days of a particularly harrowing documentary featuring the agonizing death of a mesothelioma victim, in 1982.[161] Still tighter restrictions followed in 1984, together with licensing of asbestos removal contractors.[162] This licensing regime forms one plank of the current stringent approach to asbestos, together with the complete ban on import and supply in the Asbestos (Prohibitions) (Amendment) Regulations 1999, and the various obligations on employers in the Control of Asbestos at Work Regulations 2002.

If we take the research on lung cancer in the mid-1950s as the key scientific evidence, or at latest the mesothelioma paper of 1960, then it took between forty and fifty years for the British government fully to accept the complete ban on asbestos which this new evidence logically required. While imports had been falling sharply since the 1970s, the delay is

155 Tweedale, above note 109, at pp. 206–8.

156 'Note B' to the 1969 Regulations, quoted and criticised in Editorial [1997] *JPIL* 3. See also Tweedale, above note 109, at p. 213.

157 Bartrip observes that if the destruction of the asbestos industry came through litigation in the United States, it came about in Britain through public perceptions fed by media sensation: Bartrip, above note 115, at p. 272.

158 Tweedale, above note 109, at p. 239.

159 *Ibid.* p. 212.

160 *Final Report of the Advisory Committee on Asbestos* (HSE). Turner and Newall convinced the government to wait for the EEC Directive on asbestos, expected in 1985, so as not to lose business to foreign competitors: Tweedale, above note 109, at p. 248.

161 *Ibid.* pp. 251–5.

162 Asbestos (Licensing) Regulations 1983 (SI 1983/1649).

important, not least because much of the asbestos historically imported into the country still exists in buildings throughout the United Kingdom. In that form it is inert, but demolition or maintenance work is likely to create contamination hazards, especially where work is carried out by small contractors (or indeed by amateur householders) unaware of the risks. It is arguable that, whether in failing to legislate or failing stringently to enforce the legislation enacted, the British government has been guilty of delay which has cost thousands of lives. It goes without saying, for an English lawyer, that liability of the Crown for such inaction is simply out of the question, as a matter of basic constitutional principle,[163] and indeed for elementary reasons of tort law.[164]

The question remains as to the cause of the delay. The state of scientific knowledge provides part of the solution, of course, but this cannot fully explain the failure to act in the face of evidence of the mesothelioma risk, in the third wave of victims. Tweedale argues that in the final analysis, the key actors in the asbestos tragedy all decided that there was an 'acceptable level of death' to be tolerated, in order to preserve the benefits of asbestos.[165] The extent of these benefits to society as a whole may be disputed: asbestos has undoubtedly saved lives and property as a fire retardant, but one would not only have to calculate precisely how many/how much, but also the effectiveness (and financial cost) with which alternative materials could have done so. For the asbestos companies themselves, the continued profitability of asbestos, until the crisis of the 1970s, derived from a straightforward financial cost-benefit analysis. Given the very low levels of compensation paid out throughout this period, however, one could plausibly argue that the industry was externalizing much of the cost of asbestos diseases onto the victims, and society more broadly. The belated response of tort law will be considered below. More worryingly is the failure of the government inspectors charged with enforcing the restrictions which profitability did not induce the companies to enforce upon themselves. As well as the shortcomings of the inspectorate considered above, government policy-makers were sensitive to the wider political dictates of socio-economic policy. The threat to employment in a major British industry,

[163] Although cf. for the different approach now required when the United Kingdom is in 'sufficiently serious' breach of its obligations under EU law, *R* v. *Secretary of State for Transport, ex parte Factortame Ltd (No. 5)* [2000] 1 AC 524.

[164] No general liability for omissions in English tort law: cf. *Stovin* v. *Wise* [1996] AC 923 (no duty to exercise statutory powers).

[165] Tweedale, above note 109, ch. 11.

particularly the loss of jobs to foreign competitors, was obviously one such consideration, but there may have been others.[166]

The pressure from organized labour which might have forced the government to take earlier, more decisive action on asbestos was equally lacking. The attitudes of trade unions and of workers themselves were distinctly equivocal, faced with the prospect of crippling the asbestos industry with the consequent loss of jobs. Asbestos workers in Rochdale were (initially at least) as hostile to the media assault on the asbestos industry as the Turner and Newall directors themselves.[167] Tweedale even suggests there was an 'implicit bargain' between the workforce and employers, the former accepting higher wages as the price of continuing to work in conditions which did not meet the standards laid down in legislation. Perhaps most surprisingly of all, the medical and scientific establishment performed a similar trade-off, more or less overtly. *The Lancet* argued in 1967 that it would be:

> Ludicrous to outlaw this valuable and often irreplaceable material in all circumstances [as] asbestos can save more lives than it can possibly endanger ... Workers in general are willing to take a calculated risk for the benefit of the community if they feel they have it frankly explained to them.[168]

Bartrip draws attention to the praise for the 'essential' insulating properties of asbestos, found widely in the *British Medical Journal*, *The Times*, and factory inspectors' reports. Given its widespread use by the Royal Navy, it even had a national security dimension (and asbestos manufacture had been a controlled industry during the Second World War). Outright proscription of such an important raw material was unthinkable in the 1960s, when the full implications of third-wave mesothelioma were yet to be realised.[169] Even in 1982, Doll (with Julian Peto) commented that the risks of working in the asbestos industry were comparable with those in other dangerous (but valuable) trades, like construction and deep-sea fishing.[170]

[166] It has been suggested that the delay in Mr Blair's implementation of his 1997 manifesto commitment for a total asbestos ban was influenced by pressure from the Canadian government (Quebec remains a major exporter of asbestos): *ibid.* p. 277 n. 9. The government failed to publish alarming reports about asbestos and cancer, produced for the Department of Health by Sir Richard Doll in 1968 and 1971: *ibid.* p. 208.

[167] Tweedale, above note 109, at p. 250.

[168] *The Lancet*, 17 June 1967, p. 1311. *The Lancet* later criticized asbestos campaigners for failing to weigh the benefits of asbestos against the risks created by it: Bartrip, above note 115, at pp. 276–7.

[169] Bartrip, above note 115, at pp. 205–6, 267–70.

[170] *Ibid.* p. 281.

Naturally, there are costs to be considered for every regulatory burden, and it is irresponsible for government (and other stakeholders) not to face up to these. Some have openly questioned whether the regulation of asbestos has not now gone too far, particularly as regards the high cost of engaging specialist, licensed contracts to removed textured paint such as 'Artex' which formerly contained small amounts of asbestos.[171] A support group has been set up to help businesses and members of the public to query the 'excessive' amounts demanded by such contractors.[172] Nevertheless, proposals to retrench the asbestos removal and licensing regime even by this modest amount have led to furious criticism in Parliament.[173]

More generally, there is ongoing media and parliamentary criticism about the excessively cautious approach to public life fostered by the Health and Safety Executive, or at least by overzealous (or overly timorous?) interpretations of its guidance and regulations.[174] As with the controversy over the 'compensation culture' more generally, it is difficult to decide who has the better side of the argument here, for undoubtedly there are (hard to evaluate) costs and benefits, and anecdote often prevails over evidence.[175] Not everyone will agree with the maxim, Safety First.[176] But even such people could not doubt that today's very strict controls on asbestos are fully justified, and should have been brought in much earlier than at last they were.[177]

(b) *Tort law: attitudes to compensation claims*

As mentioned, despite its toxicity, very few tort claims were brought regarding asbestos before the 1980s.[178] Turner and Newall, the largest

[171] C. Booker, 'The Big Rip-Off', *Sunday Telegraph*, 16 May 2004.

[172] www.asbestoswatchdog.co.uk.

[173] See especially Westminster Hall debate on asbestos, 31 January 2006.

[174] See, e.g., debate on health and safety at work in the House of Lords, Parl. Deb., 26 January 2006, col. 1281 *et seq.*, especially speeches of Baroness Neuberger and Lord Tunnicliffe.

[175] *Ibid.*

[176] '"This is Stanley Baldwin who got in as Prime Minister and got out again ere long," said Miss Brodie. "Miss Mackay retains him on the wall because she believes in the slogan 'Safety First'. But safety does not come first. Goodness, Truth and Beauty come first. Follow me."' Muriel Spark, *The Prime of Miss Jean Brodie* (London: Macmillan, 1961), p. 10.

[177] Bartrip, above note 115, warns us against hindsight and argues persuasively that asbestos does not deserve its 'demonized' status as the worst industrial health problem of the twentieth century: as he points out, 99 per cent of pulmonary industrial fatalities were suffered by coal-miners: *ibid.* p. 264. He alleges that other writers have been guilty of anti-asbestos zealotry, and poor historical scholarship: *ibid.* pp. 271–2.

[178] Until 1950, any claim would not be at common law but under the statutory Workmen's Compensation Act schemes.

asbestos company in Britain and indeed the world, only received fifty such claims between 1972 and 1984.[179] A study in 1988 estimated that by that time, some 10,000 claims had been brought, mainly by workers in construction and allied industries. Nevertheless, this reflected only a small fraction of the total number of sufferers, and comparing the situation with the United State, the authors commented that the British system was 'much less effective at mobilizing possible cases and there are indications of a significant pool of uncompensated victims'.[180] A comparison of litigation rates in 1994 bears this out: there was a significantly lower number of claims per capita in the United Kingdom, even though rates of mesothelioma were 3.8 times higher than in the United States.[181] The position has since changed, with a growing tide of asbestos claims,[182] mirroring what seems to be a burgeoning claims-consciousness throughout British society. This has led to considerable media concern about the 'compensation culture', which has led to recent parliamentary consideration.[183]

Undoubtedly there are some formidable legal obstacles in the way of bringing asbestos claims; these, with the courts' reaction to them, will be explored further below. However, the explanation for the long period of 'missing claims' must ultimately be sociological.[184] The Turner and Newall archive provides powerful evidence for this. Workers long proved resistant to the idea of bringing a tort claim against their employer. Partly, just as in the nineteenth century, there may have been a fear of losing employment, particularly in hard times like the 1930s, and when, as was frequently the case, many members of the same family were all employed in the same factory. In the first ever tort action brought against Turner and Newall, in 1950, many employees were unwilling to make witness statements out of fear for their jobs (although some others did make statements).[185] But the resistance seems to have been more deeply engrained than that. The 'self-reliant', stoical and uncomplaining attitude towards illness among such

179 Tweedale, above note 109, at p. 263.

180 W.L.F. Felstiner and R. Dingwall, *Asbestos Litigation in the United Kingdom* (Oxford: Centre for Socio-Legal Studies, 1988).

181 Tweedale, above note 109, at p. 252, *semble* derived from T. Durkin, 'Constructing Law: Comparing Legal Action in the United States and the United Kingdom' (unpublished Ph.D. thesis).

182 One major re-insurance group, Equitas, is reported to have set aside £56 billion to cover future asbestos claims.

183 See now Compensation Act 2006, discussed in detail below.

184 For a summary, Tweedale, above note 109, at pp. 164–7.

185 N.J. Wikeley, 'The First Common Law Claim for Asbestosis: *Kelly v. Turner & Newall* (1950)' (1998) 3 *JPI Litigation* 197.

workers was also a major factor, perhaps a more general trait amongst the British, until comparatively recently. In the case of Turner and Newall's employees, the attitude was undoubtedly encouraged by the firm's extensive use of *ex gratia* compensation payments, i.e. payments made apparently as a munificent, charitable gesture, and emphatically not as a matter of legal obligation (recipients were sometimes required to waive any claim against the company). Tweedale notes that the *ex gratia* payment was a 'classic paternalist device', assuaging any feelings of guilt in the company directors, while provoking gratitude in the recipients.[186] The strategy was certainly successful, as seen by the very low compensation payments made by Turner and Newall until the 1980s, and the (not unrelated) very small number of tort actions in this period. So low, indeed, was the level of compensation payments that Turner and Newall rejected several offers from insurers to cover its asbestos disease liability, preferring instead to self-insure purely within the group of companies.[187] This looks like a remarkably poor decision now, but the *ex gratia* and self-insurance strategy remained successful until the 1970s. The *ex gratia* payments to widows were trapped in a 1930s time-warp, the same £1 per week basic level still being paid as late as 1976. Notably, it was a media pressure-group which provided the impetus for this finally to be increased.[188]

As soon as the first common law actions were brought against it, Turner and Newall privately realized that its legal position was rather weak – limited to disputing whether asbestosis had been contracted within a 'Scheduled area' of the factory, to defend against breach of statutory duty, and focusing on causation (not, it should be noted, a defence that due care was taken) in a negligence action.[189] It was decided to settle to avoid the publicity which might spark further claims.[190] When the Limitation Act was amended in 1964 to extend the time period in which claims may be brought, the company's solicitors advised that it had lost one of its strongest negotiating weapons.[191] Nonetheless, Turner and Newall's attitude towards compensation claims was to evade liability at all costs.[192] In particular, the firm was dogged in its rejection of claims by the laggers and sprayers working in its asbestos insulation division, and outside the reach

186 Tweedale, above note 109, at pp. 115–16.
187 *Ibid.* pp. 71 and 262–3.
188 *Ibid.* p. 242.
189 Wikeley, above note 185.
190 Tweedale, above note 109, at p. 173.
191 *Ibid.* p. 181.
192 *Ibid.* p. 104. Cf. Wikeley, above note 185, who notes that the company proved unwilling to take a 'technical point' that the writ had been served on Turner and Newall Ltd instead of Turner Brothers Asbestos Ltd, the actual employer, in the *Kelly* case. The directors feared bad publicity.

of the 1931 Regulations (the 'second wave' victims). Although as seen the firm decided to underwrite the financial costs of compensation itself, in 1932 it engaged the Commercial Union Assurance Co. to handle claims on its behalf (no doubt recognizing the well-documented advantage that 'repeat players' have in tort litigation). This allowed Turner and Newall to 'pass the buck' of defending claims and adjusting down payments and settlements, helping to preserve the goodwill of injured workers. The Commercial Union proved so successful in this respect that Turner and Newall had to intervene: the policy of routinely paying 75 per cent of amounts claimed was causing hardship to injured workers.[193]

But the tenacity of defence remained, or had hardened, by the time of the *Margerson* case in the 1990s. At trial, Holland J said that Turner and Newall had 'remorselessly insisted on taking bad points', and their conduct of the litigation evinced 'a wish to contest these claims by any means possible, legitimate or otherwise, so as to wear them down by attrition'.[194] Holland J went on to contrast this hard-nosed conduct with the paternalist employment practices of an earlier period. He suggested that 'the Defendants are no longer part of this community [their Leeds factory having long since closed] and have forgotten their part in achieving an equilibrium between the conduct of an asbestos factory and the interests of a community'. With the benefit of the subsequently opened corporate archive, one might venture to doubt whether Cheeryble Brothers philanthropy ever featured prominently in Turner and Newall's business ethos. However, even when faced with less intransigent defendants, asbestos claims have had to surmount some formidable arguments of legal substance, to which we now turn.

(c) *Tort law: legal obstacles and legal development*

There are two basic causes of action upon which an asbestos claimant might rely: first, an action for breach of statutory duty. Factory safety legislation, such as the Asbestos Industry Regulations 1931, has long been held to benefit from the implied parliamentary intention that an action for damages can be brought by those injured through breaches of the statute.[195] While liability is strict, dependent simply on showing breach of the statutory requirements (here, of ventilation in Scheduled

193 Tweedale, above note 109, at p. 72.

194 Trial transcript, quoted in Steele and Wikeley in (1997) 60 *MLR* 265.

195 *Groves* v. *Lord Wimborne* [1898] 2 QB 402.

areas of asbestos factories), the obvious difficulty is that many victims, in the 'second and third waves', were not exposed during the manufacture of asbestos.[196] Therefore, they will not be able to rely upon the 1931 legislation.

Secondly, an action may be brought in the tort of negligence. It is well established that a duty of care is owed by employers to their employees, and the duty will extend to others, where there was a foreseeable risk of such persons being injured. A key question is frequently, therefore, whether the risks of causing asbestos-related diseases to the claimant were known, or ought to have been known, at the time of the exposure. Without such knowledge, the foresight of harm essential to prove negligence (careless breach of duty) will be absent. In practice, it has sometimes proved difficult to establish when and even whether a victim had been employed by the defendant company (or its predecessors), sometimes many decades previously. As ever in disease litigation, establishing the cause of a disease is often an extremely contested and difficult question for medical science. With mesothelioma, the difficulties have proved so great that the courts have created a special exception to the normal rules on proof of causation, as will be discussed below. Finally, it can be noted that defendants have raised other defences such as contributory negligence (failure to take care) on the part of the claimant, and the expiry of statutory time periods for bringing claims. It is necessary to consider all of these questions, since they have been tested in appellate litigation in England in recent years.

Before examining the development of the substantive tort law in this area, it is worth noting that it is now possible to resolve claims involving the most serious asbestos-related disease very quickly, in most cases. In May 2002, a special fast-track case list was set up in the High Court to deal with all mesothelioma claims. The aim of this procedural development was to ensure that, as far as possible, damages were paid to living mesothelioma claimants during their remaining lives – the prognosis is typically death within 12 months of diagnosis, so speed is essential. The special list has achieved these aims, with most actions by living claimants proceeding from initiation of claim to payment of damages within four months.[197] The judge (Queen's Bench Master) in charge of the list has stated that, for the most part, there is no real defence to liability in 90 per cent of the cases.[198]

[196] Cf. discussion below of *Shell Tankers* v. *Jeromson* [2001] EWCA Civ 101.

[197] Master Whitaker, 'Three Years On' [2005] *JPI Law* 172 (written by the judge in charge of the mesothelioma case list).

[198] *Ibid.* The judge notes that the speed and informality of mesothelioma list proceedings come as a culture shock to lawyers used to practising in the 'ordinary courts' in England.

The main obstacles for claims are factual and evidentiary, i.e. proving that the defendant once employed the claimant, and proving that there he is suffering from mesothelioma triggered by asbestos. Fault is straightforward to show in practice, provided the claimant was employed in one of the industries where asbestos was widely used, and where a 'persuasive' precedent finding of fault therefore exists in an earlier decision. If this is the case, the judge orders the defendant to pay a sum of up to £40,000 as an interim award of damages, while the final quantum of damages (probably much greater) is calculated. It has, however, proved necessary in the past to take several cases up to the English appellate courts, to establish the clear rules (for the most part, favourable to claimants) which make the smooth running of the mesothelioma case list possible. To those hard-fought questions of tort law we now turn our attention.

In English law, the 'clock' for the limitation of actions starts to run as soon as all of the elements of the action are present. In the tort of negligence, where 'damage is the gist of the action', the last thing to occur will normally be the damage in question (a disease may result some time after the defendant's negligence). In *Cartledge* v. *E. Jopling & Sons*, the plaintiff suffered pneumoconiosis from breathing in metal dust, in a factory. For a long period the disease was latent; only when it manifested itself could (and did) the plaintiff bring an action.[199] However, the House of Lords held that the damage had occurred (and the cause of action accrued) at the time when the plaintiff had inhaled the dust, even though at this time he was inevitably unaware of it, since pneumoconiosis presents no symptoms in its early stages. The House expressly rejected the idea that the limitation period began when the plaintiff was aware of the damage, or that a new cause of action accrued at the time of such realization. The result was that the symptoms having manifested themselves more than six years after the employment had ceased (the relevant period under the Limitation Act), the claim was time-barred, and was dismissed. Since asbestos-related diseases all have long periods of latency, such an approach to limitation would prove fatal to most if not all such claims in tort. However, following *Cartledge*, Parliament swiftly intervened to alleviate the plight of claimants suffering from latent diseases. The rules now allow a six-year period from the time of the damage, or three years from the date at which the claimant became aware, or ought to have been aware, of the damage, whichever is later.[200] There is also an overriding discretion for the court to

[199] [1963] AC 758. [200] Limitation Act 1980, ss. 11, 14 (consolidating earlier statutes).

allow an action which would otherwise be time-barred to proceed, where this would be 'just and equitable'.[201]

Another problem has related to the claims of those who have been exposed to asbestos, but who are not yet presenting symptoms of malignant disease. In some ways, these claims are in the opposite position to *Cartledge*: the claimants know of their exposure, and of the risk of disease that this causes, but the disease itself is not yet manifest, and indeed may never develop. The difficulty is that the claimant does not seem to have suffered any actionable damage, on orthodox principles. His claim is founded upon the *anxiety* of knowing that he is in a higher-risk group for the development of mesothelioma, etc. in future. English law, however, does not recognise an action in negligence for mere anxiety or other mental distress, short of a medically recognized psychiatric illness.[202] Nor does the law seem to allow an 'anticipatory action' for damage which has not yet occurred. There is a statutory power for the court to award 'provisional damages' on the basis that some future damage will *not* occur, but to allow a renewed application to the court in the event that such damage does ultimately manifest itself.[203] However, it seems that this cannot apply in the absence of a complete cause of action in the first place, in which to award the provisional damages. Furthermore, the robust 'all or nothing' approach to claims for the lost chance of avoiding some injury would seem to rule out a claim based on the future possibility of developing mesothelioma, unless the chance of developing the disease was more likely than not.[204]

Nevertheless, until recently some claims of this type had been brought successfully, when the claimant had been diagnosed with 'pleural plaques', i.e. benign (and usually asymptomatic) fibrous patches in the lung. Previously, trial courts had recognized these plaques as an 'injury' and thus actionable damage. The anxiety created by them and the risk of future diseases could then also be compensated, at the assessment of damages stage (consequential losses). In essence, the pleural plaques has provided a peg on which to hang the true motivation for the claim (de facto if not de jure) – the anxiety of developing cancer in future. In a recent test case, this approach to pleural plaques was upheld at trial.[205] However, a majority of the Court of Appeal held that in accordance with the maxim *de minimis non curat lex*, pleural plaques do not constitute an 'injury'

[201] *Ibid.* s. 33. [202] *Hinz v. Berry* [1970] 2 QB 40. [203] Supreme Court Act 1981, S. 32A.
[204] *Gregg* v. *Scott* [2005] UKHL 2, [2005] 2 AC 176.
[205] *Grieves* v. *F.T. Everard and Sons* [2005] EWHC 88 (QB).

so as to constitute actionable damage in negligence.[206] The court further refused an invitation to combine three non-actionable elements (the anxiety, the benign plaques and the risk of future disease) to come up with actionable damage, fairly describing this as illogical. The House of Lords dismissed the claimants' appeal.[207]

Many recent asbestos cases have raised the essentially factual question of breach of the duty of care. The crucial issue has normally been the state of scientific and industry knowledge at the time of the exposure. An employer can only be held not to have taken reasonable precautions when the risk was (or ought to have been) appreciated by him at the time. In *Margerson* v. *J.W. Roberts Ltd*,[208] the claimants had, as children, played in the copious dust emitted into the immediate vicinity of the defendant's asbestos factory in a residential district of Leeds. Both claimants subsequently contracted mesothelioma. The defendants conceded that during the 1930s when the claimants were children, the risk of developing asbestosis was well known. However, they pointed out that the risk of mesothelioma was not scientifically appreciated until the 1960s, and accordingly, they maintained, they could not have been careless in respect of *that disease*. The appeal court rejected this narrow approach, holding it sufficient that 'the evidence demonstrated that the defendants should reasonably have foreseen a risk of some pulmonary injury, not necessarily mesothelioma'. This test, accordingly, was satisfied.

Margerson provides an example of the manipulability of foresight, with which every English tort lawyer is all too familiar. The test for the recoverability of consequential damage is whether the type of loss in question was a reasonably foreseeable result of the accident. The definition of 'type of loss' here is all-important. The courts have sometimes required foresight

[206] *Sub nom. Re Pleural Plaques Litigation* (emphasising its test case/class action nature) [2006] EWCA Civ 27. Noted by J. Morgan, [2006] *CLJ* 269.

[207] *Johnston* v. *NEI International Combustion Ltd* [2007] UKHL 39. The effect of the decision has been reversed in Scotland by the Damages (Asbestos-related Conditions) (Scotland) Act 2009. The British government is considering whether to take similar action in England and Wales: Ministry of Justice, *Pleural Plaques*, Consultation Paper CP 14/08 (July 2008). It is pertinent to note that in the United States, 90 per cent of new asbestos claims are estimated to be filed by plaintiffs who have been exposed to asbestos but who remain entirely asymptomatic. These claims are commonly blamed for slowing the resolution of the claims by seriously ill asbestos plaintiffs. However, the former represent only 5 per cent of the total compensation bill, because the US courts have limited damages to the medical expenses incurred in monitoring for the onset of asbestos diseases. De Saulles, above note 120.

[208] [1996] PIQR P358; *The Times*, 17 April 1996. J.W. Roberts was a wholly owned subsidiary of Turner and Newall.

of the exact (sometimes unexpected, or unusual) consequential loss, and naturally such a test is often very difficult to satisfy and the relevant actions have failed.[209] On the other hand, in *Page* v. *Smith*, the House of Lords countenanced an extremely broad definition of 'kind of damage', it apparently being enough that 'personal injury' of some sort was foreseeable (the point of the case was to deny a distinction between physical injury and psychological trauma arising from a car accident, for these purposes).[210] In *Margerson*, decided in the same year as *Page* v. *Smith*, the Court of Appeal (not surprisingly) relied on the House of Lords' decision. On the facts of the case, treating all types of lung disease as one kind of damage for the purposes of foreseeability was quite sufficient to demolish the arguments of the defence.

That is not to say that any slight exposure of employees to asbestos, at any time in the twentieth century, has been found to render an employer *ipso facto* in breach of his duty of care. In *Smith* v. *P&O Bulk Shipping*,[211] the plaintiff had worked on board the defendant's ships between 1954 and 1971. He contracted mesothelioma, probably as a result to exposure to asbestos dust in the course of that employment. However, while the risks of regular exposure to asbestos (especially in confined spaces) were at the time well known, the first instance court found that 'it was not until the Notice issued by the Department of Trade in 1977 to shipowners, shipbuilders and repairers and masters, officers and seamen that there was a clear warning of a risk of developing mesothelioma from short exposure to asbestos and a requirement for precautions to be taken to protect those who had any access to the area where asbestos was being worked'.[212] Accordingly, the court held that the occasional exposure of the plaintiff (without the relevant protection) to the area of the ships where asbestos was present had not, at the relevant time, been negligent.

This case should be contrasted with *Shell Tankers UK* v. *Jeromson*,[213] which also concerned a claim by marine engineers exposed to asbestos in the 1950s. In this case, the appeal court held that there was ample evidence on which to uphold the trial judge's finding of negligence. Safety precautions should have been taken, since various warnings about the health risk to laggers and others in the ship-building industry had been published in the 1940s and 1950s.[214] The court seems to have followed a more exacting approach than in *Smith* v. *P&O*, envisaging active steps by

[209] See, e.g., *Doughty* v. *Turner Manufacturing Co. Ltd* [1964] 1 QB 518, which coincidentally concerned the foreseeability of a chemical reaction between an asbestos lid and a cauldron of molten cyanide, into which it had been accidentally dropped.

[210] [1996] AC 155. [211] [1998] 2 Lloyd's Rep. 81. [212] *Ibid.* (headnote).

[213] [2001] EWCA Civ 101, [2001] ICR 1223. [214] Quoted at paras. 49–53.

the employers to inform themselves, rather than characterizing them as a passive recipient of government warnings.[215] Curiously then, the factory inspectors' reports of this vintage, for so long ignored by government and attacked by the asbestos manufacturers, have now served to affix negligence liability to employers of 'second wave' asbestos victims.

Another issue in *Jeromson* was whether the Asbestos Industry Regulations 1931 covered the process of mixing and applying asbestos insulation during the manufacture of steam pressing machines, in an action for breach of statutory duty. The appeal court held that such a process fell within the literal terms of the Regulations,[216] and so the action would lie. This decision required Hale LJ to distinguish (and, effectively, to overrule) earlier cases which had held that the Asbestos Industry Regulations 1931 simply did not apply outside the asbestos manufacturing industry itself. The title of the Regulations and, significantly, the views of the Factory Inspectorate, were not to be decisive in this respect, and could not rebut the literal words of the statute.[217] This broader reading of the 1931 Regulations, to cover at least some 'second wave' asbestos victims, is in notable contrast to its interpretation by both industry and government at the time, although of course not all such claims can be brought within the Regulations' wording.[218]

In *Maguire* v. *Harland & Wolff plc*,[219] the Court of Appeal dismissed the claim of a woman who had contracted mesothelioma through washing the asbestos-saturated clothes of her husband, who had been employed by the defendants in the early 1960s. Reviewing the scientific literature, Judge LJ found that, at this period, the (erroneous) impression given by government circulars and regulations was that there was a 'safe limit' of asbestos exposure. It was not until rather later that the view that any exposure to asbestos was hazardous became established, and indeed the world had to wait until 1989 for a paper specifically upon 'Domestic Asbestos

[215] 'The judge was entitled to conclude that a prudent employer would have taken precautions or at the very least made inquiries about what precautions, if any, they should take. If Shell had made inquiries, the judge was in little doubt what advice they would have received' (paras. 52–53).

[216] In particular 'the mixing or sieving of asbestos, and all processes involving manipulation of asbestos incidental thereto'.

[217] At paras. 14–25.

[218] In particular, as Hale LJ admits, the Regulations might not apply to those removing old lagging (at para. 25). This might seem an ever harder distinction to justify than that between asbestos industry workers and 'second wave victims', as had previously obtained. For another example of Lady Hale reading health and safety legislation broadly (but in dissent), See *Fytche* v. *Wincanton Logistics* [2004] UKHL 31, [2004] 4 All ER 221.

[219] [2005] EWCA Civ 1.

Exposure, Lung Fibre Burden and Pleural Mesothelioma in a Housewife'. In short, the alarm on 'third wave victims' did not sound until 1965 at the earliest, with Newhouse and Thompson's paper, referred to above. In the absence of contemporaneous evidence of the danger of such secondary exposure, the defendants could not be negligent in having failed to foresee the risk to Mrs Maguire's health. As Mustill J once put it, an employer is not required to 'plough a lone furrow', taking precautions beyond what medical, scientific, governmental and industrial opinion would consider necessary.[220] Nevertheless, Mance LJ dissented. He began by stating that a failure to take safety precautions can be justified only if injury would be totally unforeseeable, 'even in the most unusual cases'. He stressed that the silence in the medical journals could not be conclusive, since the question was not what was actually foreseen, but what ought to have been foreseen. Mance LJ argued that even in 1960, it would have taken 'little imagination to see and appreciate that others outside the workforce will also be exposed to asbestos dust, [since it was] obvious that employees are taking home for cleaning clothing which is excessively and unnecessarily dust-impregnated', and that those exposed were therefore at risk of lung disease. Whether one agrees with Mance LJ or Judge LJ (and the third judge, Longmore LJ admitted to 'some hesitation' before siding with the latter), the case shows that in performing the historic 'jury role' as finders of fact, English judges today enjoy considerable decision-making flexibility. There is, however, no easily discernible trend in the recent decisions. Some, such as *Margerson* and *Jeromson*, seem notably broad, but others, like *Maguire*, rather restrictive. This suggests differing attitudes towards employers' historic conduct amongst the English judiciary, finding its expression in more or less expansive conceptions of what 'ought to have been foreseen'.

It has often been observed that causation is the main stumbling block in tort actions for disease. Mesothelioma has been a particularly difficult disease in this respect. Because it may be caused even by slight, transitory exposure, it will often be the case that out of many possible defendants, all of whom have exposed the claimant to asbestos, it is impossible to say which of them supplied the 'fatal fibre(s)' which triggered mesothelioma. This situation is especially common in the 'second wave' of victims, who may have worked with asbestos under several different employers.[221]

[220] *Thompson* v. *Smiths Shiprepairers* [1984] QB 405 at 415B.

[221] This problem, at least, is lessened for 'first wave' victims employed within the asbestos industry itself. Cf. also the geographically specific mesotheliomas in cases like *Margerson*.

Because it is impossible to say which of the employer-defendants caused the disease, the claim should, on orthodox principles, fail for want of proof of an element of the cause of action.[222] This leads to the unappealing conclusion that none of the employers caused the mesothelioma – a conclusion which is, moreover, logically incorrect.

In the well-known *Fairchild* decision, the House of Lords decided that in this situation, the dictates of justice required that each employer's contribution to the *risk* of the claimant contracting mesothelioma should be taken as an actual contribution to the disease itself.[223] Thus, the employers would all be liable, jointly and severally, for the claimant's disease. The present author has criticised the case at some length elsewhere.[224] All that need be said here is that the 'policy' cited by the House – that victims of negligence should be compensated – is most unilluminating, because it would apply with equal weight in every case where a claim is dismissed for want of a causal link, or a duty of care, or other key component of the tort. A malign and predictable consequence of such loose reasoning has been to encourage claimants to pursue ever-bolder arguments across the law of torts, which would be doomed to fail on any orthodox understanding of causation. The puzzling decision in *Chester* v. *Afshar* suggested that after *Fairchild*, the floodgates of liability (for once) really had burst,[225] although there has now been considerable retrenchment in *Gregg* v. *Scott*, the 'loss of a chance' case which was arguably more compelling than *Fairchild* itself.[226] Dismissing that claim, Lord Phillips of Worth Maltravers MR commented, significantly: 'My Lords, it seems to me that there is a danger, if special tests of causation are developed piecemeal to deal with perceived injustices in particular factual situations, that the coherence of our common law will be destroyed.'[227] Whatever the baleful effects of *Fairchild* for the coherence of causation in English law, the House of Lords' decision has, of course, been very beneficial for the victims of mesothelioma. The special mesothelioma list was set up as a direct response to that case.[228] Without *Fairchild*, most such claims would have failed completely at the hurdle of causation, rather than being resolved in the claimant's favour within a matter of months.

Even more recently, the House of Lords has returned to a question left unanswered in *Fairchild* (in fact explicitly left open). The issue was whether

[222] *Fairchild* v. *Glenhaven Funeral Services* [2001] EWCA Civ 1881, [2002] 1 WLR 1052.
[223] *Fairchild* v. *Glenhaven Funeral Services* [2002] UKHL 22, [2003] 1 AC 32.
[224] (2003) 66 *MLR* 277. [225] [2004] UKHL 41, [2005] 1 AC 134.
[226] [2005] UKHL 2, [2005] 2 AC 176. [227] At para. 172.
[228] Discussed above notes 197 and 198.

the multiple defendants in a *Fairchild* case should each be liable *in full*, on the usual principle of 'joint and several liability', or whether out of fairness, a special exception should be created to render defendants liable for a *share of the damage* only, proportionate to the exposure for which they were responsible. The House sided with the employers, and favoured proportionate liability: *Barker* v. *St Gobain Pipelines*.[229] By contrast to *Fairchild*, however, their Lordships refused to create an explicitly policy-driven exception to ordinary legal principle (i.e. joint and several liability). Instead, a principled explanation was offered for the decision. The employers were liable for a proportionate share because the damage at the heart of the action was not the outcome, mesothelioma, an 'indivisible injury', but rather the *risk* of mesothelioma, to which all the employers had contributed. Risk, of course, is inherently a divisible, proportionate concept. Hence, proportionate liability rather than liability '*in solidum*' logically follows. The main problem with *Barker* as a matter of law is that it seems inconsistent with the House of Lords' other recent decision on loss of a chance, *Gregg* v. *Scott*.[230] It is certainly curious that an employer may be liable for his 17 per cent contribution to the risk of mesothelioma, pursuant to *Fairchild* and *Barker*, but a doctor cannot be held liable for a negligent misdiagnosis which reduced a patient's chances of recovering from cancer by 17 per cent (*Gregg* v. *Scott*). Ultimately, it is *Fairchild* which is out of line with legal principle, and great uncertainty remains as to when the courts will choose to follow its lead (as in *Barker*), and when they will not (as in *Gregg* v. *Scott*).

Whatever one makes of the doctrinal controversy surrounding *Barker*, the decision was undoubtedly as much of a blow for mesothelioma claimants as *Fairchild* had been a boon. In order to recover compensation in full, it would now be necessary to trace and sue all employers who may have exposed the claimant to asbestos – clearly, a difficult task when the claimant was a jobbing builder with a number of firms, or similar situations. Simply suing one employer would no longer be good enough. The House of Lords' decision received much criticism in the media and from MPs in the House of Commons, all taking the mesothelioma victims' side. The Labour government, under pressure from its own backbenchers, promised swift action, and before Parliament rose for the summer recess it had hastily enacted the Compensation Act 2006, section 3 of which was designed to reverse the ruling in *Barker*. Thus, that decision must be one of

[229] *Sub nom. Barker* v. *Corus plc* [2006] UKHL 20, [2006] 3 All ER 785.

[230] [2005] UKHL 2, [2005] 2 AC 176. See Lord Rodger of Earlsferry's powerful dissent in *Barker*.

the shortest-lived leading cases in English legal history;[231] it is also ironic that the mesothelioma provisions, considerably more generous to claimants than the common law 'declared' in *Barker*, were tacked onto a statute whose declared purpose had been to curb tort liability and the much discussed 'compensation culture'.[232] Section 3 simply restates the law as it had been understood before *Barker*, namely that in a situation covered by the *Fairchild* principles, defendants are to be jointly and severally liable. Nevertheless, there were some calls for a mesothelioma compensation scheme, totally outside the common law of torts, in the parliamentary debates on section 3.[233] In September 2006, the Department of Work and Pensions initiated a consultation on a 'long-term solution' to compensation for mesothelioma.[234] It seems that whatever the niceties of common law principles, victims of a politically high-profile disease must receive compensation, and not be left to the *tendresse* of the welfare state.[235] Late in the day, the British government has become a staunch defender of those suffering from asbestos diseases.

[231] The decision in *Barker* was announced on 3 May 2006; the Compensation Act received Royal Assent on 25 July 2006. The mesothelioma provisions came into force retrospectively: 'Section 3 shall be treated as having always had effect' (s.16(3)); the government indeed stressed that the degree of retrospection was strictly limited: 'We are talking about barely seven or eight weeks', per Bridget Prentice MP, Parl. Deb., 17 July 2006. Cf. Lord Hunt of Wirral, Parl. Deb., 19 July 2006. See further s.16(5) and (6) (for cases determined between 3 May and 25 July).

[232] See, in this regard, Compensation Act 2006, s.1, in itself a curious provision since it seems not to change any existing rules of law. Nor perhaps was it designed to: point 10 of the 'explanatory notes' to the Act, published by the government states: 'This provision is intended to contribute to *improving awareness* of this aspect of the law; *providing reassurance* to the people and organisations who are concerned about possible litigation' (emphasis added).

[233] See Hansard debates, as cited above note 231.

[234] Although the letter notes that 'we are interested in solutions that use the *existing resources* of employers, insurers and government to deliver adequate compensation more quickly and more effectively' (emphasis added). So no new government funds seem to be on the table.

[235] A telling precedent is the Pneumoconiosis etc. (Workers' Compensation) Act 1979. Although such a scheme had only just been rejected by the ('Pearson') Royal Commission on Personal Injury (Cmnd. 7054-I, 1979), the Labour government passed the 1979 Act in a vain attempt to persuade the Welsh Nationalists to shore up its disappearing House of Commons majority. Peter Cane opines that this epitomizes the 'Gilbertian hilarity' with which the ad hoc approach can displace principle, when cutting the compensation cake: quoted N.J. Wikeley, *Compensation for Industrial Disease* (Aldershot: Dartmouth, 1993), p. 49.

3

Technological Change and the Development of Liability for Fault in France

YVONNE SALMON

1. Fires caused by sparks from railway engines

(a) *Development of the railways*

From 1837 when the first steam railway was built, until 1937 when the French rail system was nationalised, an extensive rail system was developed, albeit very slowly. Private management assisted by the public treasury was the model followed.

Railway construction in France dates back to 1823, when a line of 23 km was authorised to connect the industrial centres of Andrézieux and St Etienne in the Loire region. It was soon followed by other short lines connecting regions of iron ore and coal deposits with industrial centres on navigable waters and canals. Traction at the time was by means of stationary engines or by horse. The first passenger car drawn by a steam locomotive engine was authorised on 9 June 1826 to service the growing industrial cities of St Etienne and Lyon, but did not commence operations until 1832. The first 'railway by mechanical propulsion' to operate as a common carrier for both freight and passengers was inaugurated on 26 August 1837;[1] it connected the city of Paris with the suburban town of St Germain-en-Laye, a route of approximately 20 km.

In the same year, the government proposed several laws to the Chambre des Députés for the granting of concessions for railway lines: after long discussions as to whether the works should be entrusted to private industry as in the United Kingdom or United States, or put in the hands of the state (as in Belgium, for example) some concessions were granted to various private companies. However, these first initiatives met with very

[1] A. Dunham, 'How the First French Railways Were Planned' (1941) *J Econ. Hist.* 13.

little success, with many of the companies abandoning their concession or demanding state grants.

The true point of departure of the development of the French railways could be seen to be the Law of 11 June 1842.[2] The state was in charge of the rail infrastructure, such as the costs of acquisition and clearing of land, construction of stations, and so on. The concessionary companies were in charge of the superstructure: they were to bear the costs of the construction and installation of the rail-track itself and rolling stock, and the subsequent maintenance of the track. At the same time, the Law of 1842 established nine extensive rail-lines, seven of which would radiate out from Paris towards Belgium, the United Kingdom, Germany, the Mediterranean, the North Sea and the centre of France, with two further lines going from the Mediterranean to the Rhineland and from the North Sea to the Mediterranean.

The legislation also reserved the right for the state to have further recourse to private industry for the establishment of new lines, even burdening the concessionary companies with the costs of both infrastructure and superstructure, if such concessions were accomplished through the enactment of special laws. Numerous companies were created in the wake of the law, but the initial impetus was tempered by the financial crisis of 1847–48, the depression of 1851 and the coup d'état of 1851, which established the Second Napoleonic Empire. The state was forced to declare that certain companies had lost their concessions when they were found to be unable to fulfil their obligations, and to seize several lines and buy back the Paris–Lyon line.

By 1852, only 3,900 km of rail-track had been given by concession, of which 3,546 km were ready for use. In order to strengthen the position of the rail companies, the state decided to concentrate the lines of each particular zone in the hands of one company. The twenty-eight lines which had been established became six *grands réseaux*: North, Orléans, Paris-Lyon-Méditerranée, Est, Ouest and Midi, to which the Imperial government applied uniform general conditions, and a fixed term of ninety-nine years, imposing an upper limit on the cost of rail tickets and the obligation to construct various secondary lines, some with and some without additional state funding.[3] These innovations gave the establishment of the railway a new lease of life, and within five years, the rail network

[2] H. Bertheau, *Dictionnaire général de droit et de jurisprudence: Le Droit Civil* (Paris: J. Delamotte, 1891), p. 641.

[3] *Ibid.* p. 642.

had expanded from 3,900 km to 16,000 km of track, with the lines in use increasing from 3,546 km to 7,450 km. However, the period of uncertainty was not over: many of the companies had contracted to undertake considerable work in a short period of time, and it was quickly realised that it would be impossible for them to fulfil their obligations without some assistance. In the panic that ensued, the companies demanded government intervention and the revision of their contracts. The government decided that it would be in its own interests to respond to the request, and these negotiations resulted in the 'Franqueville contracts', named after the Conseiller d'Etat who had composed them. The contracts, ratified by a Law of 11 June 1859, assured those railways conceded since 1857 a treasury guarantee for the payment of certain fixed charges, in case of a deficit, and a treasury share in the profits, in case of a surplus. Earlier concessions were granted assistance in the creation of new lines. In return for state assistance, the companies had to give a portion of their revenue to the state from 1872.[4]

State intervention revived the railways and, by the end of 1870, more than 24,429 km of railway track had been conceded and 17,440 km of track were in use.[5] The Franco-Prussian War of 1870–71, and the overthrow of Napoleon III and the establishment of the Third Republic, did not hinder the progress of the railway system. The war had highlighted the strategic importance of railways, albeit that the war had impacted the French economy and the end of hostilities found the railways run down and in need of repairs.

On 18 May 1878, the state was authorised to repurchase and run certain small railway lines which had fallen into financial difficulty,[6] becoming a seventh '*grand réseau*', which was entirely state run.

The railways '*d'intérêt général*' formed part of the public domain.[7] The railway track itself and everything attached to it was considered to be in the public domain: station buildings, the grounds surrounding the stations, repair workshops.[8] On the other hand, the buildings occupied by the concession company did not form part of the public domain. The ground on which the railway was established, in principle remained in the public domain, unless declassified by the Minister of Public Works. In addition to the '*grands réseaux*' or '*compagnies principales d'intérêt général*', of which five were private and two publicly owned, the French railway

[4] K. A. Doukas, *The French Railroads and the State* (New York: Columbia University Press, 1945), pp. 32–8.

[5] *Ibid. p.* 36

[6] A. Picard, *Les chemins de fer français* (Paris: J. Rothschild, 1884), p. 35.

[7] Law of 15 July 1845, art. 1, on railway regulation.

[8] Cass. civ. 4 July 1866, S.1867.1.82.

system included two more classes of railway. One class comprised all local railways, '*voies ferrées d'intérêt local*', including tramways; the other comprised all secondary lines, '*réseaux secondaires d'intérêt général*'. While constituting only a very small segment of the French rail industry, both these groups were independent of the large networks, and often the two terms '*voies ferrées d'intérêt local*' and '*réseaux secondaires*' have been used interchangeably.[9]

(b) Legal regime

Given the differing levels of state involvement in the various rail companies, it must be questioned whether actions for damage caused by sparks from locomotive engines would be based in civil or administrative law. It has long been established that, in general, the principle of the inapplicability of civil law rules to public law persons is suspended when central government, local departments or communes operate in the private sphere and behave as 'private individuals'.[10] This was also reflected in a number of judgments of the Cour de cassation and the Tribunal des conflits.[11] Likewise, in the terms of article 22 of the Law of 15 July 1845 on railway regulation:

> The concessionaries or operators of a railway will be held liable either to the state, or to private individuals for damage caused by administrators, directors or employees ... the state will be submitted to the same liability towards private individuals if the railway is exploited at its expense and for its profit.

Similarly, article 9 of the Finance Law of 22 December 1878[12] reflected this approach, stipulating that 'railways exploited by the State are subject to the same regime as those operated by concession'.

In a case of 23 June 1859,[13] the Cour de cassation found that article 458 of the Code penal, which prohibited fire damage to the property of others caused by fire or sparks which have been carried or distributed with the help of machines operated by 'human hand', would also apply to

[9] R. Thévenez, M. d'Hérouville and E. Bleys, *Legislation des chemins de fer* (Paris: Rousseau, 1930), p. 839. See also report (No. 1575) of the commission of public works and remarks of the Minister of Finance in J.O. (Doc.), 27 March 1929, 584–6.

[10] P. Duez, *La responsabilité de la puissance publique (en dehors du contrat)* (Paris: Dalloz, 1927), p. 10.

[11] Tribunal des conflits, 22 June 1889, D.P.1891.3.1; Req. 8 July 1889, D.P.1889.1.353; Req. 18 November 1895, D.P.1898.1.497; Req. 22 March 1899, D.P.1899.1.452.

[12] D.P.1879.4.10. [13] D.P.1859.1.784.

locomotive engines, and this could be punished by a fine of between 50 and 500 francs.

Accordingly, special legislative enactments together with provisions of the Code civil and, on occasion, criminal law were found to be the appropriate basis for actions brought for damage caused in the operation of railways, rather than public law.

(c) *Special measures*

It would appear that no official figures were collated for the incidence of fires started by sparks from railway engines or for the value of property destroyed. However, from statements in the reported judgments, it would appear to have been a widespread problem.

Measures intended to ensure the control, operation and maintenance of railways were established in administrative law regulations, and these reflected concerns as to damage resulting from escaped sparks or coal fragments, perhaps inspired by the experience of countries with more advanced railway systems. It is interesting to note that such measures were intended to ensure the safe operation of railways, and seem to balance the public interest of an extensive railway system against the damage which could be suffered by neighbouring landowners or tenants.

The Law of 15 July 1845 on the regulation, operation and maintenance of railways made land-owners neighbouring railway tracks subject to various restrictions. Article 5 of the Law of 1845 prohibited the construction of buildings within a distance of 2 m from the outer rails of a track. Buildings existing before the promulgation of the Law of 1845, or the laying of a new railway track, were only allowed to be maintained in their pre-existing state. Article 7 of the same law prohibited thatched roofs, bales of hay, straw or other inflammable materials being placed less than 20 m from a railway, although an exception was made for the gathering of crops at harvest time.[14] Those neighbouring rail-tracks thus had to take some precautions; a failure to do so would in effect be considered a bar to compensation. The rules were strictly interpreted: a judgment of the Cour de cassation of 28 October 1903[15] held that the exception for gathering crops at harvest time was only applicable where deposit was made during the process of harvesting, not for other forms of storage (in the case at hand, moving the hay from a barn to be stored in the prohibited

[14] See art. 7(2). [15] Cass. civ. 28 October 1903, S.1904.1.93

area). In another case, bales of rye deposited less than 20 m from the rail-track caught fire after the passage of a train, due to the escape of a piece of coal or a spark. By virtue of the operation of article 7 of the Law of 15 July 1845, the railway company was exempt from liability and the farmer was unable to claim compensation for the loss of his crops.

Railway companies were also required to take certain precautions. Article 11 of the Ordonnance Royale of 15 November 1846 stipulated that rail companies had to use locomotive engines which were of such a design as to prevent burning coals falling onto the track or the surrounding land. In the event of a failure to do so, they would be found liable for the resulting fire damage. Later, article 21 of the décret of 9 August 1881 with respect to the establishment and operation of railways, provided that 'locomotive engines should not emit any odour and should not allow sparks, flaming coals, ash, smoke or water to escape into the public domain or neighbouring properties'. In this way, the owner of inflammable property, who had not taken any precautions to reduce the risk of fire which might come from sparks escaping from a passing locomotive engine, had the right to claim full compensation from the railway company, if the company failed to use machines which would not allow sparks or fragments of burning coal to escape.

(d) Civil law: fault-based liability

The duties placed on neighbouring land-owners and railway operators can be seen to operate alongside liability for fault on the basis of article 1382 *et seq.* of the Code civil,[16] which formed the main basis of early actions for compensation of damage caused by the escape of sparks from locomotive engines.

The operation of article 1382 in such cases can be seen in a judgment of 30 November 1859.[17] The Tribunal de la Seine considered a case in which it was claimed that a fire which had destroyed the plaintiff's pine trees had begun when a spark or piece of burning coal had escaped from a locomotive engine. It was found that the fact that the fire had started immediately after the passing of the train when there were strong winds, and that fires were 'frequently caused in this way',[18] was sufficient to establish

[16] Article 1382 provides that 'Any human act whatever which causes damage to another obliges him by whose fault it occurred to make reparation.' Article 1383 states: 'Every person is liable for the damage which he causes not only by his own act but also by his negligence or imprudence.'

[17] S.1860.1.41. [18] *Ibid.* 42.

fault under article 1382 C civ.: it would appear that rail companies faced a presumption of causation.

The Tribunal de la Seine noted that article 1382 applied to industries operating under concession from the state, and that the additional rules to which such concessions were subject did not remove the companies from the jurisdiction of civil law rules. 'It is therefore necessary that railway companies take a multitude of precautions to prevent their enterprise from harming third parties, and in this respect the rail companies must compensate for damage which results from a failure to take such precautions.'[19] However, in the case at hand it was not shown that the railway company had failed in its duty in any way. It would seem that such an application of article 1382 resulted in strict liability for the rail companies, unless the plaintiff had, for example, failed to take precautions stipulated by law to prevent the fire damage.

However, while article 7 of the Law of 15 July 1845 provided that thatched roofs, haystacks, and any other inflammable material must be kept at a distance of at least 20 m from rail-tracks, it did not apply to buildings existing before the construction of the track, and the Cour de cassation found in a case of 20 November 1866[20] that no restriction was placed on the storage of harvested produce in such buildings. It was claimed in that case that a fire which destroyed a barn and stables was started by sparks from a train passing near to the affected land. As a result, the railway company was held liable under article 1382 C civ. in spite of the risks taken by the land-owner.

On 30 August 1858, the Cour d'appel of Paris[21] considered that there was fault on the part of a railway company in creating a risk of fire in the exercise of its concession, and that 'when one exercises a profession which is dangerous or inconvenient for neighbours, one should compensate for resultant damage, even if the industry is operated according to all rules of ordinary care'. This approach was echoed by the Cour d'appel of Bordeaux in a judgment of 21 June 1859.[22]

It would appear from the foregoing that the rail companies were held responsible for damage caused by their locomotive engines, even where all possible precautions had been taken. It must be questioned how this approach could be justified on the basis of article 1382 C civ. in the absence of any fault being shown. The cases take place before the no-fault regime for the acts of things under one's '*garde*' had been developed under article 1384, paragraph 1 C civ.

[19] *Ibid. at* 43. [20] Cass. civ. 20 November 1866, S.1867.1.77.
[21] Reproduced in S.1905.1.104. [22] S.1860.2.41.

The cases may have been influenced by the administrative courts' approach to public works, according to which the administration is liable for damage caused in the execution of public works, without the need to prove fault.[23] The difficulty here would be that the fire would be considered a consequence of the operation of the resource, rather than its construction. However, it would appear that during this period the administrative courts and the Tribunal des conflits increasingly assimilated exploitation to construction, where the consequences of operation were considered to be the necessary and inevitable consequence of the ordinary functioning of the public work, such as vibration caused by the passage of trains,[24] or the damage resulting from the smoke emitted by locomotive engines.[25]

Nonetheless, the Bordeaux and Paris cases expressly assimilate the rail companies to industrial companies, and they refer to the provisions of article 1382 C civ. The case of 21 June 1859 expressly states that 'the companies which operate the railways are, like all industrial companies, subject to civil law and thus according to article 1382 C civ, liable to repair the damage which they cause'.

A second possible influence is the case law relating to neighbour issues, which reflects a similar approach to liability as that seen in cases relating to fire caused by sparks from locomotives.

While the existence of fault was considered to be essential for liability, the ongoing development of industry made it difficult to maintain the same conception of fault. It was considered inequitable that neighbours would have to bear the numerous inconveniences which the activities of large companies might give rise to. The noise, vibration, smoke, and so on, harmed the neighbouring buildings whatever precautions the factories took. In order to award compensation for such harm, the courts departed from a strict interpretation of article 1382 and its fault requirement. They considered that the mere fact of having established a site which, by its nature, inconvenienced neighbours more than traditional neighbour activities, constituted a fault which must be compensated. It did not matter that the law or administration had authorised the construction and use of such establishments: such a legal or administrative authorisation only concerned the general interest and not the individual interests of neighbours.[26]

[23] Tribunal des conflits, 17 June 1899, S.1901.3.131; Conseil d'État, 18 January 1901, S.1903.3.79.

[24] Conseil d'État, 16 May 1879, S.1880.2.342.

[25] Tribunal des conflits, 28 July 1894, S.1896.3.121, 2 March 1901, S.1903.3.111.

[26] Cass. 18 November 1884, S.1885.1.69; Paris, 19 April 1893, S.1893.2.124.

In 1820, the Cour d'appel of Metz developed the doctrine that 'it would be unjust if a proprietor found himself suddenly harmed and inconvenienced by the use to which his neighbour puts his property'.[27] In a case of 3 December 1860, it was stated that 'the exercise of the right to property ceases to be legitimate and becomes a fault, from the moment that it seriously harms the right of a neighbour'; where 'the inconveniences resulting from the proximity of the establishment exceed the tolerance threshold that neighbour relations impose' it 'necessarily gives rise to a finding of fault'.[28]

It is clear that in neighbour matters, the 'fault' which gives rise to the duty to compensate is neither negligence nor recklessness: it is natural and even desirable that certain industries operate, and in so doing the establishment of such industries is not in itself a '*faute*'. However, the case law would suggest that the effects of the operation of such industries can be considered to be a fault and affected neighbours may claim damages for the resulting inconvenience.

It would seem that the judges in the cases at hand conceived of fault in this way. What took place was a balancing of the public interest in allowing railways to operate, against the possible harm to neighbouring land-owners: even if the company had taken all possible measures to prevent harm, the railway company had to compensate for burned crops or property.[29]

Later, Capitant put forward the theory that neighbour relations should not be considered as merely simple relations between individuals: each has an obligation to the other which forbids all acts which might cause damage.[30] It is thus unnecessary to call upon article 1382: 'the neighbour only has to establish the damage which the owner has caused in exercising his right, and the latter is held to repair the damage even if all possible precautions were taken, even if he had used all possible means at his disposal to avoid the damage'.[31] It would seem that the case law on railways reached the same conclusion by a broad interpretation of article 1382.

[27] Metz, 16 August 1820, affirmed by Cass. 27 November 1844, S.1844.1.811. This principle was followed more or less explicitly in subsequent cases: Cass. civ. 17 July 1845, S.1845.1.825; 20 February 1849, S.1849.1.346; 8 June 1857, S.1858.1.305.

[28] Cass. 3 December 1860, S.1861.1.840. In a similar way, Cass.civ. 14 July 1875, S.1875.1.352: 'In exceeding the normal relations between neighbours, a neighbour exceeds the limits of his rights. As such this constitutes a fault and gives rise to the need to compensate damage.'

[29] In a similar way, Grenoble, 21 June 1863, *Jurisprudence générale des assurances* 1863, Part II, 533.

[30] H. Capitant, 'Des obligations de voisinage' *Revue critique de législation et de jurisprudence* 1900, 156.

[31] *Ibid.* 168.

(e) *Emergence of no-fault liability for the act of things*

In response to industrial accidents, as will be examined in the context of exploding boilers, article 1384, paragraph 1 C civ. came to be interpreted as meaning that the guardian of a thing is held liable for damage caused by it, without the need to prove fault.[32] This had the result that it was not necessary to establish fault and, with respect to liability for fire caused by sparks from locomotive engines, it allowed claimants to rely upon article 1384, paragraph 1, rather than the presumption of causation established under article 1382.

Thus, in a case of 6 May 1902,[33] the Cour de Toulouse found that the locomotive had been 'furnished with equipment to prevent the escape of burning material from the grille and to prevent the escape of burning coals through the chimney', in accordance with article 11 of the Ordonnance of 15 November 1846 and no negligence or recklessness had been shown on the part of the railway company; however, the Cour de Toulouse found the railway company liable under article 1384, paragraph 1 C civ. for damage caused to a forest by a fire started by small pieces of burning coal or sparks.

As the custodian of a locomotive engine, the railway company would also be liable for a fire started by sparks escaping from the wheels of the engine.[34] It was still necessary to prove that the locomotive had emitted the sparks or other material capable of starting a fire;[35] if the cause of the fire was unknown, no one having seen the escape of sparks or burning coal, third parties who suffered harm had to prove the fault of

[32] See Note of Labbé to Cass. 19 July 1870, S.1871.1.9; R. Saleilles, *Les accidents du travail et la responsabilité civile: Essai d'une théorie objective de la responsabilité individuelle* (Paris: A. Rousseau, 1897), p. 26; L. Josserand, *De la responsabilité du fait des choses inanimées* (Paris: A. Rousseau, 1897), p. 53 and A. Esmein, note on Cass. civ. 16 June 1896, S.1897.1.17, and on Cass. civ. 23 February, 29 and 30 March 1897, S.1898.1.65.

[33] Toulouse, 6 May 1902, S.1905.2.105, note Appert.

[34] See also, Bordeaux, 28 March 1904, Req. 24 July 1905, D.P.1908.1.183; Req. 17 July 1922, D.P.1923.1.172; Trib. civ. Saint-Nazaire, 23 May 1924, *Gaz. Trib.* 30 November 1924; *Rev. trim.* 1925, 109 and 110; note Josserand, D.P.1926.1.130; Bordeaux, 28 February 1927, *Gaz. Pal.* 2 April 1927; Trib. civ. Issoudun, 25 November 1930, *Gaz. Pal.* 9 January 1931; Paris, 8 January 1932, D.P.1932.2.69, note H. Lalou; Aix, 12 May 1934, *Gaz. Trib.* 1 September 1934; Trib. civ. Verdun, 11 March 1936, *Rec. pér. assur.* 1936, 284; Trib. civ. Orange, 21 February 1939, *Rev. trim* 1939, 750, D.H.1939. 382; Req. 1 December 1942, D.A.1943. som. 8; *Rev. trim* 1943, 115; *Rev. trim* 1945, 114; Cass. civ. 20 January 1948, D.1948.201; Cass. civ. 30 January 1950, D.1950.390; Tribunal de paix Bessèges, 29 May 1952, *Gaz. Pal.* 10 October 1952.

[35] Alger, 8 March 1926, D.P.1928.2.112.

the *gardien* or those for whom he was responsible.[36] Again, in the absence of eye-witnesses, circumstantial evidence was adduced to establish a presumption of causation. Sufficient proof that the cause of the fire had issued from a locomotive could be made by 'substantial and precise presumptions', such as when a fire immediately followed the passage of a train, the railway bank was too low, there were strong winds and no other likely perpetrator was observed in the area.[37]

Special measures with respect to railway regulation continued to play a role in combination with article 1384, paragraph 1. Thus, the Tribunal civil de Laval[38] found a railway company liable, by virtue of article 1384 C. civ., for a fire which broke out in a farm next to a railway line. It was claimed that under the terms of article 7 of the Law of 1845, the haystacks were within a distance of 20 m from the railway line, and so the railway company could avoid liability. However, it was held that where a permanent storage site for crops had been established before the construction of the railways it could continue to exist within the protected zone, unless it could be shown that at the time of the expropriation, the land-owner had been compensated for the loss of use of the zone. In the absence of such compensation, the railway company was held liable under article 1384, paragraph 1.

(f) *Special regime for damage caused by fire*

In 1920, the Cour de cassation appeared to apply article 1384, paragraph 1 to buildings, in a case in which the damage consisted in the communication of a fire to neighbouring properties.[39] The case concerned a fire starting in a railway station and fed by barrels of resin stored there, which then spread to cause damage to the property of a third party. Thus, it did not in fact concern damage caused by a building, but damage caused by barrels of resin, moveables which were under the supervision of the railway company. The application of article 1384, paragraph 1 was thus in perfect conformity with the case law on the article at this time.

However, subsequent judgments used this case as a basis to apply article 1384, paragraph 1 to damage caused by a fire which started in a building and spread to neighbouring buildings or chattels, to the effect that the victims did not need to prove the fault of the owner of the property

[36] Trib. civ. Saint Girons, 24 January 1951, somm. 72.

[37] Trib. civ. Orange, 21 February 1939, D.H.1939.382.

[38] 12 December 1921, *Gaz Pal.*, 1922.1.604.

[39] Cass. civ. 16 November 1920, D. P.1920.1.169, note Savatier.

where the fire had started in order to obtain compensation.[40] Insurance companies demanded legislative intervention to rein in the expansive interpretation given to article 1384, paragraph 1, and the response was the Law of 7 November 1922,[41] which inserted a second paragraph:

> However, he who possesses by whatever right all or part of a building or of personal property in which a fire occurs is liable vis-à-vis third persons for damage caused by such fire only if it is proved that it should be attributed to his fault or to the fault of persons for whom he is responsible.

The law only covered instances of accidental fire. He who had the *garde* of the inanimate object from which a non-accidental fire had escaped was strictly liable without the need for fault to be proved. However, the rapporteur of the Law of 1922 to the Chambre des Députés expressly stated that 'Cases in which fire has been communicated by an explosion, sparks or fragments of burning coal from a locomotive are not covered by the law.'[42] This conclusion is explained in part by the fact that it was necessary that the fire 'started' in the property of the *gardien*, which implied that the thing had itself fallen prey to fire: one could not say that the fire had accidentally started in the locomotive. The Law of 1922 thus effected no change with respect to liability for fire caused by locomotive engines and article 1384, paragraph 1, continued to apply.

2. Liability for exploding boilers in French law

(a) *Introduction: the historical context*

'Steam power today constitutes almost the only source of power for industry',[43] wrote Armand Béhic, the Minister for Public Works, in 1865. Indeed, in 1850 there were 6,832 steam boilers in use, but this had risen to 22,516 by 1863.[44] However, steam was recognised as a 'form of power which has its dangers'.[45]

40 Toulouse, 8 June 1921, D.P.1922.2.20; Trib. civ. Seine, 25 February 1921, *Gaz. Trib.* 2 June 1921.

41 R. Dufourmantelle, 'Les principes de responsabilité en matière d'incendie et les modifications législatives proposées aux articles 1384 et 1386 du Code civil' *Journal des assurances* 1922, 99.

42 Reproduced in H. Lalou and P. Azard, *Traité pratique de la responsabilité civile* (Paris: Dalloz, 1962), p. 725.

43 A. Béhic, 'Rapport a l'Empereur sur la fabrication et l'établissement des machines et chaudières à vapeur' in V. Vidal, *Legislation des machines à vapeur: Décret du 25 janvier 1865* (Paris: De Noblet and Baudry 1865), p. 11.

44 *Ibid.* p. 12. 45 *Ibid.*

A number of distinct constituencies of victim may be affected by a boiler explosion: (1) a neighbour may complain of nuisance, which can be considered when examining the regulation of steam boilers and planning law; (2) an employee of a public or private body may suffer injury,[46] (3) the owner may suffer damage to his factory. Each of these has distinct mechanisms of compensation.

There are relatively few reported cases on liability for exploding boilers, with the majority of reported cases on boilers concerning nuisance, and it must be concluded that civil liability was not a major feature in dealing with the issues to which boiler explosions gave rise. The key feature would appear to be the great capacity of the French administration of the time to engage in regulation and inspection. Significant boiler explosions in the 1820s led to a series of rather piecemeal legislative initiatives until an overhaul of the law resulted in major regulation in 1843.[47] It is clear that attempts are made to balance public safety against the interests of industry, which often results in a tug of war between interventionist and *laissez-faire* policies. Particular attention must also be paid to the role of insurance, which developed in parallel with advances in steam technology.

(b) Regulation of steam boilers: authorisation to install and safety measures

Planning measures and administrative rules seem to have formed the major method to handle matters of safety and installation, in particular in the early years of the age of steam, which may explain the paucity of reported cases on the matter.

(i) The Decree of 1810 on 'unhealthy and inconvenient' establishments

From a very early stage the installation and use of steam-powered machines was considered a matter of industrial nuisance, and planning law was the chosen vehicle to manage issues arising from the new technology. The use of steam-powered machines inflicted hitherto unknown problems of smoke, polluted water, noise and vibration on those neighbouring the factories. New tensions developed between industrialists and

[46] Public bodies became subject to a distinct set of legal rules after 1873: Tribunal des conflicts, 8 February 1873, *Blanco* S.1873.2.153. As the owners of boilers could either be public bodies (such as armaments factories, tobacco factories or the armed forces) or private bodies, this must also be borne in mind.

[47] Ordonnance of 22 May 1843.

property owners: an 'urban crisis'[48] which resulted in a Decree of 1810 on 'unhealthy and inconvenient establishments'.[49] The 1810 Decree provided that all factories which might emit noxious or disagreeable fumes would have to seek authorisation from the administration. The text was only concerned with fumes or odours: no other form of nuisance was considered.

Factories were grouped into three categories: the first concerned establishments which should be placed far from domestic residences, although no distance was specified; the second, those which did not necessarily have to be far from domestic residences, but which had to ensure that the activities pursued did not inconvenience neighbours. The third category concerned those establishments which could neighbour residential areas but would be monitored. While a table stipulating the industries to be grouped under the different categories was annexed to the décret, no administrative or criminal sanction was imposed on industrialists who contravened the law. All decisions concerning factories of the first and second categories were undertaken by the préfet of the département.

The fact that the décret spoke of the protection of 'owners of neighbouring property', and not the protection of neighbours per se, may be significant. Rather than a measure to protect public health or to combat industrial nuisance, it might be considered to be a planning measure: to manage urban areas and mediate between industrialists and private property owners. For example, article 9 applied the principle of anteriority: 'any individual who builds in the area of such authorised factories or workshops, may not demand that the factories be moved further away'. While a strict interpretation of the 1810 décret would seem pro-industrialist, its application had consequences which had not been foreseen by the legislature.

The speed and uniformity of application of the décret varied from département to département and authorisation remained largely a formality, but as the nineteenth century progressed, complaints became more numerous, not only against companies who had failed to seek authorisation, but against companies who already had permission to operate in a particular area.[50] Such actions were brought with respect not only to odours but also to smoke, dust and the pollution of groundwater (and towards the end of the century, noise). These claims were based on

[48] G. Massard-Guilbaud, 'La régulation des nuisances industrielles urbaines (1800–1940)' (1999) *Vingtième Siècle. Revue d'histoire, No. 64, Numéro spécial: Villes en crise?* 53 at 54.

[49] Décret impérial of 15 October 1810 relatif aux Manufactures et Ateliers qui répandent une odeur insalubre ou incommode, *S. lois annotées* 1789–1830, 834.

[50] Massard-Guilbaud, above note 48, at 56.

the 1810 décret which was given an extended scope in the case law.[51] Such actions often demanded that a factory be relocated some distance away from a domestic dwelling. By the last quarter of the nineteenth century, the decree was used to challenge industrialists who found themselves in an increasingly defensive position.

In addition to the Decree of 1810, steam boilers became regulated by increasingly numerous safety measures. Such regulations were rather interventionist in style and imposed duties to monitor and test boilers. So, for example, an Ordonnance Royale of 2 April 1823 regulated the use of steam-operated water pumps following a cluster of accidents. Then later the same year,[52] and in 1828,[53] 1829,[54] 1830[55] and 1842, the body of regulations was added to in a rather piecemeal fashion, so, for example, the Ordonnance of 28 January 1842 noted the dangers of the use of cast iron and a later Ordonnance demanded the replacement of flexible disc couplings with valves.[56]

(ii) The Ordonnance of 1843: the legislature intervenes with respect to steam boilers

General legislation was enacted in 1843[57] and it constituted real progress from the earlier rules as it recognised technological advances and attempted to unify the many different regulations to which steam boilers were subject. All parts of the boiler, including any accessory parts and safety mechanisms,[58] were individually regulated (articles 22–34) and tested (articles 14–21), and any changes or repairs would be subject to new tests (article 72). The materials used and the methods of construction were also strictly controlled. Every detail of the use and installation of boilers was regulated by administrative rules.

The conditions to which each steam boiler was subject differed according to the capacity of the boiler and the level of steam pressure. According to these considerations, steam boilers were separated into four categories,

[51] G. Massard-Guilbaud, 'Die Unannehmlichkeit zum Ausdrück bringen' in C. Bernhardt (ed.), *Umwelt europäischer Städte im 19 und 20 Jahrhundert/Environmental Problems of European Cities in the 19th and the 20th Century* (Munster: Waxman Verlag, 2001).

[52] 29 October 1823 Ordonnance Royale relative aux machines à vapeur à haute pression.

[53] 7 May 1828 Ordonnance Royale relative aux machines à vapeur à haute pression.

[54] 23 September 1829 Ordonnance Royale relative aux machines à vapeur à haute pression.

[55] 25 March 1830 Ordonnance Royale portant règlement sur les appareils à vapeur.

[56] 22 May 1843.

[57] Ordonnance Royale of 22 May 1843 relative aux machines et chaudières à vapeur autres que celles qui sont placées à bord des bateaux, *S. lois annotées* 1831–1848, 746.

[58] Safety mechanisms included safety valves (arts. 22–24), manometers (arts. 25–27), water level indicators (arts. 28–31) and separate feeders for multiple boilers (art. 32).

with the most powerful boilers only allowed to be established outside of residential houses and workshops (article 34 of the Ordonnance), and 10 metres from a residential house or a public highway, with a defensive wall 1 metre thick (article 36). Second category boilers could be placed inside a workshop as long as it did not form part of a residential house or a factory of more than one storey (article 38). These boilers had to be 5 metres from residential houses or the public highway and again a 1 metre thick defensive wall must be built (article 39). Boilers of the third category might be placed inside a workshop as long as the latter did not form part of a residential house and no defensive wall was required (article 42). Boilers of the fourth category might be placed inside a workshop even if this was part of a residential house.[59] Separate rules existed for the installation of boilers inside mines (article 46), mobile boilers (articles 47–51) , locomotive boilers (articles 52–60) and steam boilers installed on ships or boats.[60] The professional qualifications necessary for boiler operators were stipulated, in addition to the maintenance procedures to be followed (article 69).

Article 4 of the Ordonnance of 1843 expressly recognised the role of the law of 1810: 'steam boilers ... can only be installed if authorisation has been given by the préfet of the department conforming to that prescribed by the decree of 15 October 1810 on unhealthy and inconvenient establishments of the second category'. In this way, steam boilers could only be installed or authorised after an enquiry in which all interested parties were heard (articles 9–13), after which engineers would meet on site to establish that the conditions of installation and the distance from neighbouring buildings or public highways were observed. Authorisation would then be granted by a decision of the préfet, who would determine the conditions to be observed by the permit-holder.

However, article 11 provided that the decisions of the préfet could be challenged by industrialists to whom authorisation was denied,[61] or third parties who contested the granting of permission,[62] causing great loss of time to industry. The complexity and stringency of the rules and procedures led some to consider it a regime 'prejudiced against industry'.[63]

With respect to liability, in practice almost all difficulties which could arise were dealt with by préfets, monitoring commissions, the Central

[59] Article 43. Those boilers in the third and fourth categories must be 50 cm from third party residential homes: art. 44.

[60] Steam boilers on ships were regulated by a special ordonnance: art. 1.

[61] By way of a challenge before the Conseil d'État.

[62] In the first instance before the Conseil de Préfecture and then the Conseil d'État.

[63] Béhic, above note 43, at p. 16.

Commission for Steam-Powered Machines[64] and the Ministry for Agriculture, Public Works and Commerce. In the event of an accident, the local police and an authorised engineer would go immediately to the scene, which had to remain undisturbed, and then transmit their findings to the préfet and if necessary the public prosecutor. The causes of the accident could thus be established at the public expense, which would assist the claim of any victim (employee, neighbour or owner). In practice the courts were only called upon to determine compensation in the aftermath of accidents which had resulted in death or serious injury. There is very little case law on the matter.[65]

The Ordonnance of 1843 noted that factory owners would be responsible for any accidents or damage resulting from the negligence or incapacity of their agents, in accordance with the dispositions of article 1384.[66] Penal sanctions for the infringement of preventive measures were introduced by article 69 of the Law of 21 July 1856.

However, due to the speed of industrial progress, the stringent controls placed on the materials and methods of boiler construction were increasingly derogated from, albeit in a piecemeal way, to avoid 'industrial paralysis'.[67] But these limited concessions became increasingly inadequate. By 1865, a great deal of progress had been made with respect to the production and design of boilers, and steam power had found new and unforeseen applications. Boilers had changed both in design and in substance, as the materials from which they were made were of higher quality; in addition workers charged with the control of boilers were more experienced and more numerous. Following a far-reaching enquiry, the Central Commission for Steam-Powered Machines issued a report on the matter, which was then taken up by the Minister for Agriculture, Public Works and Commerce, Armand Béhic.[68]

(iii) The Decree of 1865 brings a degree of deregulation

The report of the Minister recognised that by 1865, steam boilers had an integral and ubiquitous role in industry, and that the evolution of technology meant that many of the preventive measures imposed by earlier legislation could be attenuated without harming the greater public interest. The resulting Law of 25 January 1865 deregulated the installation of steam

[64] Commission centrale des machines à vapeur, which was attached to the Ministry for Agriculture, Public Works and Commerce.

[65] V. Vidal, *Legislation des machines à vapeur: Décret du 25 janvier 1865* (Paris: De Noblet et Baudry, 1865), p. 8.

[66] Article 69. [67] See Béhic, above note 43, at p. 15. [68] *Ibid.* p. 16.

boilers and relaxed certain safety regulations. The role of the administration was reduced, its intervention replaced by the liability of the owners of steam boilers and by the monitoring of third parties who aimed to protect their own interests. Boilers no longer required prior authorisation: their regulation was divorced from the Law of 1810, so that they would no longer fall under the rules on 'unhealthy and inconvenient establishments': allowing the lengthy administrative procedures to be replaced with a simple declaration to be made to the préfet of the département (article 10). This seems to be the most significant aspect of the new rule: it replaced case-by-case examination by the préfet with general rules. In addition it allowed third party neighbours to release boiler owners from the need for the fulfilment of the installation conditions (article 18).

The law contained a classification of types of boilers (closed boilers, fixed boilers and mobile boilers),[69] with each subject to different forms of prior inspection and testing. As in the 1843 regulations, different rules applied according to the power and capacity of the steam boiler, in particular with respect to location, such as the distance from residential dwellings (article 46). In the new rule, the conditions were significantly relaxed. There were three categories of steam boiler instead of four. Boilers of the first (most powerful) category could be installed in single-storey workshops, for example, and need only be 3 metres from a domestic residence or public highway, with a defensive wall only required in certain cases (article 13); boilers of the second category could be placed inside a workshop without any requirement for a defensive wall,[70] and boilers of the third category might be installed in a workshop even when it formed part of a residential dwelling inhabited by third parties.[71]

The various safety mechanisms required for a boiler were stipulated in articles 5 to 9, but in contrast to the original 1843 rules, only general conditions were set and the industrialist could construct and use the boiler as he wished: the nature and quality of the materials used were left within the authority and responsibility of the engineer. The new rules of 1865 maintained testing for the boilers but did not require the testing of

[69] It was recognised that rules with respect to locomotives were now regulated predominantly by rules of public administration, as has been explored in the previous section: arts. 22–27. In addition, there was a separate ordonnance of the same year with respect to steam boilers on ships and boats.

[70] Article 15; and needed only to be 1 metre from third party residential areas (art. 17).

[71] Article 16; and needed only to be 50 cm from third party residential areas (art. 17).

accessory parts, and reduced the rigorousness of such tests.[72] A regime of inspection was also retained (article 28).

Recognising that smoke had become 'the greatest inconvenience for neighbours',[73] the law introduced the requirement that all boilers have a smoke filter or smoke-reducing mechanism of sufficient efficiency (article 19), The obligation to filter smoke was absolute, but the 'courts have great latitude; the rights of neighbours will not be the same in a town and in an industrial centre which is built around a significant factory and which only exists by virtue of it'.[74]

With respect to liability, article 30 retained the inquiry into any accident or explosion: where there is an accident resulting in death or serious injury, the owner or manager of the establishment must immediately notify the local authorities and an inspecting engineer who would in turn report to the préfet: again providing a method to ascertain the causes of an accident at the expense of the public purse. The discontinuance of the system of administrative authorisation means that henceforth any problems which might emerge were handled by the ordinary courts, in contrast to the regime of the *ordonnance* of 1843, which only allowed decisions of the préfet to be challenged by recourse to the Conseil d'État. Those who suffer damage due to the infringement of safety measures are also allowed to obtain civil remedies (article 57). Furthermore, article 6 of the law of 1856 continued to apply, so that the failure to observe safety measures would incur criminal liability.[75]

In practice, it seems to have been hard to establish liability. Difficulties of proof arose in showing that clear breaches of the planning legislation relating to boilers caused the loss in question.[76] Given the paucity of reported cases, it would appear that the regulatory regime proved adequate for identifying those responsible for explosions and for punishing them through the criminal law. Reported cases would seem to suggest that defective systems were often in place for many years before an explosion, implying that the inspection regime was not strictly or efficiently implemented, possibly due to the overwhelming increase in the number

[72] The dimensions of various accessory parts were not stipulated at all, the only requirement being that these were 'adequate'.

[73] Vidal, above note 65, at p. 74. [74] *Ibid.* p. 75.

[75] There are very few reported criminal cases on the matter, however, perhaps because few would be considered by the higher courts.

[76] Req. 7 March 1892, DP 1892.1.502: it was not shown that a breach of regulations with respect to the wooden construction surrounding the boiler caused the loss; also Aix 4 July 1894, DP 1895.2.167.

of boilers in use. It may also be that the spread of enterprise and property insurance meant that fewer actions were being brought.[77]

In the wake of the law of 1865, an association of the owners of steam boilers was formed with its own statute to control the selection of engineers for their own machinery.[78] A later Decree of 24 March 1919 allowed associations of steam boiler owners to have authority over the control and inspection of their members' machines.

(c) *Employees and industrial accidents*

(i) Public law

In 1873, the Tribunal des conflits decided that public authorities were not governed by private law legislation, but rather the 'legislation of justice'.[79] The basis of liability is the benefit-burden principle:[80] the state benefits from the cooperation of the employee who acts in the public service and so ought to compensate him for the unequal burden that he suffers. The *Cames* decision of 1895 was taken against the background of this evolution and marked a change for public law factories. Cames had 'injured his left hand ... when a fragment of metal was thrown out by a power hammer',[81] and was unable to work. It was impossible to find a *faute*, error or negligence, either on the part of Cames or on the part of his state employer.

Facing a legal void, the Conseil d'État applied a negatively formulated rule that 'in the circumstances it ... is not in a position to find that the state has not incurred any liability', and so found the public company liable. Commissaire du gouvernement Romieu based his argument more specifically on liability for public works, which is strict.[82] That principle is also supported by a pragmatic argument that the state is well placed to assume the burden and use parliamentary credits to pay it (drawn from taxation, rather than from insurance, i.e. the state is a self-insurer).[83] The Conseil d'État followed the reasoning of Romieu and no further controversy existed in the administrative courts.[84]

77 G. Hamon, *Histoire générale de l'assurance en France et à l'étranger* (Paris: 1895), p. 242 *et seq.*

78 M. Amar, 'Legislazione sugli stabilimenti insalubri, pericolosi e incomodi, sulle macchine e caldaie a vapore, regolamenti d'igiene, edilità e polizia locale' (1878) *Archivio giuridico* 21 at 164.

79 "*Equité*": Tribunal des conflits, 8 February 1873, *Blanco* S.1873.2.153; see also Tribunal des conflits, 30 July 1873, *Pelletier* Lebon, 117.

80 Declaration of the Rights of Man, art. 13.

81 CE 21 July 1895, S.1897.3.33, note Hauriou.

82 Under the Law of 12 Pluviôse An VII (1798).

83 M. Hauriou, R.D.P.1896.2.51 *et seq.*

84 L. Duguit, *Traité de droit constitutionnel*, Tome III (Paris: Fontemoing, 1930), p. 469.

(ii) Civil law

Fault-based liability under article 1382 of the Code civil French civil law remained wedded to fault, and liability for an exploding boiler would generally rest on proof of fault under article 1382.[85] However there were clear problems with respect to the burden of proof and an injured employee would often face problems in establishing the *faute* of his employer and causation. Furthermore an employee would have to show a construction defect in order to bring an action against the manufacturer under article 1382.[86] Liability might well be split between the owner, manufacturer and inventor in proportion to their fault.[87] If an accident was caused solely by the employee, and no *faute* could be attributed to the employer, the worker would be entirely without recourse.[88] It would appear that the overriding implications of the rigours of article 1382 were that 'in accepting the job, the employee freely accepts the professional risks of the work and may not then bring a claim against his employer for the consequences of his work'.[89] Statistics from 1888 would suggest that 68 per cent of industrial accident cases were considered to be instances of *cas fortuit* or *force majeure*, 20 per cent due to the *faute* of the worker himself, and only 12 per cent were found to be due to the *faute* of the employer: the only situation in which the worker could obtain compensation.[90] Various means were sought to assuage the situation of the employee.

In Belgium, the liability of employers was established in contract under article 1710 of the Civil Code governing the hire of services, supplemented by article 1135, which imposes equitable principles in addition to the express terms of an agreement, so that the employer would owe a duty to provide a safe system of work.[91] The employer would be liable for failure to foresee dangers to employees, but the burden of proof remained on the employee to show fault.[92] In France an attempt was also made to base the liability of the employer on violation of a contractual obligation. Some claimed that a tacit guarantee clause existed within the contract of employment (what would come to be recognised as the contemporary

85 Civ. 19 July 1870, D.P.1870.1.361.

86 Trib. com. Seine, 21 September 1905, D.P.1906.5.15.

87 Req. 17 July 1872, D.P.1873.1.206.

88 Req. 9 March 1880, D.1880.1.296; Req. 17 November 1894, D.1895.1.368.

89 A. Vassart and A. Nouvion-Jacquet, *La loi du 9 avril 1898 sur les accidents industriels* (Paris: Larose, 1899), p. 14.

90 Figures presented in parliamentary discussions: JO, Chambre, séance of 28 June 1888, 1926.

91 Vassart and Nouvion-Jacquet, above note 89 at p. 8.

92 Cass. Belgique, 8 January 1886, D.P.1886.2.153; Cass. Belgique, 28 March 1889, S.1890.4.17.

obligation de sécurité) but its time had not yet come and this solution was rejected.[93]

There was a movement in the case law to appreciate *faute* in a particular way in the case of industrial accidents, so that the fault, and liability, of the employer could be found with greater ease.[94] The fault of the employer was, on occasion, established in rather controversial circumstances: in a case of 27 April 1877, having stated that it would have been absolutely impossible to prevent the accident, the court added that it was nevertheless unacceptable that the worker should be left without protection, and censured the employer for failing to ascertain what no one could have discovered.[95] Employers were also held at fault if they failed to take all necessary precautions to protect their workers, even against their own lack of skill.[96]

Likewise, in a judgment of 7 August 1895, the Cour de cassation stretched the causal link between fault and damage to award compensation to an injured worker.[97] An employer had retained a 15-year-old worker beyond the legally prescribed working hours, during which time he had a disagreement with a colleague and fell into a machine, crushing his skull. The Cour d'appel had refused to find a link between the violation of the rules on working hours and the damage. However this was criticised by the Cour de cassation, which found that the employer had committed a fault which had the consequence of the prolongation of the presence of the young man at the printing press, without which it would have been impossible for the boy to have committed the act of indiscipline: the employer had 'contributed by his fault to the resultant accident suffered by his apprentice'. Contributory negligence was found and apportionment stipulated. In his commentary on the judgment, Planiol considered this to be very significant 'given the generalisation to which it is susceptible'.[98] With regard to these decisions, Saleilles concludes that '*faute*, which rests on a voluntary act, is henceforth to be seen from an objective perspective: what one could call objective fault'.[99]

In Luxembourg, the legal basis also remained article 1382 of its Civil Code, but the view was taken that the use of machines was dangerous per

[93] Rennes, 20 March 1893, D.P.1893.2.526. Also note Saleilles (on *Teffaine*): critical of the attempt to base liability on a tacit guarantee clause: D.P.1897.1.433.

[94] Vassart and Nouvion-Jacquet, above note 89, at p. 14.

[95] Dijon, 27 April 1877, D.P.1878.1.383.

[96] Nancy, 18 November 1897, Gaz. Pal. 1898.1.314.

[97] Cass. civ. 7 August 1895, *Jousset c. Aubert*, D.P.1896.1.81.

[98] Cass. civ. 7 August 1895, *Jousset c. Aubert*, D.P.1896.1.81, note Planiol.

[99] D.P.1897.1.433, note on *Teffaine*.

se, and so a presumption of fault was established. The employer had to show that there was *force majeure* in order to exonerate himself, but this could not be shown simply by demonstrating that the machine had malfunctioned, if this was a normal risk in the industry in question.[100]

Insurance also seems to have been of some influence here. For example, when considering which employer could be sued when an employee was loaned to another company and was then injured, the choice of which employer to sue would appear to turn on which was paying remuneration and contributing to the insurance fund to provide relief for the injury suffered to the employee.[101]

By the middle of the nineteenth century, the role of insurance was increasing.[102] In 1861, an insurance company was founded[103] which issued a collective insurance policy with two functions: to guarantee compensation to injured workers, regardless of their fault or otherwise; and to provide liability insurance for the employer where fault was established. Such insurance was encouraged by the government and a second specialised insurance company, the Soleil-Sécurité générale, was founded in 1865. A public insurance fund for industrial accidents[104] was created in 1868 to provide greater insurance coverage,[105] but this met with great hostility from employers, who preferred the conditions offered by private companies, and had little success.[106] Nonetheless, by 1898 it is estimated that half of all industrial workers active in France were already protected by private insurance.[107]

Evolution of article 1384(1) of the Code civil: liability for the act of things In his conclusions on the *Cames* case, commissaire de gouvernement Romieu had suggested that article 1384(1) of the Code civil could form the basis for no-fault liability of private employers for harm suffered by workers,[108] because such injuries were caused by a machine

[100] Cour supérieur Lux., 27 November 1884, S.1885.4.25.

[101] Paris, 27 February 1892, D.P.1892.2.465; Req. 7 June 1896, D.P.1896.1.207.

[102] P. J. Richard, *Histoire des institutions d'assurance en France* (Paris: Editions de L'Argus, 1956).

[103] *La Préservatrice*, see Hamon, above note 77, at p. 618.

[104] Caisse d'assurance contre les accidents du travail.

[105] Law of 11 July 1868. [106] Hamon, above note 77, at pp. 164–5, 644.

[107] Richard, above note 102, at p. 107; albeit the lack of a public insurance fund allowed private companies to set allegedly high premiums: P. Pic, *Traité élémentaire de législation industrielle: les lois ouvrières* (Paris: Arthur Rousseau, 1909), p. 825.

[108] His analysis of art. 1384(1) seems to fluctuate between objective liability and presumed fault: CE 21 July 1895, S.1897.3.33, note Hauriou. On this case law, see in general F. Leduc *et al.*, *La responsabilité du fait des choses, réflexions autour d'un centenaire* (Paris: Economica, 1997).

and not by the individual, and in recognition of the availability of insurance. He noted that an employer need only pay a small premium to 'insure the lives of his employees as he insures his home or his business'.

A year after Romieu's comments, the Cour de cassation reacted. In *Teffaine*, an employee of a private company was killed when a steam boiler exploded on a tug. The circumstances were such that it was impossible to find *faute*, error or negligence on the part of the employee, or *faute* on the part of his employer.[109]

The Cour d'appel de Paris had awarded compensation on two grounds: first, on the basis of the labour contract, which it held to be supplemented by a *clause de sécurité* on the basis of article 1160 C civ., which allows the court to read implied terms into an agreement; and secondly, under article 1386 by virtue of the fact that the situation in question was 'analogous' to damage caused by a building in disrepair. The court explained that its reasoning was based on the fact that it was 'equitable to decide' to hold the owner of the machine liable.[110]

On appeal, the Cour de cassation developed liability for the act of things (*responsabilité du fait des choses*) on the basis of article 1384(1).[111] Hitherto, this subsection had been considered to serve a merely introductory purpose so that 'the scope of the disposition of article 1384 relative to things is limited by articles 1385 [which governs liability for harm caused by animals] and 1386 [liability for harm caused by buildings]'.[112] Certain judgments rendered by lower courts had already challenged this view, and had awarded compensation on the basis of article 1384(1),[113] but this line of case law was far from constant or universal.[114]

However, in *Teffaine* the Cour de cassation recognised for the first time that article 1384(1) could be considered an autonomous provision.

109 Cass. civ. 16 June 1896, *Oriolle, Guissez et Cousin c. Teffaine*, S.1897.1.17, note Esmein; D.1897.1.433, concl. Sarrut, note R. Saleilles. The workmen's compensation Bill was already in Parliament, so it might be said that the Conseil d'Etat and the Cour de cassation were merely bowing to the inevitable.

110 Cour d'appel de Paris, 19 May 1893, D.1897.1.435–439.

111 See W. Ernst (ed.), *The Development of Traffic Liability* (Cambridge University Press, 2010), pp. 58–9.

112 Avocat général Sarrut in *Teffaine*: D.1897.1.439–440, who had proposed the quashing of the judgment of the Cour d'appel; the judgment of the Cour de cassation went against his findings.

113 The first example of such an application would appear to be that of the judgment of 26 April 1888 of the Tribunal de Gien, S.1890.2.221, which found in art. 1384(1) a general principle which allowed the presumption that the guardian is responsible for damage caused by things in his custody. This reasoning was followed by the Tribunal de Bourgoin of 10 June 1891 and was confirmed by the Cour d'appel de Grenoble on 10 February 1892 (Tribunal de Bourgoin, 10 June 1891, Cour de Grenoble, 10 February 1892, S.1893.2.205).

114 Leduc *et al.*, above note 108.

Given that the Cour d'appel had accepted that the boiler had suffered from a construction defect, 'which excludes *cas fortuit* and *force majeure*', the Cour de cassation found that there existed in the 'terms of article 1384 of the Code civil' the basis of the liability of the owner of the tug 'excluding any recourse to proof of fault on the part of the producer of the boiler, or to the hidden nature of the defect'. The judgment would appear to suggest that the only possible means of exoneration would be the rupture of the chain of causation by an extraneous event, unrelated to the machine.

The issue arose again in a judgment rendered by another chamber of the Cour de cassation on 30 March 1897.[115] Grange was killed when a boiler exploded on a ship. Compensation was sought on the basis of article 1384 C civ., claiming that it was for the defendant shipping company to prove a means of exonerating their liability. The court of first instance rejected the claim because 'the existence of a legally imputable fault is an essential element in an action for liability' and so the burden of proving fault lay with the plaintiff. The Cour de cassation followed the same reasoning, stating that 'it is absolutely impossible to determine the cause of the accident, which should be considered a *cas fortuit* ... which cannot incur any liability' and upheld the judgment of the lower court, rejecting the claim brought by Grange's widow.

A new reading was imposed by the *Grange* decision, as the relative autonomy of article 1384(1) seemed to be limited to transferring the liability of the producer to the guardian of the machine, but without imposing true objective liability.[116] It was argued that rather than being contradictory to the judgment in *Teffaine*, the Cour de cassation had introduced unforeseeability and unavoidability as extraneous causes which could exonerate the defendant from liability, whereas *Teffaine* had seemed to limit any exoneration to circumstances of *exteriorité* (external acts).[117]

While the presumption of *faute* became harder for the guardian to rebut from 1919,[118] the two divergent lines of case law which emerged from *Teffaine* and *Grange* would not be resolved until the *Jand'heur* case of 13 February 1930,[119] which confirmed that the guardian of a thing may only be exonerated from the presumption of liability under article 1384(1)

[115] Req. 30 March 1897, *Veuve Grange c. Compagnie générale Transatlantique*, S.1897.1.440.

[116] The Chambre des requêtes followed its traditional position: Req. 5 April 1894, *Judet c. Compagnie de Chatillon-Commentry*, D.1894.1.479.

[117] D.1897.1.433, concl. Sarrut, note R. Saleilles.

[118] Civ. 21 January 1919 and Civ. 15 March 1921, D.1922.1.25, note Ripert; Civ. 16 November 1920, D.1920.1.169, note Savatier.

[119] *Jand'heur c. Les Galeries Belfortaises*, 13 February 1930, Cass. Ch. réun., S.1930.1.121.

'if there is proof of a *cas fortuit*, *force majeure* or an external cause which is not imputable to him: it is not sufficient to prove that no *faute* was committed' by him.

In spite of the new reading of article 1384(1) C civ., an employee could still be found contributorily negligent, which accounted for a large number of cases.[120] Furthermore, civil procedures were often lengthy and costly, with judges apparently awarding somewhat 'arbitrary' and unpredictable levels of compensation.[121] Ultimately, the influence of article 1384(1) had a brief reign with respect to industrial accidents, as a Law of 9 April 1898 instituted a new regime.

(iii) Legislative intervention in the industrial accident context

The medical profession and concerned philanthropists began to draw attention to the seriousness of the problem of industrial accidents in the second half of the nineteenth century. However, industrial accidents were for many years consistently attributed to worker negligence until the 1850s, when it was realised that human capital was a resource to be conserved. Statistical research conducted by the Conseil du Travail[122] highlighted the extent of the problem, which was further aggravated as mechanisation spread.[123] Under the Second Empire, funds were established to assist the victims of industrial accidents, financed by means of fines, employee contributions and employer donations. The evolution of this practice relied a great deal on the will of business to avoid recourse to the courts. In the face of employer-led initiatives, industrial action remained rare: in 1897, only 1.6 per cent of strikes concerned the issue of industrial accidents.[124] The recognition of freedom of association on 21 March 1864 instigated a new wave of legislation to protect the interests of workers in a newly industrialised society.

In 1880, the Chambre des députés voted for a law which had been proposed by Martin Nadaud, substituting the principle of professional risk and a system of compulsory insurance for private law liability. Due to

[120] P. Pic and J. Kréher, *Le nouveau droit ouvrier français* (Paris: Pichon, 1943), p. 254.

[121] *Ibid.* [122] Which had been instituted by a Law of 19 May 1874.

[123] More specifically a 'bulletin on accidents caused by the use of steam-driven machines' was published in 1894 and 1895: 'Bulletin des accidents arrivés dans l'emploi des appareils à la vapeur pendant l'année 1894', JO 1895, 5411 and 'Bulletin des accidents arrivés dans l'emploi des appareils à la vapeur pendant l'année 1895', JO 1896, 616.

[124] V. Brumeaux, *La question des accidents du travail à la fin du XIXe siècle* (Paris: Université Paris X-Nanterre, 1979).

opposition in the Sénat, the law did not succeed. However, in response to the proposed law, an Association des Industriels de France was created in 1883, to give workers greater protection from industrial accidents. The Association was developed by powerful groups of industrialists[125] and, in return for a modest premium, it disseminated regulations and new safety measures throughout the companies of its 2,287 members. Linking the prevention of accidents to greater regulatory discipline and greater profitability, it offered a technical solution rather than the social solution proposed by Nadaud, and sought to assuage concerns about worker safety.

Further proposed laws failed in the Sénat in 1884 and 1887, as did a new text in 1893, based on the principle of professional risk and obligatory insurance.[126] But the 1890s eventually brought a legislative solution. The observation of initiatives abroad, socialist schemes and the willingness of moderate Republicans to extend social rights to the working class conspired to allow the unanimous enactment of the Law of 9 April 1898 concerning liability for industrial accidents.[127] After eighteen years of parliamentary debate the Law of 9 April 1898[128] instituted no-fault liability for industrial injury.

The Law replaced the notion of employer liability with that of professional risk: on the grounds that 'from the moment that industry entails inevitable risks, the worker should not be forced to bear the consequences'.[129] No fault had to be proved on the part of the employer, who was compelled, but not obliged, to insure himself. Total incapacity gave absolute and permanent right to an income of two-thirds of the salary lost; temporary incapacity to an income of 50 per cent of the salary; death entitled successors in title to an income of 20 per cent of the salary and any surviving children were entitled to between 15 and 40 per cent, according to their number (article 3). It was estimated that a worker injured in an industrial accident falling under the new law would obtain compensation in more than 80 per cent of cases.[130] Under article 20, only

[125] Such as the Association des Industriels de l'Est and the Union des Houillères du Pas-de-Calais.

[126] See also 'Proposal for a law concerning the liability for accidents to which workers fall victim, during their work, and the organisation of compulsory insurance', Documents parlementaires, Sénat, 1895, 268 and 297; on the many legislative initiatives, see Hamon, above note 77 at pp. 639–41.

[127] JO 10 April 1898, 2209.

[128] Loi 1898 of 9 April concernant les responsabilités des accidents dont les ouvriers sont victimes dans leur travail, Bull. off. 1977, n.34, 652.

[129] JO, Chambre, séance of 18 May 1888, 1431, speech by M. Ricard.

[130] Vassart and Nouvion-Jacquet, above note 89 at p. 93.

an intentional act or *faute inexcusable* on the part of a worker would remove or diminish the right to compensation respectively.

Insurance remained voluntary[131] and could be sought from private insurance companies or underwriting consortiums.[132] A Law of 24 May 1899[133] reorganised the public *Caisse d'assurance contre les accidents du travail*,[134] to provide insurance coverage albeit only against death or permanent incapacity: and it remained a voluntary matter. In case of non-payment by an employer, compensation could be awarded by the Caisse nationale des retraites,[135] which could then pursue the claim against the company owner or their insurer.

The Law faced opposition from industrialists who feared that it would prove costly to them, in spite of the success of similar legislative initiatives abroad, and their resistance succeeded in limiting the benefits accorded to those who fell victim to industrial accidents.[136] The new regime did not cover all industrial accidents: it concerned only certain industries enumerated in article 1 and companies using explosives or machines driven by neither human nor animal force. The concept of 'professional risk' on which the Law of 1898 was based eventually found broad acceptance and was extended, in several legislative steps, to almost every worker who was party to a labour contract.[137] Employers were generally satisfied: accidents which caused workers to stop work for less than five days were not compensated (article 1) and insurance was not compulsory. If compensation was awarded, the worker could not seek further recourse, which ultimately led to a decrease in the total amount of compensation paid by employers.[138] The Law was criticised for 'legalising fatal accidents' and encouraging industrialists to employ foreign and unmarried workers.[139]

131 In contrast to other legislative initiatives such as that in Germany.

132 Syndicats de garantie; P. Pic, *Traité élémentaire de législation industrielle: les lois ouvrières* (Paris: Arthur Rousseau, 1903), p. 826.

133 D.1899.4.830.

134 Which had been established in 1868 but had met with little success.

135 Decree of 28 February 1899, Rec. 1899, 184.

136 Pic and Kréher, above note 120, at p. 254.

137 For example, Law of 30 June 1899 for those injured by agricultural machinery; Law of 12 April 1906 which extended the 1898 law on industrial accidents to all commercial companies; Law of 15 July 1914 for those employed in the forestry sector; Law of 15 December 1922 which extended protection to anyone injured who was employed in the general agriculture sector.

138 Pic and Kréher, above note 120, at p. 255.

139 Who would receive reduced compensation in the terms of art. 3. Foreign workers injured abroad would fall outside the scope of the law and it is notable that it did not apply to Algeria. Brumeaux, above note 124.

As all accidents had to be reported under article 11, the Law resulted in the collection of more comprehensive statistics, but the number of accidents continued to increase. One of the most significant consequences of the regime of 1898 was the improvement and increased enforcement of safety regulations and supervision.[140] Later, the institution of the social security system by the Law of 30 October 1946 meant that the individual liability of the employer was replaced by collective liability on the basis of obligatory insurance under the auspices of social security funds, which replaced the system of private insurance companies.[141]

(iv) Summary

It would appear that private law liability had a limited and rather short-lived role in providing redress for injured employees. Changing ideas during the 1880s led to legislative initiatives in the 1890s, as industrialisation grew apace and ideas of social solidarity came to the fore. It would appear that judicial responses were taken almost to pre-empt proposed legislation, rather than to be as proactive as the Belgian and Luxembourgeois judiciary. By 1898, most industrial workers were protected by workers' compensation for injuries at work (within the limits of the regime), and either in private law by a presumption of fault for the act of things, or in public law by virtue of a principle of equality before public burdens in the face of dangerous public activities.

(d) *The owner*

The owner faced three possible causes of an explosion: a defect in the boiler construction, deficiencies in the way employees operated it, or just bad luck. However, it must be borne in mind that enterprise and property insurance became increasingly widespread and owners would be able to insure themselves against such risks.[142]

With respect to defects, the rule of *caveat emptor* would normally exclude the liability of the vendor of a boiler (usually the manufacturer) for its defects and their consequences, under article 1641 of the Code

[140] *Ibid.*

[141] An injured employee benefits from a presumption of causation: under art. L.411-1 of the Code of Social Security, 'any accident which occurs during work, whatever the cause, is considered to be an industrial accident'.

[142] M. Bellom, *De la Responsabilité en matière d'Accidents du Travail. Commentaire de la loi du 9 avril, 1898, et des lois et decrets subsequents* (Paris: Rousseau, 1899), p. 25.

civil.[143] Most of the hidden defect (*vice caché*) cases reported in the nineteenth century concern agricultural products, where the purchaser is in the trade and should be able to inspect.[144] However, the courts seemed more indulgent to the purchaser where a complex mechanism was involved and where expertise was needed to conduct an inspection and identify the existence of a defect. In such cases, the defect may be considered hidden.[145]

It was accepted from 1847 that damages paid to the ultimate purchaser could be recovered as 'costs incurred through the sale' under article 1646 C civ.[146] These combined to create quite a strict liability on the purchaser. While there is relatively little recorded case law on boilers in this regard, it is clearly governed in much the same way. A *vice caché* case on exploding boilers turned on the thickness of the boiler wall, whether there was an evident risk and whether it could be discovered.[147] However, *vice caché* cases faced a short time limit and the issue of transmissibility. This was settled in 1884, so that the subpurchaser was protected by the device of implied assignment of the guarantee against hidden defects.[148] It would appear that breach of legislative construction regulations would trigger liability towards the purchaser under article 1382 as a way round these limits.[149] But this would be limited to situations in which the owner could show a defect in construction.[150]

Courts began to find that employers were under a duty to protect employees from the effects of their own imprudence,[151] and while a master might be able to sue his employees, as there is a clear contractual nexus, there was usually little point, as the employee would often be penniless or dead. Instead, the employer would often have to rely on enterprise

[143] Article 1641: 'The seller is held to a guarantee against hidden defects in the thing sold, which render it unsuitable for the use it is intended for, or which so diminish such use that the buyer would not have purchased it, or would have given only a lesser price for it, had he known of them.'

[144] For example, CA Nantes, 19 January 1859, D.P.1859.3.69; CA Nantes, 22 March 1862, D.P.1863.3.15.

[145] Lyon, 18 August 1858, D.P.1859.2.102, where the machine had been assembled from parts from elsewhere. Liability for a defective motor is clearly a technical matter: Req. 26 December 1906, D.P.1907.1.278; Req. 31 July 1905, D.1908.1.148. Also Com. 11 June 1858; Bourges, 27 June 1893.

[146] 'Frais occasionnées par la vente': Req. 29 June 1847, D.P.1848.1.187. This was confirmed as applying to damages paid to the victim of a road accident in 1925, Req. 21 October 1925, D.1926.1.9 observations Célice.

[147] Req. 9 February 1857, D.P.1859.1.259.

[148] Cass. civ. 12 November 1884, D.1885.1.357; Paris, 24 February 1882, D.P.1883.2.78.

[149] Civ. 16 June 1879, D.P.1880.1.36.

[150] Besançon, 7 June 1898, D.P.1900.2.178.

[151] Caen, 19 March 1880, D.P.1881.2.79.

and property insurance, whether the damage was caused by the act of an employee or just bad luck.

(e) *Concluding remarks*

In the early years of the age of steam, particular attention was paid to balancing the needs of industrialists with those of private property owners. Planning regulations and rules on the construction and inspection of boilers had some success with respect to pre-empting problems of nuisance and safety issues. However, with growing numbers of steam boilers in use, stringent controls had to be relaxed due to fears of 'industrial paralysis'. The breach of regulations would incur criminal liability and, where causation could be established, liability for regulatory breach would arise under article 1382 of the Code civil.

It would appear that civil law did not play a major role in offering solutions to the victims of exploding boilers. An owner would only rarely be able to use contract law to recover against the manufacturer, because most cases would involve issues of operation, rather than construction. In the event of an industrial accident, the injured employee would often struggle to prove *faute* and causation and, confronted by increasing numbers of accident victims, a way was sought to assuage the position of injured workers. The courts had to negotiate a somewhat discordant doctrinal minefield, with contractual solutions being rejected in France, in favour of a particular appreciation of *faute* specific to the circumstances of industrial accident. Ultimately, the courts had to have recourse to a case-by-case approach, which led to a rather incoherent system. The civil and administrative laws provided some protection in the period 1895 to 1898, before their practical effect was largely superseded by the institution of a workers' compensation regime in 1898. Insurance would appear to be of considerable influence, with the growth of insurance protection operating in the background, both influencing and reacting to the evolution of new rules.

3. Injuries from asbestos

(a) *Introduction: the history of asbestos*

A report issued by the French Senate has termed the effects of exposure to asbestos a 'health catastrophe',[152] which was further exacerbated by the

[152] G. Dériot and J.P. Godefroy, *Rapport d'information fait au nom de la mission commune d'information sur le bilan et les conséquences de la contamination par l'amiante* (2005, No. 37 Sénat), p. 9 ('Senate report').

delay in promulgation of safety measures and the later prohibition on the use of asbestos.[153] As a result, illnesses resulting from exposure to asbestos (notably lung cancer and mesothelioma) are still on the rise, while the number is declining in other countries such as the United States, where companies took preventive measures some twenty years before those in France.

While 35,000 deaths have been imputed to asbestos exposure during the period between 1965 and 1995, because of the thirty to forty year latency period of mesothelioma, between 50,000 and 100,000 deaths are expected in the coming twenty years, to which should be added approximately 10 per cent of the 25,000 cases of lung cancer diagnosed each year.[154] The scientific community believes the 'epidemic' to come is unavoidable and irreversible.

Before its tardy prohibition in 1997,[155] asbestos was in widespread use, meaning that the risk of secondary exposure to asbestos still remains for builders and others who are exposed to residual or environmental asbestos.[156]

In 1906, a report was issued documenting the high mortality rate of textile workers employed by a factory using asbestos in Condé-sur-Noireau, but no action was taken.[157] Some twelve years before, when commenting on the application of the Law of 12 June 1893 with respect to dust in workshops, the Cour de cassation had already noted that 'the presence of dust of whatever kind may always constitute a danger'.[158] The dangers of asbestos in the industrial context were in some ways admitted by the inclusion of certain illnesses caused by asbestos dust in tables of industrial diseases: inclusion in such tables created a presumption that the illness had resulted from the exercise of the affected worker's profession.[159] On 3 August 1945, pulmonary fibrosis resulting from the inhalation of silicon or asbestos dust was included in the tables of industrial diseases, followed in 1950 by the recognition

[153] Yves Saint-Jours, 'L'amiante: de la prévention négligée aux conséquences induites' *Droit ouvrier* 1999, 486.

[154] Senate report, above note 152, at p. 48.

[155] 1 January 1997, Decree 96–1133 of 24 December 1996 (J.O. 26 December 1996).

[156] Senate report, above note 152, at pp. 195, 203.

[157] *Ibid.* p. 43; *Bulletin de l'inspection du travail,* 'Note sur l'hygiène et la sécurité des ouvriers dans les filatures et tissages d'amiante, M. Auribault, inspecteur départemental du travail à Caen', republished in Supplément 1082, Semaine sociale Lamy (1 July 2002).

[158] Commenting on Decree of 10 March 1894 (with respect to the application of the Law of 12 June 1893), *Bulletin de l'inspection du travail* 1894, 626.

[159] M. Babin, 'Les logiques de reconnaissance des maladies professionnelles' *Droit social* 1998, 673.

of asbestosis with the creation of table 30.[160] However, it was only in 1976 that lung cancer and mesothelioma were included in table 30,[161] followed in 1985 by the inclusion of bronchial-pulmonary cancer where there was strong medical evidence that it had been caused by asbestos.[162] In 1996, table 30*bis* was added, which included primary bronchial-pulmonary cancer where the worker had been exposed to asbestos for ten years.[163] Allowances had to be made for the long latency periods of many asbestos-related diseases. The legislature intervened to modify the starting point for the period of prescription[164] and a law was passed to permit the reopening of cases for a further three years, commencing from its entry into force.[165]

Only in 1977 was the use of asbestos in domestic housing prohibited, and the concentration of asbestos to which workers could be exposed in the workplace restricted. France thus took the first precautionary measures some forty-six years after the United Kingdom, thirty-one years after the United States and thirteen years after the meeting of the International Conference on Asbestos of the New York Academy of Sciences of 1964 documented the risks associated with asbestos, the reports of which were readily available in the library of the Paris Faculty of Medicine 'some few hundred metres from the offices of the relevant ministers'.[166] While the carcinogenic nature of asbestos had been highlighted by a resolution of the European Parliament in January 1978,[167] France only banned asbestos production some nineteen years later.

Several reasons have been traditionally invoked to explain France's inertia in reaction to the situation. First, the long latency periods of illnesses connected with asbestos resulted in an underestimation of the real consequences. Secondly, there was scientific uncertainty as to the effects of exposure to small amounts of asbestos. Thirdly, the absence of systems of supervision and alert

[160] Decree of 31 August 1950. [161] Decree of 5 January 1976. [162] Decree of 19 June 1985.

[163] Decree of 22 May 1996.

[164] Under art. L.431-2 C séc. soc., the starting point for the limitation period of two years to bring a claim for the recognition of an industrial disease corresponded to the date of the first medical diagnosis of the disease; since the enactment of art. L.461-1 C séc. soc., it henceforth starts to run from the date upon which the victim is informed, by medical certificate, of the possible link between his employment and illness.

[165] Article 40 of Law 98-1194 of 23 December 1998 concerning the social security budget for 1999 (J.O. 27 December 1999). 'Prescription et victimes de l'amiante' *Semaine sociale LAMY*, no. 967 (February 2000), 2.

[166] Senate report, above note 70, at pp. 12 and 92, illustrating the delay when compared with other countries.

[167] O.J. C006, 138, 9 January 1978, Résolution sur les risques sanitaires de l'amiante (in French only).

and the ambiguous role of the Institut national de recherche et de sécurité reduced the flow of information on the growing crisis.[168] Even the political environment of the 1980s and consequential reduction in trade union strength[169] were raised to explain the inaction of the authorities. However, it would appear that between 1982 and 1995 the issues surrounding asbestos had effectively been delegated to an informal body, the Comité permanent amiante (CPA), which was in fact merely a lobby of the industry, on which scientists and representatives of the relevant ministries sat.[170] A 2005 Senate inquiry found that in the absence of any ministerial control, the CPA had benefited from the caution of eminent scientists and inexperienced researchers, who were led to believe that no suitable substitutes for asbestos existed. As 'industrialists with false noses',[171] the CPA was a model of lobbying and manipulation which, in the absence of governmental control, exploited and capitalised on any scientific uncertainty,[172] and attempted to perpetuate the 'myth of the safety of controlled asbestos use',[173] in order to ensure 'continued asbestos use and delay its prohibition for as long as possible'.[174]

While the risks associated with asbestos exposure were known, documented and accessible to scientists, doctors and public authorities, for forty years it would appear that a 'culture of lies' allowed the continued use of asbestos in France.[175] Some actors certainly knew more than others – notably the asbestos industry: large multinational groups and their professional organisations (which were in close contact with Anglo-Saxon multinationals) had access to scientific studies on these dangers, and until the early 1970s were forced to actively construct a strategy to allow them to continue using asbestos.

In attempting to gain compensation for the consequences of asbestos exposure, several means of recourse are available to the affected parties, and from the foregoing analysis, it is perhaps unsurprising that both employers and the state have been pursued. Likewise, the various means of recourse reflect the diverse means of exposure to asbestos. These could be professional, as when one works with asbestos, of 'a para-professional nature (workers working in buildings which contain asbestos), environmental (inhabitants living near to a factory using asbestos) or domestic (people having used asbestos in their own homes)'.[176]

[168] Senate report, above note 152, at p. 62. [169] *Ibid.* p. 69.

[170] F. Chateauraynaud and D. Torny, *Les sombres précurseurs: une sociologie pragmatique de l'alerte et du risqué:* (Paris: Éditions de l'École des hautes études en sciences sociales, 1999).

[171] Senate report, above note 152, at p. 83. [172] *Ibid.* p. 75. [173] *Ibid.* p. 79. [174] *Ibid.* p. 80.

[175] F. Malye, *Amiante: 100.000 morts à venir* (Paris: Éditions le cherche midi, 2004).

[176] C. Guettier, 'Le fonds d'indemnisation des victimes de l'amiante' *Resp. civ. et assur.* 2002, chr. 19.

In the present study, the liability of employers, both as a consequence of the new case law of the Cour de cassation concerning the concept of *faute inexcusable* under social security law and also under civil law, will be examined, as will the liability of the state for failure to enact regulations specific to asbestos use before 1977 and for late and insufficient regulation after 1977, as confirmed in 2004 by the Conseil d'État.[177]

In 1997, the Institut national de la santé et de la recherche médicale (INSERM) published a report entitled *Effets sur la santé des principaux types d'exposition à l'amiante*,[178] at the request of the Ministers of Health and Employment. The health consequences of asbestos were for the first time put in terms of an epidemic, which was met with 'great surprise'.[179] With increased public awareness and outrage,[180] following the ban on asbestos production in 1997, the compensation of victims of asbestos exposure became an urgent matter.[181] As will be explored later, the concern for compensation was translated into a system of early retirement and pension for the victims of asbestos exposure (FCAATA)[182] and a special compensation regime (FIVA),[183] which has been the subject of some controversy.[184] It is estimated that the compensation of victims may reach between 27 and 37 billion euros in the coming twenty years.[185]

However, it should be noted that claimants may be faced with the insolvency of the liable companies, even if they overcome the considerable obstacles to fulfilling the conditions of causation and the prescription period. Moreover, the burdens of proof and compensation available

177 C. Guettier, 'L'État face aux contaminations liées à l'amiante' *AJDA* 2001, 529. While it will not be examined here, actions may also be brought against the state as employer: TA Lille, 7 July 1998, *Bulcourt, Petites Affiches* 1999, no. 133, 26, note V. Mutelet.

178 M. Goldberg and D. Hemon (INSERM), 'Les effets sur la santé des principaux types d'exposition à l'amiante. Résultats d'une expertise collective de l'I.N.S.E.R.M.: Surveillance médicale des salariés exposés à l'amiante' *Archives des maladies professionnelles de médecine du travail* 1997, vol. 58, no. 7, 569.

179 Senate report, above note 152, at p. 58; the INSERM report was followed by C. Got, *Rapport sur la gestion politique et administrative du problème de santé publique posé par l'amiante en France* (Paris: Ministère de la Santé, 1998), which revealed the extent of damage caused by asbestos, and gave similarly alarming predictions as to the extent of future cases.

180 Termed 'a new collective drama': C. Guettier, 'L'État face aux contaminations liées à l'amiante' *AJDA* 2001, 529 at 530.

181 Senate report, above note 152, at p. 14.

182 Le Fonds de cessation anticipée d'activité des travailleurs de l'amiante.

183 Le Fonds d'indemnisation des victimes de l'amiante.

184 A. Guégan-Lecuyer, 'À propos de la confrontation des offres d'indemnisation du Fonds d'indemnisation des victimes de l'amiante au pouvoir judiciaire' D.2005, chr. 531.

185 Senate report, above note 152, at p. 15.

to victims of asbestos exposure vary widely according to the means of recourse chosen.[186]

(b) State liability

In 2000, the state's failure to react to the threat posed by asbestos was brought into question when four victims of asbestos exposure, or their successors in title, brought an action against the state before the administrative courts for failure to exercise its powers of regulation and control.[187]

The actions were successful, with the state being held liable on two accounts: for having only imposed the first restrictions on asbestos use in 1977,[188] and then for having adopted insufficient and tardy regulations for the protection of employees after 1977.[189]

In establishing the liability of the state pre-1977, the court found, citing the INSERM report and the relative dates that precautionary measures were taken by other countries, that 'the existence of a significant risk could not have been ignored' and that the public authorities had 'failed to obtain a precise assessment of the risk from scientific authorities' in order to 'evaluate the risk, understand the danger and subsequently take the necessary measures to limit if not prevent the danger from materialising'.[190]

With respect to liability after 1977, which concerned the regulation of the safe limits of inhalation of asbestos fibres in an industrial environment, the court considered that the state had 'seriously damaged public health' by transposing the 1983 European Directive[191] four years too late, which had reduced by half the safe limit which had been established by the *décret* of 1977, and that the state had been a year too late in transposing the 1991 European Directive,[192] which had again lowered the safe limit.

The mesothelioma from which two of the victims had died was found to be an industrial disease contracted following exposure to asbestos at their

[186] A. Schneider, 'Lorsque le mieux à travers les systèmes d'indemnisation apparaît comme étant l'ennemi du bien' D.2001, 1979.

[187] TA Marseille, 30 May 2000, *Bourdignon, Thomas, Xueref, Botella*, *Gaz. Pal.* 2001, 40–1, note B. Pauvert, *Droit de l'environnement* 2000, no. 80, 5, concl. C. Fedi; Guettier, above note 98, comparing these cases to the contaminated blood products cases (at 531).

[188] In the *Bourdignon* and *Thomas* cases, above note 187.

[189] In the *Xueref* and *Botella* cases, above note 187.

[190] *Ibid.*

[191] Council Directive 83/477/EEC of 19 September 1983 on the protection of workers from the risks related to exposure to asbestos at work.

[192] Council Directive 91/382/EEC of 25 June 1991 amending Directive 83/477/EEC on the protection of workers from the risks related to exposure to asbestos at work.

place of work,[193] and the same was found for lung cancer in the other two cases.[194] However, such conditions may develop independently of exposure to asbestos and in considering that causation was established 'the court was far from thorough'.[195] As the Commissaire du gouvernement noted, the court relied on the views of the Collège de médecins and the INSERM report of 1996, both of which suggest that while such diseases are characteristic of asbestos exposure, such a link is 'probable and not certain'.[196] Nonetheless, the mere fact that the victims had been exposed to asbestos in an industrial context, and that they had developed illnesses which asbestos may cause, was considered sufficient to establish causation. Likewise, the fact that the government had failed to promulgate stricter norms to regulate the inhalation of asbestos fibres in good time was considered the cause of death. It has been suggested that the wish to give the victims better compensation than that available under claims for industrial illness was the main motivating factor,[197] albeit that proceedings in the administrative courts would take many years, which might well be too late for those already suffering from asbestos-related diseases.[198]

These cases were confirmed on appeal[199] and again by the Conseil d'État on 3 March 2004[200] which noted that:

> the public authorities ... are under a duty to keep workers informed of the dangers to which they may be exposed in the exercise of their profession, particularly taking into account products or substances which they handle or with which they come into contact and they are under a duty to take the most appropriate measures with the help of additional studies or enquiries to reduce and if possible eliminate any danger.

Likewise, the causal link between the fault of the public authorities and the harm suffered by the victims was confirmed. However, in asbestos

[193] *Xueref* and *Botella* cases, above note 187.

[194] *Bourdignon* and *Thomas*, above note 187.

[195] Guettier, above note 180, at 533, noting that the court did not question the victims' consumption of tobacco or way of life.

[196] TA Marseille, 30 May 2000, above note 187. [197] Guettier, above note 180, at 533.

[198] H. Arbousset, 'Amiante: la responsabilité de l'État est, enfin, reconnue par le Conseil d'État' D.2004, 973 at 976.

[199] CAA Marseille, 18 October 2001, *Ministre de l'emploi et de la solidarité contre Thomas*, obs. C. Guettier, 'Responsabilité de la puissance publique, panorama de jurisprudence' *Resp. civ. et assur.* 2002, comm. 4.

[200] C.E. Ass. 3 March 2004, *Ministre de l'emploi et de la Solidarité c/ Cts Bourdignon, no.* 241150; *Ministre de l'emploi et de la solidarité c/ Cts Xueref; Ministre de l'emploi et de la solidarité c/ Cts Thomas; Ministre de l'emploi et de la solidarité c/ Cts Botella*. Note C. Guettier, 'Panorama de la jurisprudence administrative' *Resp. civ. et assur.* 2004, comm. 234; G. Viney, 'Chronique de responsabilité civile' JCP 2004.I.163, no. 6.

cases the temporal factor is crucial due to the long latency periods which may pass before the diagnosis of a disease. It has been suggested that industrial exposure to asbestos before its use in France was regulated should give rise to the presumption that it is the cause of the illness,[201] and that this presumption should remain if the exposure took place at a time relatively early in regulation,[202] but that a more recent exposure (until December 1996 when asbestos was prohibited) might demand more convincing proof.[203]

(c) *Employer's liability*

The law of 9 April 1898[204] based the liability of the employer on the notion of professional or industrial risk, together with a fixed compensation scheme, and avoiding contractual liability. The employer would thus be held strictly liable, but would benefit from immunity from other civil actions being brought against him. In the case of *faute inexcusable* on the part of the employer, increased compensation would be awarded. The *faute inexcusable* of an employer is a jurisprudential notion which was defined by the Cour de cassation in the *Veuve Villa* case of 15 July 1941, as 'a fault of an exceptional seriousness, arising from a voluntary act or omission, the danger of which the author should have known, in the absence of any justificatory defence, distinguished from intentional fault by the lack of an intentional element'.[205] The result of this restrictive definition had been that 'during several decades, the number of *fautes inexcusables* found had been small, if not insignificant'.[206]

Social security laws substituted the concept of industrial risk with 'social risk', and in 1946 and 1947 replaced the 1898 scheme with social security funds ensuring the compensation of the victim of an industrial accident or illness, while the concept and role of *faute inexcusable* remained the same. Pierre Sargos, president of the Chambre sociale of the Cour de cassation, wrote that such compensation schemes:

> prevented the putting into place of effective accident prevention schemes … The certainty of compensation … contributed to the perverse idea – which formed part of a general consensus for a long time which

201 As was the case in *Thomas* and *Bourdignon*, above note 187.
202 As in *Botella* (1982) and *Xueref* (1985).
203 Guettier, above note 180, at 533. 204 J.O. 10 April 1898, 2209.
205 Ch. Réunies, 16 July 1941, JCP 1941.II.1705, note J. Mihura.
206 P. Sargos, 'L'évolution du concept de sécurité au travail et ses conséquences en matière de responsabilité' JCP 2003.I.121.

> in turn led to the 'social crime' of asbestos – that it was not necessary to incur costs or put in place safety measures since the human damage was compensated.[207]

Law 76–1106 of 6 December 1976 relating to the prevention of industrial accidents provided for payment of lost wages and other harm, and employers were for the first time allowed to insure against the risk of *faute inexcusable*.

With regard to asbestos-related conditions, the concept of *faute inexcusable* was discussed with regard to the knowledge of the danger, considered objectively given business practices, scientific knowledge and regulatory measures.[208] In their defence, employers cited the late inclusion of mesothelioma in the table of industrial diseases and the lack of regulations, stating that they could not have had knowledge of the danger until the end of the 1970s.

Although the Cour de cassation had frequently held that, since asbestosis had been considered an industrial disease since 1951, the employer must have known of the dangers to which his workers were exposed,[209] the case law was ambiguous.[210]

However, 'a revolution in the law of industrial accident and industrial disease'[211] occurred when the Cour de cassation gave a new definition of *faute inexcusable* based on the contract of employment rather than a fault of exceptional gravity. In a special hearing examining thirty appeals brought by employers who had been held liable for asbestos-related conditions on the basis of a *faute inexcusable,* the Cour de cassation made reference to an '*obligation de sécurité de résultat*' based on the contract of employment.[212] Any failure in the exercise of this obligation would constitute a *faute inexcusable* under article L.452–1 of the Code de la sécurité sociale 'if the employer has not taken the measures necessary to prevent the employee's exposure to a danger of which he had, or should have had, knowledge'. In this way, the Cour de cassation effectively put an end to employers attempting to rely on the defence of absence of knowledge of danger, in effect reversing the burden of proof.[213]

[207] *Ibid.*

[208] J.-J. Dupeyroux, M. Borgetto, R. Lafore and R. Ruellan, *Droit de la sécurité sociale* (14th edn, Paris: Dalloz, 2001), nos. 927 *et seq.*

[209] See Orléans, 23 February 1994, *Droit ouvrier* 1994, 288, observations Y. Saint-Jours; Dijon, 18 December 1997, *Gaz. Pal.* 1998.

[210] X. Prétot, 'La nouvelle définition de la faute inexcusable de l'employeur: une jurisprudence contra legem?' D.2002, 2696.

[211] A. Lyon-Caen, *Droit social* no. 4 (April 2002).

[212] Soc. 28 February 2002, Bull.; JCP 2002.II.10053, concl. A. Benmaklouf; Prétot, above note 210; Senate report, above note 152, at p. 93.

[213] Prétot, above note 210, at 2700.

Since the judgments of 2002, there has been 'a quasi-automatic finding of *faute inexcusable*' in cases of industrial disease linked to asbestos-related illnesses.[214] In 2005, it was estimated that 1,500 judgments had been delivered involving a claim on the basis of the redefined notion of *faute inexcusable* in the context of asbestos-related conditions.[215] As was predicted,[216] rather than a corrective mechanism, actions brought on the basis of *faute inexcusable* have arguably become the norm.[217]

In the event of a claim on the basis of a *faute inexcusable,* the victim or those successors in title entitled under the Code de la sécurité sociale[218] may receive an increase in social security benefits[219] and claim for other harm suffered, such as physical and emotional harm and non-material damage, in addition to harm resulting from the loss or reduction of opportunities for professional advancement.[220] However, this has come at a great cost to companies[221] and many victims may be confronted with the insolvency of their employer.

(d) *Exposure to asbestos outside the industrial context*

Those successors in title who do not have the right to claim benefits under the provisions of the Code de la sécurité sociale[222] and *victimes par ricochet,* may have recourse to civil liability.[223] As third parties to the employment contract, such victims can establish delictual liability by relying on the *faute inexcusable* of the employer in failing to fulfil the *obligation de sécurité de résultat.*[224] Where victims are exposed to asbestos outside of the context of a professional activity, civil liability may play a significant role.

Damages for asbestosis were awarded to the wife of an employee who had been recognised as having contracted asbestosis in the industrial context.[225] It was established that her husband had brought asbestos into the marital

214 Senate report, above note 152, at p. 96. 215 *Ibid.*

216 Prétot, above note 128, at 2701. 217 Senate report, above note 152, at p. 95.

218 Enumerated under art. L. 434–7 *et seq.* C séc. soc. 219 Article L. 452–2 C séc. soc.

220 Article L.452–3 C séc. soc.

221 Senate report, above note 152, at p. 97, detailing the resultant insolvency of various companies.

222 That is to say the successors in title who are not listed in art. L.434–7 *et seq.* and may not be prevented from claiming by the principle of employer immunity. Civ. 2, 15 January 1997, Bull. no. 14, D.1997, 317, note Y. Saint-Jours.

223 Y. Lambert-Faivre, *Droit du dommage corporel: Systèmes d'indemnisation* (4th edn, Paris: Dalloz, 2000), no. 295.

224 Cass. civ. (1) 18 July 2000, Bull. Civ. no. 221, JCP 2000, II, 415 rapp. P. Sargos.

225 Caen, 20 November 2001, *Environnement*, July 2002, 19, no. 96, note A. Faro; JCP 2003. II.10045, note F.G. Trébulle.

home on his skin and hair, and until 1977, on his work clothes, which she would wash and leave to dry in their bedroom. The causal link between the wife's illness and the domestic exposure generated by the working conditions of the husband was considered to have been established.

The Cour d'appel noted that 'the spread of dust outside of the factory though the intermediary of its workers and the inhalation of this dust by third parties, originates from the company's failure to exercise its power of control over the dangerous material', and then found the husband's employer liable for the wife's asbestosis on the basis of article 1384, paragraph 1 C civ. as *gardien* of the asbestos.

In this way, it would seem that companies may be held liable for environmental contamination where a company has failed to prevent asbestos fibres from escaping the confines of the factory, on the basis of liability for the acts of things. Likewise, liability may also be found in nuisance.[226]

'Residual but omnipresent'[227] asbestos as a consequence of its 'massive usage' in construction and a wide range of products before its prohibition,[228] may also lead to contamination. The Association de défense des victimes de l'amiante (ANDEVA) has indicated that, in contrast to asbestos exposure in the professional context, the diagnosis of illnesses resulting from this passive or secondary form of exposure are on the increase.[229] Moreover, given certain geological conditions, passive exposure to asbestos may occur in a purely environmental context.[230]

Given the vast number of potential claimants and the disparate means of recourse, the ease, speed and quantum of compensation became the primary concern. It is perhaps unsurprising that compensation schemes were considered as a matter of urgency.

(e) *Need for timely and full compensation creates unlikely bedfellows*

While efforts to pursue employers for criminal liability have met with little success,[231] victims seeking redress attempted to obtain

226 *L'assurance française* no. 719, June 1996, 7 and case law referred to there.

227 Senate report, above note 152, at p. 194. 228 *Ibid.* 36.

229 In his deposition to the Senate inquiry, F. Malye estimated that 15 to 20 per cent of newly diagnosed sufferers of asbestos-related diseases have never been in contact with asbestos in an industrial context: Senate report, above note 152 at p. 195.

230 Where serpentine asbestos is present: in particular in Nouvelle-Calédonie (*Ibid.* p. 241), and Haute-Corse: Direction Départementale des Affaires Sanitaires et Sociales de Haute-Corse, *L'amiante environnemental en Haute-Corse* (2004).

231 Senate report, above note 152, at pp. 102–11.

compensation from the Fonds de garantie des victimes des actes de terrorisme et d'autres infractions on the basis of provisions of the Code de procédure pénale to 'obtain full compensation for the damage resulting from personal injury' where caused by 'an offence'.[232] This possibility was confirmed in a judgment of the Cour de cassation of 18 June 1997,[233] and has since been followed in cases of asbestos-related diseases,[234] with victims benefiting from faster procedures, more flexible limitation periods and arguably larger sums of compensation than conventional methods of recourse.[235]

Given the original purpose of the fund and the large number of potential victims, the success of these cases has met with criticism.[236] However, comparisons were made with the contaminated blood products crisis and the compensation fund which followed,[237] which reflected increasing concern for compensation and the realisation of the possible extent of the problem.[238]

The first explicit governmental initiative for the compensation of victims of asbestos exposure was the creation of an early retirement fund for affected workers in 1999.[239] The government went further a year later, taking the initiative to set up a compensation fund under public administrative law,[240] for victims of both industrial and environmental asbestos exposure, and their successors in title: the Fonds

232 Articles 706–3–706–15 of the Code de procédure pénale. See also A. Guégan-Lecuyer, *Dommages de masse et responsabilité civile* (Paris: LGDJ, 2006), p. 163. Recourse may be had to this fund independent of criminal proceedings, see in general, F. Labrousse, 'Amiante et responsabilité' *Petites Affiches* 1996, no. 42, 4.

233 Cass. civ. (2), 18 June 1997, reproduced in H. Groutel, 'L'application aux accidents du travail de la législation sur l'indemnisation des victimes d'infraction, Un arrêt de la Cour de cassation du 18 juin 1997 risque de "faire sauter la banquet"' *Resp. civ. et assur.* 1997, chr. 21.

234 Cass. civ. (2), 30 November 2000, A. Schneider, 'Lorsque le mieux à travers les systèmes d'indemnisation apparaît comme étant l'ennemi du bien' D. 2001, 1979.

235 Critical of the disparities in result according to the jurisdiction chosen: Schneider, above note 234, at 1981.

236 *Ibid.*

237 Tribunal de Grande Instance Cherbourg, 25 March 1999, *Droit ouvrier* 1999, 338, note F. Meyer.

238 H. Arbousset, 'L'amiante: la création d'un fonds d'indemnisation pour canaliser une "bombe à retardement"' *Revue de la recherche juridique* 2001, 873.

239 Fonds de cessation anticipée d'activité des travailleurs de l'amiante (FCAATA), art. 41 of Law 98–1194 of 23 December 1998 concerning social security budget for 1999; Senate report, above note 152, at pp. 125–9 and 172–2.

240 It is under the supervision of the finance and social security ministers and is presided over by a senior judge. See Senate report, above note 152, at pp. 132–3 for details of its organisational structure.

d'indemnisation des victimes de l'amiante (FIVA).[241] The transfer of claims from the Fonds de garantie des victimes des actes de terrorisme et d'autres infractions to the new fund was instituted and this method of recourse was effectively closed.[242] However, the fund became operational much later than intended, with the first compensation awards only being made in 2003.[243] It is predominantly state funded,[244] although it may also bring a subrogatory action against those responsible for the harm.[245]

The FIVA aims to give the victim of asbestos exposure (or his successors in title) *réparation intégrale*,[246] to return him to the state in which he was, or would have been, if the damage had never occurred, be it of a material or non-material nature.[247] In order to shorten the length of the compensation process, where the victim has been recognised to have contracted an industrial disease as a consequence of asbestos exposure, or where the victim has contracted an asbestos-related disease which features on a specific interdepartmental list, the victim benefits from a presumption that his illness was caused by asbestos.[248] Within six months of receipt of the claim, the fund offers an indication of the quantum of damages and the amount to be awarded.[249] The legislation provides that the acceptance of the compensation offered by the fund would entail the giving up of any other claim for compensation,[250] but where the claim is

[241] Article 53 of Law 2000–1257 of 23 December 2000 concerning social security budget for 2001, J.O. 24 December 2000, 20558 and Decree 2001–963 of 23 October 2001 concerning a compensation fund for the victims of asbestos exposure, J.O. 24 October 2001. Comparing the fund for victims of contaminated blood products: Guettier, above note 180, at 533–4. See Guégan-Lecuyer, above note 232, at p. 159 *et seq*., for general overview and analysis.

[242] Article 53-IX of Law 2000–1257 of 23 December 2000 concerning social security budget for 2001, J.O. 24 December 2000, 20558 and art. 42 of Decree 2001–963 of 23 October 2001 concerning a compensation fund for the victims of asbestos exposure, J.O. 24 October 2001; Schneider, above note 234, at 1982.

[243] Senate report, above note 152, at p. 131.

[244] Article 53-VII of the Law of 23 December 2000.

[245] As well as against persons or institutions which are charged with giving partial or full compensation: art. 53-VI of the Law of 23 December 2000.

[246] Article 53 of Law 2000–1257 of 23 December 2000 concerning social security budget for 2001, J.O. 24 December 2000, 20558.

[247] See further Senate report, above note 152, at pp. 133–6.

[248] Article 53 III of the Law of 23 December 2000. Guettier, above note 180, at 535.

[249] Allowance is made for the sums already awarded to the victim through social security and any damages received or to be received from other claims on the basis of the same harm.

[250] Article 53-III of Law 2000–1257 of 23 December 2000.

rejected, the procedure takes longer than six months, or the fund's offer is not accepted,[251] the victims of asbestos-related diseases benefit from a special procedure before the appeal courts.[252]

While 86 per cent of asbestos-related cases were resolved by the FIVA in 2004,[253] the legislature appears to have failed to avoid a re-enactment of the jurisdictional confusion which had arisen with the establishment of the compensation fund for contaminated blood products.[254] While compensation cannot be obtained twice for the same harm, where the victim has not received a satisfactory offer from the fund, a case may be brought before a court of appeal,[255] with some appeal courts awarding significantly higher sums.[256] Likewise, it is still possible to bring an action for *faute inexcusable* and since its redefinition by the Cour de cassation in 2002 there has been a marked increase in such cases: from 300 in 2002 to over 1,500 in 2004, of which less than 2 per cent of claims for *faute inexcusable* were unsuccessful.[257] This state of affairs has been strongly criticised by the Conseil constitutionnel,[258] which noted that in addition it may still be possible for victims to bring an action before the criminal or civil courts. It should be recalled that it is still possible to bring an action against the state for failure to exercise its powers of regulation and control. The 2005 Senate inquiry also noted the heterogeneous levels of award made by the various jurisdictions.[259]

[251] Where quantum is disputed, an action must be brought within two months of the offer.

[252] Article 26 *et seq.* of the Decree of 23 October 2001; this is comparable to the procedure of the fund for the victims of contaminated blood products, but it should be noted that appeals relating to the FIVA have to be made before local courts, unlike appeals concerning the contaminated blood products fund, which have to be brought before the Cour d'appel de Paris.

[253] Senate report, above note 152, at p. 166.

[254] C.E. 16 June 1997, D.1999, somm. 7, note P. Bon and D. de Béchillon; France has also been condemned several times before the European Court of Human Rights for the violation of its Article 6 ECHR obligations in this respect: see M. Dreifuss, 'L'indemnisation des victimes du sida à l'épreuve du dualisme juridictionnel' *Revue française de droit administratif* 1996, 561.

[255] Five per cent of offers made by the FIVA were appealed in this way in 2004: FIVA, *Troisième rapport d'activité au Parlement et au gouvernement* (June 2003–May 2004), p. 24.

[256] See A. Guégan-Lecuyer, 'À propos de la confrontation des offres d'indemnisation du Fonds d'indemnisation des victimes de l'amiante au pouvoir judiciaire' D.2005, chr. 531 at 532–3.

[257] Senate report, above note 152, at p. 166.

[258] Conseil constitutionnel décision no 2000–437 D.C. of 19 December 2000, Law concerning social security budget for 2001.

[259] Senate report, above note 152, at p. 168.

While it is certain that those suffering from the effects of asbestos-related diseases, faced with long procedural delays and compromised life expectancy, may ultimately find it most advantageous to bring claims and accept the compensation offered by the FIVA, the continued use of other means of recourse is also an indication that victims are dissatisfied with the levels of compensation offered,[260] to the extent that they are willing to brave the vagaries of procedures outside of the fund with the attendant difficulties of proof of causation or the risk of an insolvent employer. The credibility of the FIVA may ultimately 'depend on the tables of compensation',[261] but the predicted scale and longevity of the epidemic will place hitherto unknown burdens on both the judicial and compensation systems.

[260] Guégan-Lecuyer, above note 256; Guettier, above note 180, at 537.
[261] Guettier, above note 180.

4

Technological Change and the Development of Liability for Fault in Germany

JENS M. SCHERPE

1. Dominance of the fault principle in Germany

The central element of the German[1] *Deliktsrecht* (law of torts) always has been that liability in principle requires fault.[2] Strict liability or prima facie cases of negligence only exist in specialised areas and even today are seen as concepts alien to the general private law and as exceptions. One of the first of these exceptions was §25 of the Prussian Railway Act (Preußisches Gesetz über die Eisenbahnunternehmungen), dating from as early as November 1838; it was held that such an exception was necessary because of the risks the railways presented to the general public. However, before looking at the problem of escaping sparks, etc. from railway engines and how it was dealt with in Germany, it is necessary very briefly to outline in which legal and social context these developments took place, namely the Pandectist roots of, and liberal thinking in, German law in this area.

Theorists of the nineteenth century perceived the law of contract to be a tool for the use of freedom, while the law of torts was to deal with the abuse of freedom.[3] Already here we find the concept of fault deeply engrained; liability for accidents without fault was seen as exceptional and even inappropriate for a systematic classification.[4] More specifically,

[1] In some parts of Germany, namely Baden and territories west of the Rhine, the French Code Civil was applicable for a long part of the nineteenth century; this, however, will not be dealt with in this chapter.

[2] The same, of course, is true for the other European jurisdictions, cf. H. Kötz and G. Wagner, *Deliktsrecht* (10th edn, Munich: Beck, 2006), p. 11 and K. Zweigert and H. Kötz, *Introduction to Comparative Law* (3rd edn, Oxford: Oxford University Press, 1998), pp. 596, 647.

[3] F. C. von Savigny, *Das Obligationenrecht als Theil des heutigen römischen Rechts* (Berlin: Veit, 1853), vol. II, p. 330. Cf. J. Esser, *Grundlagen und Entwicklung der Gefährdungshaftung* (Munich and Berlin: Beck, 1941), p. 50.

[4] Cf. H.-P. Benöhr, 'Zur außervertraglichen Haftung im gemeinen Recht' in D. Medicus and H. H. Seiler (eds.), *Festschrift für Max Kaser zum 70. Geburtstag* (Munich: Beck, 1976), p. 690.

it was generally felt that the law had to be pure and devoid of social and political considerations.[5] So, in the absence of fault, the loss lay where it fell – *casum sentit dominus.* Savigny and his contemporaries focused on Roman law and its sources and had no intention of adapting the solutions of this system to the problems created by growing industrialisation. Interestingly, the existing types of liability without fault in Roman law were mostly overlooked or seen as exceptions that did not merit further examination.[6] If seen in their historical context, those exceptions actually have much in common with the exceptions created later by courts and legislators: they dealt with risks created by economic growth and urban development, where strict standards of fault liability and proof simply would not be able to reach practical and fair results.[7] The same can be said (and indeed has been said), for example, about sparks escaping from railways, exploding boilers and asbestos-related diseases.

However, the significance of the exceptions in their historical context seems to have escaped the commentators at the time, who either explicitly refused to draw analogies with the social and technological situation of their time or simply discarded those exceptions as irrelevant. They focused on fault liability and the damaging act alone and did not accept the social function of law in this area.[8] Social justice was to be achieved

[5] F. C. von Savigny, *System des heutigen Römischen Rechts* (Berlin: Veit, 1840), vol. I, p. 54; see also R. Ogorek, *Untersuchungen zur Gefährdungshaftung im 19. Jahrhundert* (Cologne: Böhlau, 1975), p. 11 and Esser, above note 3, at pp. 55, 62.

[6] E.g. *actio de pauperie; actio aquae pluviae arcandae; actio de deiectis vel effusis*. Savigny, Obligationenrecht above note 3, at p. 330, wrote that the '*variae causarum figurae*' were not suitable to derive general principles from. See further Lohsse, in W. Ernst, *The Development of Liability for Traffic* (Cambridge: Cambridge University Press, 2009).

Despite that, one form of strict liability which stems from Roman law even found its way into the German Bürgerliches Gesetzbuch (BGB) in 1900, the liability for animals (§833-Tierhalterhaftung). This provision originally covered all animals but in 1908 was amended due to political pressure by the agrarian sector, so that from then on only liability for damage caused by 'luxury' animals, i.e. animals not used for commercial purposes, was strict. For all other animals the owner could now exculpate himself by showing that he acted with due care or that the damage would have occurred even if he had acted with due care; cf. R. Zimmermann, *The Law of Obligations: Roman Foundations of the Civilian Tradition* (Oxford: Oxford University Press, 1996), p. 1116, and further J. Scherpe, in P. Mitchell, *The Role of Institutions in Legal Development* (Cambridge: Cambridge University Press, 2009). This created the rather absurd result that commercial livestock owners were privileged. For criticism of that rule see Bundesgerichtshof 27 May 1986, NJW 1986, 2501 and Kötz and Wagner, above note 2, at p. 191. See also Ogorek, above note 5, at p.14.

[7] In other words: those exceptions were created to deal with 'industrial' issues of their time!

[8] For an overview and summary see Ogorek, above note 5, at pp. 9–47 and Esser, above note 3, at p. 49. See also Lohsse, above note 6.

by freedom alone and the state was not to interfere. Influential writers such as Egid von Löhr,[9] and especially Johann Christian Hasse,[10] paved the way to what was later to become the maxim of German tort law as formulated by Rudolph von Jhering: '*Kein Übel ohne Schuld*' ('No wrong without fault')[11] and '*Nicht der Schaden verpflichtet zum Schadensersatz, sondern die Schuld*' ('It is not the occurrence of the harm which obliges one to make compensation, but fault').[12] To Jhering and his contemporaries, this was 'as simple as the chemical fact that what burns is not the light but the oxygen in the air'.[13] The initial concept of *culpa* was eventually narrowed down to its subjective meaning and thus equated with fault, explicitly rejecting the earlier Germanic rule of '*Wer Schaden tut, muß Schaden bessern*' ('He who causes damage, must pay damages').[14] *Culpa* understood this way was seen as a means to promote adequate behaviour while at the same time allowing industries to flourish.[15]

Roman law (as outlined above) already contained exceptions from the strict approach that tortious liability should be based on fault alone. Increasingly, modern developments could not be dealt with satisfactorily under the existing rules and soon more and more exceptions evolved, both in case law and in legislation.[16]

2. Law concerning the respective interests of occupiers of adjoining property

One of the areas that generated exceptions from fault liability is the law concerning the respective interests of occupiers of adjoining property. While principally regarded as property law problems and dealt with accordingly, some interesting arguments can be seen in the decisions by

[9] E. von Löhr, *Theorie der Culpa* (Gießen, 1806–08), Vols. I and II.

[10] J. C. Hasse, *Die Culpa des Römischen Rechts* (Kiel, 1815).

[11] R. von Jhering, 'Das Schuldmoment im römischen Privatrecht' in R. von Jhering (ed.), *Vermischte Schriften juristischen Inhalts* (Leipzig: Bretikopf and Härtel, 1879), p. 163.

[12] *Ibid* p. 199 (translation by Tony Weir in Zweigert and Kötz, above note 2, at p. 630).

[13] *Ibid.*

[14] Hasse, above note 10, at p. 16. See also Esser, above note 3, at p. 46 and R. Bienenfeld, *Die Haftung ohne Verschulden* (Berlin and Vienna: Springer, 1933), p. 14 for further references to Germanic law.

[15] Ogorek, above note 5, at pp. 32–6, 46–7; Esser, above note 3, at p. 54 (referring among other sources to the *travaux préparatoires* for a Bavarian codification of 1861 and an Austrian codification by von Zeiller of 1812), see p. 58 *et seq.* See also the *travaux préparatoires* for the BGB, *Protokolle der Kommission für die Zweite Lesung des Entwurfes eines Bürgerlichen Gesetzbuches* (Berlin, 1898), vol. II, pp. 569, 585.

[16] Cf. Benöhr, above note 4, at p. 694.

the courts, despite the fact that they varied enormously.[17] Those arguments dominated the discussions in Germany where damages to property were concerned, including damage caused by fires ignited by escaping sparks from railways. Some of these arguments were then later picked up in the debate as to whether there should be liability without fault or proof of fault in 'industrial' cases.

While legal disputes between neighbours have occurred since Roman times,[18] the problem intensified with the growing industrial use of land. Claims for damages and injunctions, however, were systematically categorised as restrictions on the use of property, and no real attempt was made to generalise the ideas contained therein.[19]

When balancing the rights of the land-owner and entrepreneurial freedom, the courts had to answer two key questions, often with conflicting results. First, could there be liability for damages or room for an injunction in the absence of fault? As the legislator did not provide any help in this conflict, it was mostly up to the judges to deal with the clash between the right to the unrestricted use of one's property, on the one hand, and the fault principle, on the other. If the courts did not want to leave the affected land-owner without any remedy, they had to find a solution that provided for one. Very often this led to decisions allowing injunctions but not awarding damages; many of the decisions lacked clear reasoning and were obviously result-oriented. The Oberappellationsgericht Celle, for example, decided on 22 June 1837, referring to D.8.5.8.5, that a land-owner did not have to suffer the smoke and vapour emissions from a neighbouring factory; the court did not require any fault by the factory owner.[20] This decision was followed by a number of courts.[21] In 1840, the Oberappellationsgericht Lübeck held, relying on D.8.5.8.5–7, that there was a general principle in the

[17] For a detailed account of this problem area see R. Ogorek, 'Actio negatoria und industrielle Beeinträchtigung des Grundeigentums' in H. Coing and W. Wilhelm (eds.), *Wissenschaft und Kodifikation des Privatrechts im 19. Jahrhundert*, vol. IV, *Eigentum und industrielle Entwicklung, Wettbewerbsordnung und Wettbewerbsrecht* (Frankfurt/Main: Klostermann, 1979), pp. 40–78.

[18] E.g. Ulpian D.8.5.8.5.–7; D.8.2.18 and Ulpian D.43.27.1.7.

[19] Cf. Ogorek, above note 5, at p. 52; Ogorek, above note 17, at p. 41 and J. Gordley, *The Development of Liability Between Neighbours* (Cambridge: Cambridge University Press, 2009), ch. 4.

[20] OAG-Celle, Seufferts Archiv, vol. 11, 18, no. 14 (22 June 1837). See also OAG-Celle, Seufferts Archiv, vol. 8, 452, no. 346 (1848, no further date specification).

[21] See, e.g., OAG-Lübeck, Seufferts Archiv, vol. 9, 292, no. 218 (29 October 1840); OAG-Lübeck, Seufferts Archiv, vol. 11, 166, no. 114 (28 June 1853); AG-Dresden, Seufferts Archiv, vol. 3, 8, no. 7 (17 December 1843, confirmed by OAG-Dresden on 11 May 1844).

digests that no one could be required to suffer emissions, etc., without just cause.[22]

Secondly, if the use of the land was within legal limits, i.e. within the concession or permission for the use of land, could there nevertheless be room for an injunction or even damages? There was general concern that if interpreted widely, property rights could seriously impede industrial and economic progress.[23] For example, when a land-owner in Pforzheim claimed damages from the neighbouring chemical factory for the noxious vapours and smoke emanating from the factory and the consequential pollution of his land, the Obergerichtshof Mannheim decided on 10 December 1860 not to award damages, as the use of the factory was within the concession held by its operators.[24]

Similarly, the First Senate of the Preußisches Obertribunal refused damages in a decision on 18 September 1848, stating explicitly that there was no room for liability in the absence of fault; since there was a concession allowing for the factory to be run, it was held that therefore there could not be any fault.[25] Interestingly, only two years later the second Senate of the same court allowed damages, arguing that the smoke and vapours would infringe the property rights of the neighbour; no one should derive an advantage from another's affliction and infringement of rights.[26] As the two decisions of the Senates were clearly contradictory, the Plenum had to decide between them, and essentially followed the second Senate's reasoning in a decision on 7 June 1852.[27] While a valid concession would in principle prevent the factory operators from being liable for damages without being at fault in tort, this would not affect property rights. The key argument remained that someone had infringed another's right and gained an advantage from it, which could not be permitted. The same line of argument can also be seen in cases and legislative debates regarding railways and the damage created by sparks (see below p. 140 *et seq.*).

[22] OAG-Lübeck, Seufferts Archiv, vol. 9, 292, no. 218 (29 October 1840).

[23] Cf. Ogorek, above note 17, at p. 44; C.U. Hesse, *Die Rechtsverhältnisse zwischen Grundstücksnachbarn (Nachbarrecht)* (2nd edn, Jena: Fischer, 1880), p. 311; W. Werenberg, 'Über die Collision der Rechte Verschiedener Grundeigentümer' in C.F. von Gerber and R. von Jhering (eds.), *Jahrbücher für die Dogmatik des heutigen römischen und deutschen Privatrechts (Jherings Jahrbücher)* (Jena: Fischer, 1863), vol. VI, p. 1 at p. 79.

[24] OHofG-Mannheim, Seufferts Archiv, vol. 14, 177, no. 113 (10 December 1860).

[25] Excerpts of the decision can be found in Entscheidungen des königlichen Ober-Tribunals, vol. 23 (1852), 252. Cf. also Benöhr, above note 4, at p. 700.

[26] Entscheidungen des königlichen Ober-Tribunals, vol. 23 (1852), 255.

[27] *Ibid.* 256.

While this can be seen as a tentative first step towards liability without fault, this was not expanded by the courts beyond the realm of property rights and the interests of occupiers of adjoining property, although there were no obvious restrictions preventing the courts from doing so.[28] This can be explained by the importance of and focus on property and especially land, which enjoyed special protection.[29] The criteria applied, however, often seemed to be based not on principle but on the facts of the individual case; frequently used arguments were danger of recurrence (*Wiederholungsgefahr*), substantial interference (*nicht unbeträchtlicher Eingriff*), and permanence of the facilities (*bleibende Vorrichtung*).[30]

The general tendency was to hold that, even where emissions could normally be expected from an undertaking, they would nevertheless constitute a basis for a claim by owners of neighbouring land. To a certain extent, therefore, the balance seemed to tip in favour of the land-owners, as is also shown by the codifications of the time, particularly the Saxonian Civil Code (Sächsisches Bürgerliches Gesetzbuch) which incorporated this view in §§321–323, 358, and draft Bills in Hesse and Bavaria.[31]

The focus of the problem then shifted to the remedy: obviously, an injunction would terminate the industrial activity whereas claims for damages would allow it to continue and could be included in the business calculations. Unsurprisingly for the time, the courts tended to prefer the latter.[32] This was finally confirmed by legislation: §26 of the Trade, Commerce and Industry Regulation Act (Gewerbeordnung (GewO)) of 21 June 1869,[33] which expressly stated that there could not be a claim for a

[28] See Ogorek, above note 5, at pp. 56–61.

[29] Cf. e.g. R. von Jhering, 'Zur Lehre von den Beschränkungen der Grundeigentümer im Interesse der Nachbarn' in C.F. von Gerber and R. von Jhering (eds.), *Jahrbücher für die Dogmatik des heutigen römischen und deutschen Privatrechts (Jherings Jahrbücher)*, vol. VI, pp. 81–130.

[30] See, e.g., AG-Dresden, Seufferts Archiv, vol. 3, 8, no. 7 (17 December 1843, confirmed by OAG-Dresden on 11 May 1844); Obertribunal zu Berlin, Seufferts Archiv, vol. 31, 148, no. 116 (12 October 1875); Ogorek, above note 17, at p. 53; E. Spangenberg, 'Einige Bemerkungen über das Nachbarrecht' (1826) 9 *Archiv für die civilistische Praxis* (*AcP*) 265 at 270; Benöhr, above note 4, at p. 704.

[31] See Ogorek, above note 17, at p. 53.

[32] Cf. *ibid.* p. 58, referring to an 1862 decision by the Oberappellationsgerichts Rostock, Entscheidungen des großhergzoglich Mecklenburgischen Oberappellationsgerichts zur Rostock (von Bucha/Budee (eds.)), vol. 4 (1862) 108, and OLG zu Braunschweig, Seufferts Archiv, vol. 46, 391, no. 249 (22 October 1890), and von Bucha/Budee, p. 77. Cf. also N. Jansen, *Die Struktur des Haftungsrechts* (Tübingen: Mohr Siebeck, 2003), p. 66 and H. Kötz, 'Haftung für besondere Gefahr' (1970) 170 *AcP* 3 at 4.

[33] In force until 1974, when it was replaced by §68 of the Federal Emissions Control Act (Bundesimmissionschuzgesetz), BGBl. 1974 I, 721.

cessation of the undertaking, but only measures to end the interferences or, if that was not possible, damages could be demanded.[34] This provision reflected the spirit of the decisions: that once an industrial undertaking had been granted the relevant permission or concession, its operation as such was guaranteed and injunctions expressly excluded.[35] Due to the practical problems in enforcing a claim for measures to minimise or end emissions etc., the law suits soon focused on damages alone. The courts limited this claim to damages for expected future interferences, hence leaving land-owners without a remedy for damages already suffered; for those they still had to rely on the law of torts. So, essentially, the codification limited the rights of land-owners (as they had evolved) for the benefit of entrepreneurs, in effect a return to fault liability.[36] However, this was soon circumvented by the courts, and later articles 105 and 125 EGBGB[37] allowed for stricter rules to be implemented by the German states.[38]

Thus, interestingly, in the law concerning the interests of occupiers of adjoining property (independently of the law of torts) an action for damages *not* based on fault evolved, as a result of the weighing of the interests of the parties involved. The ideas and arguments of this discussion, however, also influenced the law of torts. More importantly, cases dealing with fires caused by sparks escaping from railways were mostly argued on the basis of property law rather than tort law.[39]

3. Special liability of railway operators

(a) *Legislation: §25 of the Prussian Railway Act*

The most remarkable feature of the German law regarding liability for damage caused by railways is that it was dealt with by legislation at a very early stage.[40] Three years after the first German railway line on German

[34] Cf. Ogorek, above note 17, at p. 63.

[35] *Ibid.* p. 60 See also protocols of the deliberations, *Stenographische Bereichte des Reichstages 1869*, vol. 1, p. 273, especially the account by the members of Parliament Bähr, p. 280 *et seq.*, and Weigel, p. 295; in the deliberations references were made to the existing local legislation, namely §321 of the Saxonian Civil Code (Sächsisches Bürgerliches Gesetzbuch), §12 Lübisches Stadtrecht and §§3, 12, 14, 16 Rostocker Stadtrecht.

[36] For the limiting effect of the new provision see Ogorek, above note 17, at p. 62.

[37] Introductory Act for the BGB (Einführungsgesetz zum BGB).

[38] Cf. Prinz, 'Der Ausgleichsanspruch bei Funkenflugschäden', *Zeitung des Vereins Deutscher Eisenbahnverwaltungen* 1929, 473–81 at 480–1. See also below p. 171 *et seq.*

[39] See *ibid.*, 473 and the cases discussed below p. 171 *et seq.*

[40] Cf. Zimmermann, above note 6, at p. 1130; F.H. Lawson, *Negligence in the Civil Law* (Oxford: Oxford University Press, 1950), p. 45. On railway liability in Germany see also Lohsse, above note 6, and Scherpe, above note 6.

territory, leading from Nürnberg to Fürth,[41] was opened, the Prussian Railway Act (Preußisches Gesetz über die Eisenbahnunternehmungen) (PGE) of 3 November 1838[42] was enacted – four days after the first Prussian railway line, Berlin-Potsdam, which was privately owned, was opened on 29 October 1838.[43] In its §25 PGE, it contained a far-reaching provision, allowing for recovery of damages without fault. This provision is the gamete of all later provisions stipulating strict liability. Interestingly, the driving force and mastermind behind this section was von Savigny, the very same man who in his major works gave little or no regard to any tortious liability without fault (see above). Indeed, the regulation was solely based on an initiative by the Prussian Ministry (of which von Savigny was a part) and there was little input from the legal profession or academia.[44] Even after it entered into force it was widely ignored by academic authors and even specialist literature referred to it as a singular exception.[45]

Initially, the provision proposed (then §19) by the Staatsministerium allowed recovery from the railway company only if an employee of the company was at fault and no compensation could be gained from that employee.[46] The proposal was presented to the Council of State (Staatsrat), of which Savigny was a member. In his written comments[47] Savigny pointed out that it would be overly burdensome if one first had to sue the employee. The better way would be to allow claims against the company,

[41] 7 December 1835. For a description of the historical development of railways in Germany, see G. Eger, *Eisenbahnrecht* (Leipzig: Gloeckner, 1910), pp. 26–60.

[42] Gesetzessammlung für die königlichen preußischen Staaten 1838, no. 35, 505–16.

[43] At that point Prussia possessed some 34 km of railways, whereas there were some 141 km in all of Germany. See Table 1 in M. Eckardt, *Technischer Wandel und Rechtsevolution – Ein Beitrag zur ökonomischen Theorie der Rechtsentwicklung am Beispiel des deutschen Unfallschadensrechts im 19. Jahrhundert* (Dissertation, Ph.D. thesis, 1998, also published, Tübingen: Mohr Siebeck, 2001), p. 402; M. Eckardt, 'The Evolution of the German Tort Law in the Nineteenth Century: An Economic Analysis of the Evolution of Law' in O. Volckart (ed.), *The Institutional Analysis of History* (Munich: Accedo, 2004), p. 83 at p. 99; and Benöhr, above note 4, at p. 707. As Zimmermann, above note 6, at p. 1130, n. 261, reports, a pastor in Berlin on that day implored his congregation to keep away, for the sake of their eternal salvation, from these 'infernal dragons'.

[44] On the genesis of this provision see T. Baums, 'Die Einführung der Gefährdungshaftung durch F.C. von Savigny' (1987) 104 *Zeitschrift der Savigny-Stiftung für Rechtsgeschichte, Germanistische Abteilung* 277. Cf. also Esser, above note 3, at pp. 49 and 62; Eckardt, *Technischer Wandel*, above note 43, at p. 292; Eckardt, '*Evolution*', above note 43, at p. 99; Jansen, above note 32, at p. 369; Kötz, above note 32, at p. 14.

[45] Cf. Ogorek, above note 5, at p. 62.

[46] Cf. Eckardt, *Technischer Wandel*, above note 43, at p. 298.

[47] The text is reprinted in Baums, above note 44, at pp. 278–9. The original files can be found in the Staatsarchiv Merseburg.

which then could have recourse from the employee, particularly since it would be much easier for the company to identify the employee who might be at fault. Further, he explicitly referred to the problem of fires caused by escaping sparks, which 'had caused great damage several times in England'; often nobody would be at fault for such events. As those damages could not be prevented, the company should be liable for them unless the claimant was responsible for the damage himself or the damage came about through an inevitable event (*unabwendbares Ereignis*). He considered it unfair that the owners of the neighbouring land or passengers should suffer the damages which were the inevitable result of the operation of a railway.[48] Savigny's statement was adopted for the official papers and reports. In the ensuing discussions criticism was raised, particularly as the railways were singled out and other hazardous activities, including the operation of steam engines etc., which essentially presented the same risk, were not to be liable in the same way. These concerns were taken into account, yet it was found that the railways, unlike the other industrial activities mentioned, by their very nature were not suitable for preventative measures, as the tracks covered vast distances. Throughout the entire rail network this presented a risk to land-owners and people living close to the railway. Making them carry the risk and the damages would amount to an expropriation to the benefit of the railway operators. Therefore, the state was under a duty to take protective measures for the potential victims, 'just as has been done in England'.[49] The state was obliged to recognise the special risk created by the operation of railways and stipulate the remedies. This argument prevailed and §25 PGE was accepted by a majority of twenty-nine votes to four.[50]

The wording of §25 PGE as adopted reads as follows:

> The company is liable for all damages occurring during train transport to transported goods and persons as well as to other persons and goods, and the company can only exonerate itself from this liability by proving that the damage was caused through fault of the claimant or by an inevitable

[48] Cf. the text in Baums, above note 44, at p. 281.

[49] Ibid. See also C.J. Mittermeier, 'Die Rechtsprechung in England und Nordamerika über die Haftungspflicht der Eisenbahnverwaltungen wegen des durch Anlegung oder durch Betrieb der Bahn an benachbarten Grundstücken, oder Personen oder Waaren zugefügten Schadens, mit Rücksicht auf die deutsche Rechtsprechung' (1858) 41 *AcP* 410 and G. Lehmann, *Körperverletzungen und Tödtungen auf deutschen Eisenbahnen und die Unzulänglichkeit des Rechtsschutzes* (Erlangen: Enke, 1869), pp. 65 and 84, with data comparing Prussia and England.

[50] Baums, above note 44, at p. 281.

event. The dangerous nature of the undertaking as such is not such an event exonerating from liability.

This strict rule that allowed claimants to sue for damages arising from personal or property injury in the ordinary courts was quickly adopted by other German states.[51] It was applicable regardless of whether the railways were privately or publicly owned[52] and applied regardless of whether the claims were based on contract or tort.[53] While primarily created for personal injury cases (and personal injury of employees of the companies in particular), the provision also covered claims to transported goods and 'any other goods', the latter including neighbouring land and premises. The only requirement was that the damage must have occurred 'during train transport' ('*bei der Beförderung auf der Bahn*'). This was generally interpreted widely, but the wording of §25 PGE makes clear that even the tiniest form of contributory negligence by the victim would immediately rule out the entire claim. Therefore, railway operators often tried to prove contributory negligence, and apparently quite successfully. Statistics show that only 52 per cent of the passengers injured 'during train transport' were counted as having been injured without being at fault themselves, which means that only about half of them were eligible for damages under §25 PGE.[54] The situation for employees of railways seems to have been even worse, with only 12 per cent of them receiving compensation.[55]

Importantly, the fact that the operation of the railway was a permitted activity did not in any way leave room for arguments excluding liability (arguments that had successfully been raised in cases concerning the respective interests of occupiers of adjoining property, see above p. 136 *et seq.*); this was made clear by the second sentence of §25 PGE.

[51] *Cf.* Lehmann, above note 49, at p. 17. The Duchies and Grand Duchies of Saxony and Anhalt brought §25 PGE into force with regard to the Thuringian and Anhaltian railways. Holstein adopted a comparable provision in a declaration of 19 May 1840 (§22), as did Mecklenburg by an Act dated 23 May 1855. Austria also implemented similar rules: *Eisenbahnbetriebsordnung*, 16 November 1851 and 5 March 1869 (cf. the account by Lehmann, above note 49, at, p. 18), as did Switzerland in 1875 (cf. Benöhr, above note 4, at p. 707, n. 87).

[52] Cf. Preußisches Obertribunal, Entscheidungen des königlichen Ober-Tribunals, vol. 37 (1858), 42, no. 3 (14 December 1857); Eckardt, *Technischer Wandel*, above note 43, p. 288.

[53] Eger, above note 41, at p. 131.

[54] See Eckardt, *Technischer Wandel*, above note 43, at pp. 289, 316, 333.

[55] See Eckardt, 'Evolution', above note 43, at p. 105; Eckardt, *Technischer Wandel*, above note 43, Table 12 on p. 411.

However, in some respect §25 PGE was interpreted narrowly by Prussia's courts: while there was no doubt that the railways were to pay damages for physical injury to a person, it was held that there was no liability for consequential damages, particularly loss of present or future income because of the injury.[56]

(b) Case law on liability in parts of Germany where §25 was not applicable

Interestingly, even in those regions where §25 PGE was not in force, the courts in in some cases seemed one way or the other to follow a similar path.[57] The Oberappellationsgericht Braunschweig in 1856,[58] in a case where a cart was damaged by a train when crossing the tracks, held that the operation of a railway necessarily was dangerous and awarded damages; the court sidestepped engaging with the question of whether there should be liability in the absence of proof of fault by interpreting the statement of the defendant company that its employees acted in accordance with regulations to mean that, as there had been an accident, the instructions must have been faulty.

Because of the 'dangerousness of operating such a means of transport' the Oberappellationsgericht Lübeck held that the operator of a railway was liable to pay damages to an injured passenger even beyond the scope of contractual claims. Because of the special nature of railways, the railway operator was found to be under such a special duty of care that the smallest mistake by one of its employees would lead to liability.[59] As regards the proof of such a minor mistake, the court (referring to the reasoning of §25 PGE although the Act was not applicable in the case) – in complete reversal of established principles – held that for such claims both under contract and tort it was not for the claimant to prove fault but the onus was on the operator to prove the absence of fault! The court found support for this in the rules on liability deriving from the *receptum* and in the legislation of other German states and the court decisions following

[56] Preußisches Obertribunal, Entscheidungen des königlichen Ober-Tribunals, vol. 28 (1854), 270, no. 33 (24 April 1854); Preußisches Obertribunal, Entscheidungen des königlichen Ober-Tribunals, vol. 37 (1858), 42, no. 3 (14 December 1857).

[57] For a description of the development see Ogorek, above note 5, at p. 63.

[58] OAG-Braunschweig, Seufferts Archiv, vol. 10, 221, no. 165 (10 January 1856).

[59] OAG-Lübeck, Sammlung der Entscheidung des Ober-Appellationsgerichtes zu Lübeck in Frankfurter Rechtssachen, vol. 6, 116, no. 259 (27 September 1858). Cf. Zimmermann, above note 6, at pp. 397 and 1124.

from them. The reversal of the burden of proof in essence amounted to a presumption of fault, thus in practice rendering the fault requirement almost meaningless.

The struggle of the courts in the regions where §25 PGE was not applicable to uphold (or at least pay lip-service to) the fault requirement, while at the same time allowing damages, is apparent in a number of cases concerning fires caused by sparks emanating from a locomotive. The Oberappellationsgericht München on 11 May 1853[60] held that the railway company was liable for the damages because the 'sparks could only be explained by the employees' negligence or insufficient measures to prevent such sparks'. This was deemed to be sufficient for a claim based on fault under the *lex Aquilia*.[61] The defence that the company's activity was lawful and within the concession obtained (*qui iure suo utitur, neminem laedit*) was denied because this did not entail the right to damage another's property; this again was is in line with the property-based arguments concerning the respective interests of occupiers of adjoining property as outlined above. In 1859, the Oberhofgericht Mannheim awarded damages in a case concerning a fire caused by sparks, explicitly stating that neither the concession of the railway operator nor the fact that he used state-of-the-art technical equipment to prevent the escape of sparks would allow him a defence. The fact that a dangerous activity caused the harm was seen as sufficient and the operator could not claim exoneration for policy reasons. The court compared the position of the claimant to that of a person who would be compensated had he been expropriated.[62] The reduction of the fault requirement reached its peak in 1861 in a decision by the Oberappellationsgericht München:[63] while in theory upholding the requirement of a culpable act ('*culpose Handlung*'), this was held to be fulfilled if the defendant knew that his acts could lead to the damage and he acted nonetheless; as this was inevitable in the operation of a locomotive, the operation of a locomotive as such was seen as *necessarily* constituting a culpable act![64] This was a remarkable and rather strained intellectual exercise to uphold the fault principle. The presumption of fault triggered solely by the operation of a locomotive essentially removed the fault requirement and leads to a form of strict liability very much like that under §25 PGE.

[60] OAG-München, Seufferts Archiv, vol. 10, 220, no. 164 (11 May 1853).

[61] The court referred to Gaius D.19.2.25.7.

[62] OHofG-Mannheim, Seufferts Archiv, vol. 13, 329, no. 235 (20 December 1859).

[63] OAG-München, Seufferts Archiv, vol. 14, 354, no. 208 (16 April 1861).

[64] *Ibid.* 358. See also OLG für Bayern, Seufferts Archiv, vol. 41, 403, no. 271 (1 April 1886). See also Lohsse, above note 6.

Despite the development of the case law, the situation of claimants in regions where §25 PGE or similar Acts were in force was far from certain. Still, the court decisions reflect the rising awareness that the strict adherence to the fault principle in many ways was unsuitable for the problems arising out of the growing industrialisation. Hence, the courts often resorted to watering down the fault principle as just described or, more frequently, to allowing claims based on property arguments, as shown above. The latter tendency continued even after the enactment of the German Civil Code (BGB) (see below p. 171 *et seq.*).

Nevertheless, §25 PGE, enacted in 1838, was a remarkable piece of legislation for its time, the spirit of which not only influenced the legislation and jurisprudence in many other German states but also future developments.

Despite the increasing convergence of the law regarding liability for railways (and other industrial enterprises), the lack of certainty was much lamented[65] and was one of the reasons for the enactment of the Reichhaftpflichtgesetz in 1871 (see below p. 160 *et seq.*).

(c) *Attempts to limit tortious liability by contract*

While owners of neighbouring land and premises and injured persons without contractual relations to the railway companies in theory enjoyed the full protection of §25 PGE, the railway companies and their associations soon started excluding their liability under §25 PGE through contract and works council agreements. As mentioned earlier, the general jurisprudence still regarded any liability not based on fault as anomalous and a singular exception and thus neither academia nor the courts raised principled objections to contracts excluding liability.[66]

It did not take long until railway companies would no longer employ people who had not waived their claims under §25 PGE, thus thwarting one of the main objectives of the legislation, safeguarding the interests of the employees.[67]

[65] See, e.g., Lehmann, above note 49, and G. Lehmann, *Der Nothstand der Schädensprocesse und der Entwurf der Königl. Sächsischen Civilproceßordnung* (Leipzig: Fischer, 1865); cf. Ogorek, above note 5, at p. 100 and Lohsse, above note 6.

[66] §25 PGE was not seen as part of the mandatory legal rules as it 'clearly' established an exception. *Cf.* Jansen, above note 32, at p. 369; Eckardt, 'Evolution', above note 43, at pp. 98 and 105; Prinz, above note 38, at 478.

[67] Cf. Ogorek, above note 5, at p. 62; Eckardt, *Technischer Wandel*, above note 43, at p. 314; I. vom Feld, *Kontrollierte Staatsentlastung im Technikrecht – Dampfkesselgesetzgebung und-Dampfkesselüberwachung in Preußen 1831–1914* (Frankfurt: Mohr Siebeck, 2007), p. 132.

It was even argued by some that by buying land next to a railway and being aware of the danger of fires caused by escaping sparks, the purchaser had tacitly waived his claims for compensation, but this apparently was not followed by the courts.[68] It was also submitted that when the owner of a piece of land had already received compensation for the decrease in value of his property because of the risk the adjacent railway posed, he could not make any further claims if the risk actually materialised. This, however, was rejected, as any further loss still amounted to quasi-expropriation and thus was not covered by the original compensation payment.[69]

The attempts to limit liability under the PGE by contract and other measures in the end (after thirty years!) led to the Prussian Act of 2 May 1869[70] which barred such contractual exclusions and declared them void.[71]

Regarding the transport of goods, the General German Commercial Code (Allgemeines Deutsches Handelsgesetzbuch) of 1861 contained a provision (article 423) which prohibited the exclusion of liability for damages to goods transported by rail; here, apparently, the associations of the industry dependent on railway transport had lobbied successfully.[72]

(d) *Procedural position of the claimant*[73]

Before the enactment of §25 PGE and similar Acts in other German states and the development of the case law, the main obstacle for a claimant was the need to prove the fault of the defendant.[74] For the majority of the cases, i.e. those not dealing with railway liability, this still remained the case, and this included industrial accidents in particular.

[68] Cf. Prinz, above note 38, at 478, who argues for such a tacit waiver.

[69] *Ibid.* 479 and especially a decision by the Reichsgericht in 1892, Entscheidungen des Reichsgerichts in Zivilsachen (RGZ) 29, 268, no. 66 (27 April 1892): the seller of a piece of land will have to suffer the emissions that under normal circumstances can be expected from a railway or a factory if at the time of the sale, he was aware that the property would be used for that purpose. This, however, does not apply to fires caused by escaping sparks and other unusual or unexpected events.

[70] Gesetzessammlung für die königlichen preußischen Staaten 1869, no. 38, 665.

[71] Eckardt, 'Evolution', above note 43, at p. 111.

[72] Eckardt, *Technischer Wandel*, above note 43, at p. 316; Jansen, above note 32, at p. 369.

[73] See the account by Ogorek, above note 5, at p. 87 and for the consequences of the onus of proof on liability, H. Stoll, 'Haftungsverlagerung durch beweisrechtliche Mittel' (1976) 176 *AcP* 145.

[74] Cf. Eckardt, *Technischer Wandel*, above note 43, at p. 314; Eckardt, 'Evolution', above note 43, at p. 105; vom Feld, above note 67, at p. 132.

It was not the onus of proof as such, but often the procedural laws concerning evidence that were very burdensome for the claimant, and it was said by some that it indeed negated claims and made successful claims for damages impossible. One of the most influential critics, Lehmann, wrote that the courts were strangled by the strict and illogical rules concerning evidence and in some cases needed to ignore facts that were commonly known; in essence, the burden of proof determined the outcome of claims for damages, particularly when the means for proof had perished in the accident and the exact damage needed to be stated with utmost accuracy.[75] In essence, he wrote, 'one might as well advise the claimant to climb to the moon to get the damages awaiting him there'.[76] Counsel should advise their clients to spend their money on a nice holiday instead of wasting it in a futile attempt to recover damages.[77]

This can be exemplified by the two following cases. In a decision from 1836, the Oberappellationsgericht Kassel[78] refused to award damages to a woman who had been injured by a dog under an *actio de lege Aquilia*. The court argued that walking a dog was an activity generally allowed by the law, and the fact that the dog was known to be wild and had caused injury before was deemed irrelevant because the claimant could not show that earlier (undisputed) incidents had not been caused by taunting or similar acts. Therefore, there was no reason to assume that the owner of the dog was under a duty to take special care when walking his dog, because the claimant had failed to establish the dangerousness of the dog. The proof and knowledge of earlier incidents was held to be insufficient to establish such a duty.

In another case, a land-owner was not awarded damages by the Oberappellationsgericht Berlin.[79] The damage had undisputedly been caused by flooding from a municipal drainage system but the claimant was denied damages simply because he could not prove that the damage would not have occurred naturally even without the flooding caused by the drainage system.

The practical impossibility of obtaining damages, particularly in an industrial environment, caused growing concern and became a

[75] Lehmann, above note 49, at p. 74.

[76] Lehmann, *Nothstand*, above note 65, at p. 8.

[77] *Ibid.* p. 1. See also the report by W.M. Schaffrath, in *Verhandlungen des 4. deutschen Juristentages*, vol. 3, p. 252.

[78] OAG-Kassel, 9 September 1836, *Annalen der Justiz und Verwaltung in der Provinz Hessen, hrsg. von O.L. Heuser* (Heuser-Annalen), vol. 1, 1854, p. 543.

[79] OAG-Berlin, 10 October 1868, *Annalen der Justiz und Verwaltung in der Provinz Hessen, hrsg. von O.L. Heuser* (Heuser-Annalen), vol. 15, 1869, p. 413.

major argument in the deliberations on later legislation, namely the Reichshaftpflichtgesetz of 1871 (see below, p. 160 *et seq.*).[80]

(e) *German railways as the leading sector in industrial growth*

In the nineteenth century, Germany experienced a rapid economic growth, particularly after the beginning of railway construction.[81] As can be seen from Table 4.1, the track length expanded rapidly from only 6 km in 1835/36 (and none in Prussia) to almost 43,000 km in 1879 (almost 27,000 km of which were in Prussia).

The importance of the railways for the general economy of Germany and Prussia can also be seen in Table 4.2, showing the growth in passenger and goods traffic, and Table 4.3, giving data on the labour force employed by railway operators and the revenue, expenditure and capital income for these companies.

While the coal mining and iron industries were quite dependent upon the development of the railway, the growth of the railway as such was relatively independent from other industries. Nevertheless, as Table 4.4 shows, the average transportation costs, according to the research done by Fremdling, declined steadily.[82]

The fact that Germany, and particularly Prussia, adopted comparatively strict rules for liability of railway operators seems not to have had a negative impact on this industrial sector. No other sector experienced growth on this scale and railways were the leading and vital factor in German economic development in the nineteenth century. In 1885, the railway sector accounted for almost 12 per cent of Germany's capital stock, and in 1875–1879 the net investment in this sector even amounted to 25 per cent of the net investment of the entire German economy.[83] At this time, the liability rules as described above were long established and did not seem to deter investors.

[80] *Cf.* Ogorek, above note 5, at p. 94.

[81] For a comprehensive and detailed description of the development of railways in Germany from 1835–1989, see R. Fremdling, R. Federspiel and A. Kunz, 'Statistik der Eisenbahnen in Deutschland 1835–1989' in W. Fischer, F. Irsigler, K.H. Kaufhold and H. Ott (eds,). *Quellen und Forschungen zur historischen Statistik von Deutschland* (St. Katharinen: Scripta Mercaturae, 1995), vol. XVII.

[82] R. Fremdling, *Eisenbahnen und deutsches Wirtschaftswachstum 1840–1879* (St Katharinen: Scripta Mecaturae, 1995), p. 194.

[83] *Ibid.* p. 194; Fremdling, Federspiel and Kunz, above note 81, at p. 2.

Table 4.1. Length of railway lines in Germany[a] and Prussia[b] in km, 1835–1879

	Germany		Prussia	
Year	*Single track*	*Double track*	*Single track*	*Double track*
1835	6			
1836	6			
1837	20			
1838	141		34	
1839	242		69	
1840	462	117	185	
1841	677	135	395	18
1842	922	248	588	113
1843	1,310	264	815	136
1844	1,758	264	924	136
1845	2,151	286	1,106	158
1846	3,291	426	1,972	188
1847	4,317	650	2,424	320
1848	5,002	1,139	2,718	567
1849	5,458	1,229	2,880	632
1850	5,875	1,248	2,967	365
1851	6,162	1,334	3,153	731
1852	6,649	1,394	3,487	735
1853	7,177	1,557	3,697	803
1854	7,608	1,808	3,697	992
1855	7,862		3,822	1,040
1856	8,672		4,373	1,184
1857	9,055		4,652	1,426
1858	9,721	2,873	4,901	1,550
1859	10,648		5,452	1,600
1860	11,157		5,762	1,742
1861	11,567		5,951	1,800
1862	12,150		6,113	1,821
1863	12,773	3,710	6,416	1,960
1864	13,240	4,048	6,560	2,110
1865	14,034	4,400	6,895	2,265
1866	14,941	4,618	7,133	2,418
1867	15,793	4,929	7,425	2,640
1868	16,442	5,275	10,125	3,645
1869	17,322	5,673	10,457	3,796
1870	18,810	5,959	11,460	4,067

Table 4.1. (*cont.*)

	Germany		Prussia	
Year	*Single track*	*Double track*	*Single track*	*Double track*
1871	20,405	6,337	12,474	4,377
1872	22,522	7,454	13,632	4,971
1873	23,853	8,061	14,461	5,398
1874	25,495	8,290	15,493	5,575
1875	27,931	9,152	16,940	6,358
1876	29,088	9,186	17,556	6,391
1877	30,288	9,476	18,346	6,514
1878	31,337	9,601	18,974	6,634
1879	33,094	9,782	20,134	6,795

Source: R. Fremdling, *Eisenbahnen und deutsches Wirtschaftswachstum 1840–1879* (St Katharinen: Scripta Mecaturae, 1995), p. 48 (also reproduced in Eckardt, *Technischer Wandel*, above note 43, at p. 402).

[a] From 1872 on, the figures include Alsace-Lorraine.
[b] From 1868 on, the figures include the territories annexed in 1866.

Table 4.2. Passengers and goods traffic on German[a] and Prussian[b] railways, 1840–1879

	Persons (million km)		Goods (million km, incl. transport of livestock)	
Year	Germany	Prussia	Germany	Prussia
1840	623	20.1	3.2	0.5
1841	97.2	42.5	8.1	3.9
1842	145.3	84.9	14.5	8.6
1843	185.0	113.0	25.6	18.3
1844	255.7	153.0	33.2	23.0
1845	308.5	161.7	50.8	28.4
1846	427.8	224.5	81.9	46.6
1847	546.7	318.8	159.7	105.2
1848	614.9	346.4	168.4	112.2
1849	664.3	359.4	233.8	149.5
1850	782.7	421.7	302.7	194.2

Table 4.2 (*cont.*)

	Persons (million km)		Goods (million km, incl. transport of livestock)	
Year	Germany	Prussia	Germany	Prussia
1851	865.3	440.5	393.8	239.0
1852	914.2	450.2	527.4	330.7
1853	935.1	477.8	620.8	412.5
1854	1040.8	505.2	897.7	551.5
1855	1090.4	531.9	1094.9	663.0
1856	1263.0	627.4	1241.8	752.5
1857	1457.1	740.5	1531.2	907.8
1858	1490.6	760.6	1505.0	876.6
1859	1637.2	810.1	1475.2	816.9
1860	1732.9	874.6	1675.2	963.7
1861	1900.9	958.6	1997.6	1157.8
1862	2063.7	1042.4	2431.4	1452.4
1863	2359.8	1201.8	2776.8	1637.5
1864	2571.2	1384.0	3219.8	1934.9
1865	2676.0	1385.9	3671.8	2244.0
1866	3131.8	1772.7	3777.0	2291.1
1867	2977.9	1518.9	4527.1	2707.6
1868	3236.1	2073.1	5184.0	3590.4
1869	3533.8	2271.7	5520.4	3848.1
1870	4446.8	3029.5	5875.9	4176.6
1871	5031.3	3374.3	7072.0	4918.4
1872	5019.9	3157.9	8361.3	5724.6
1873	5692.7	3553.8	10060.0	7016.8
1874	5843.7	3651.6	10295.6	7193.5
1875	5994.3	3784.1	10624.6	7443.6
1876	6109.4	3826.0	10978.7	7737.3
1877	6111.9	3750.3	11224.7	7953.9
1878	6151.2	3777.6	11497.1	8035.4
1879	6147.5	3825.3	12244.3	8644.6

Source: R. Fremdling, *Eisenbahnen und deutsches Wirtschaftswachstum 1840–1879* (St Katharinen: Scripta Mecaturae, 1995), p. 17 (also reproduced in Eckardt, *Technischer Wandel*, above note 43, at p. 403.

[a] From 1868 on, the figures include the territories annexed in 1866.
[b] From 1872 on, the figures include Alsace-Lorraine.

Table 4.3. Labour force, expenditure and capital income of German[a] and Prussian[b] railways, 1840–1879 (in millon marks)

Year	Germany				Prussia			
	Labour force	Revenue	Expenditure	Capital income	Labour force	Revenue	Expenditure	Capital income
1840	1,648	3.6	2.0	1,539	648	1.1	0.8	0,314
1841	2,776	5.8	3.4	2,420		2.6	1.7	0,972
1842	4,114	9.1	5.1	4,022	2,335	5.3	3.0	2,304
1843	5,719	13.0	6.5	6,456	3,299	8.3	3.9	4,353
1844	8,475	17.4	9.3	8,018	4,884	10.7	5.6	5,094
1845	9,541	20.6	11.2	9,462	4,976	11.5	6.1	5,348
1846	12,505	30.0	16.3	13,675	7,065	17.4	9.6	7,776
1847	17,059	42.4	23.3	19,027	9,674	26.2	13.7	12,441
1848	20,910	46.0	27.7	18,336	11,449	27.5	15.8	11,739
1849	23,345	55.9	29.8	26,168	12,534	32.6	16.6	16,014
1850	26,084	67.5	34.6	32,916	13,706	39.3	18.8	20,478
1851	32,487	77.7	37.8	39,984	17,163	43.8	20.6	23,162
1852	38,625	92.0	44.6	47,509	20,633	51.5	24.6	26,924
1853	44,287	105.9	53.6	52,318	23,687	59.6	30.1	29,451
1854	47,734	125.1	62.3	60,708	26,326	70.3	36.5	33,799
1855	51,480	145.9	74.7	69,047	27,380	81.0	43.0	37,997
1856	55,427	164.1	82.7	79,262	30,276	92.1	49.5	42,517
1857	64,975	191.8	90.3	96,913	34,357	106.9	53.8	53,193

1858	70,145	191.0	94.4	93,474	36,181	105.7	53.6	52,083
1859	80,102	196.1	95.6	99,127	41,467	103.2	50.1	53,119
1860	85,608	218.2	100.2	117,607	44,852	117.4	52.1	65,35
1861	90,159	243.6	110.0	133,608	48,420	132.3	58.5	73,774
1862	93,782	275.8	122.3	153,494	51,502	151.5	66.0	85,493
1863	101,859	292.5	135.2	157,284	56,462	160.6	72.6	88,022
1864	105,740	322.2	145.7	176,510	58,005	180.0	77.9	102,067
1865	113,570	353.6	160.2	193,355	62,294	196.5	85.5	110,973
1866	121,630	374.5	182.9	191,613	64,859	208.9	98.1	110,767
1867	131,867	414.7	205.5	209,154	73,190	222.3	109.7	112,658
1868	143,562	450.5	222.2	228,348	96,987	290.8	140.9	149,869
1869	153,778	472.2	231.4	240,868	101,388	310.3	146.9	163,393
1870	161,014	510.7	253.3	257,429	106,542	314.8	161.2	180,595
1871	178,461	608.0	305.0	302,985	122,506	397.8	193.0	204,816
1872	193,506	670.7	381.7	288,958	135,875	434.5	238.8	195,678
1873	234,114	750.7	464.0	286,693	162,114	490.5	297.4	193,155
1874	263,554	788.9	503.6	285,266	178,350	516.1	326.6	189,485
1875	274,018	836.9	492.4	344,580	182,476	552.1	313.5	238,648
1876	278,107	843.5	484.7	358,776	182,724	553.8	303.6	250,182
1877	262,490	839.3	476.1	363,188	179,091	551.3	310.1	241,167
1878	269,185	800.3	441.6	358,669	179,203	550.9	277.2	273,671
1879	272,831	814.8	436.3	378,527	182,143	565.9	276.8	289,133

Source: Eckardt, *Technischer Wandel*, above note 43, at p. 404, based on the data in R. Fremdling, *Eisenbahnen und deutsches Wirtschaftswachtum 1840–1879* (St Katharinen: Scripta Mecaturae, 1995), pp. 24 *et seq.*, 141, 143.

[a] From 1872 on, the figures include Alsace-Lorraine.

[b] From 1868 on, the figures include the territories annexed in 1866.

Table 4.4 Average transportation cost per ton-kilometre (assuming that 1840 = 100)

1840	100
1850	60
1860	47
1870	33
1880	26
1913	21

4. Steam boilers

The advent of steam boilers and steam engines had a deep impact on German industrial development, but they caught on later than in other parts of Europe. Again, there was legislation at a very early stage when there were almost no steam engines to be found. However, as the legislator was well aware of the potential danger as accidents had been occurring in other countries, it was felt necessary to intervene, albeit at first not by establishing stricter liability rules (as had been done for railways), but by installing a licensing/concession system.[84] Rules on liability were first implemented in the Reichshaftpflichtgesetz of 1871 (see below p. 160 *et seq.*). Exploding steam boilers never reached the prominence in public debate and importance in case law and legislation that they did in other countries (see Figures 4.1 to 4.4 below).

(a) Use of steam power in Prussia

The first steam engine in Prussia was apparently used in a mine in 1785; outside of a mine, the first steam engine was utilised in a porcelain manufactory in Berlin in 1799.[85] However, in 1812 there apparently were still only two steam engines in use in all of Prussia.[86] They were then increasingly utilised by the mining industry and gradually found use in manufacturing businesses, as well as for locomotives and ships. Smaller businesses often rented or shared machines. The Prussian government was indeed so worried about the lack of industrial progress that they even provided steam engines to businesses so that Germany did not fall too far behind countries

[84] vom Feld, above note 67, at p. 193. [85] *Ibid.* p. 15. [86] *Ibid.*

such as England in its industrial development.[87] Figures 4.1, 4.2 and 4.3[88] show the late growth in numbers of steam engines in Prussia.

The increasing number of steam engines and boilers in Germany inevitably led to an increase in accidents and explosions. Apparently, between 1857 and 1869 there were 107 explosions with 155 persons killed in Prussia; therefore it was assumed that every year approximately 11 or 12 people were killed in the operation of steam boilers.[89] However, reliable data can only be found from the late 1870s onward.[90]

As can be seen from Figure 4.4, the number of explosions remained relatively stable over the years; the drastic increase after 1894 was due to a (temporary) redefinition and extension of the term 'explosion' by the Reichskanzler in a decree; the more extensive definition was 'corrected' in 1897.[91] The number of persons killed varies somewhat, but this can be explained by the fact that just one or two major accidents could cause a significant increase. Interestingly, the number of persons killed drops somewhat after 1901, which seems to coincide with an increase in regular controls of steam boilers.[92]

Exploding steam boilers only accounted for a small portion of industrial accidents in Germany, many more being caused by other machines and machine parts like propeller or drive shafts and drive lines.[93] According to Eckardt,[94] after 1889 there were usually around 200 reported accidents per year involving steam boilers (including explosions), compared to 1,500–3,300 involving railways, 5,000–15,000 involving motors in general and 500–2,000 involving lifts!

While explosions of steam boilers and engines occurred, and certainly the larger incidents in England and the United States were also reported in newspapers, there does not seem to have been a wider concern. Mostly, politicians and academics seemed to assume that

[87] *Ibid.* p. 17. [88] All taken from *ibid.* p. 17 *et seq.*

[89] Data given by the Department of Commerce (Handelsministerium) in the parliamentary discussion of the 1861 Act regulating the operation of steam boilers, see *Anlagen zu den Stenographischen Berichten über die Verhandlungen des Hauses der Abgeordneten während der 2. Session der 11. Legislaturperiode (1871–1872)*, vol. I, no. 34, pp. 190–2.

[90] Chart taken from vom Feld, above note 67, at p. 23, who based it on the data in *Badische Gesellschaft zur Überwachung von Dampfkesseln* (Mannheim: Badische Gesellschaft, 1906), p. 48.

[91] vom Feld, above note 67, at p. 23.

[92] But vom Feld, *ibid.* p. 24, concedes that there is no sufficient proof for a link between the lower number of deaths and the increase in controls.

[93] *Ibid.* p. 24.

[94] Eckardt, *Technischer Wandel*, above note 43, at p. 412, Table 13, based on *Amtliche Nachrichten der Reichsversicherungsanstalt* (Berlin, 1885).

workers were compensated for dangerous work by having an increased salary.[95] Where steam engines were installed in industrial undertakings, the disputes more often concerned smoke emission and noise; where workers opposed the introduction of new technology, they mostly did so because of the fear of losing their livelihood, not because of the danger it might pose.[96] Therefore, explosions of steam boilers do not feature prominently in the public or parliamentary debates or in the case law.

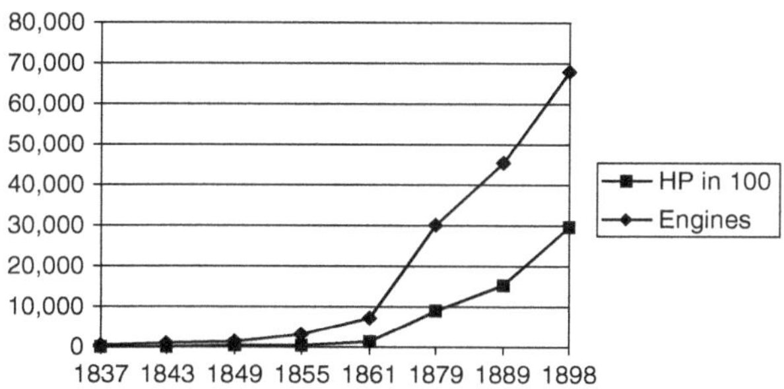

Figure 4.1 Stationary steam engines in Prussia 1837–1898
Source: vom Feld, above note 67, at pp. 17–18.

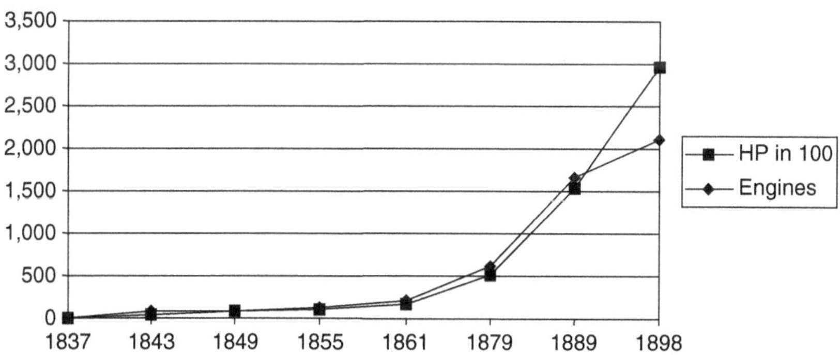

Figure 4.2 Steam engines on vessels in Prussia 1837–1898
Source: vom Feld, above note 67, at pp. 17–18.

[95] Eckardt, *Technischer Wandel*, above note 43, at p. 249.
[96] Cf. vom Feld, above note 67, at pp. 28–31.

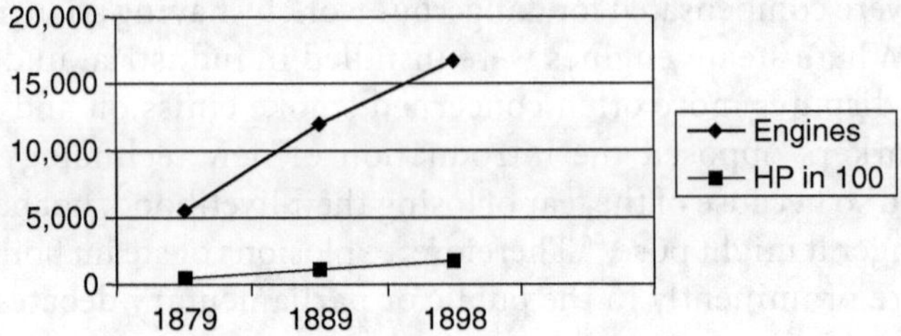

Figure 4.3 Mobile steam engines in Prussia: 1879–1898
Source: vom Feld, above note 67, at pp. 17–18.

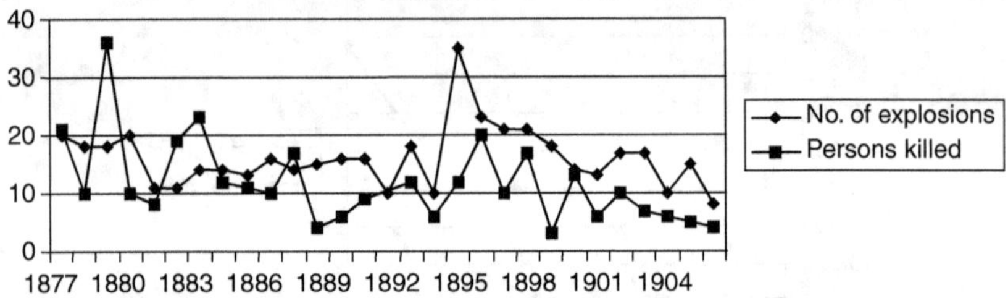

Figure 4.4 Steam boiler explosions in the Deutsche Reich 1877–1905
Source: vom Feld, above note 67, at pp. 17–18.

(b) *Regulation and control of steam boilers and engines before the Reichshaftpflichtgesetz of 1871*

With the exception of those German territories where French law remained applicable until 1831, there was no general duty to obtain a licence or concession in order to operate steam boilers or engines. The reluctance of the Prussian authorities to implement such a duty despite the fact that the potential dangers of such machinery were well known can be explained by the strong desire to increase industrial activity. In the territories where French law was applicable there was such a duty, and the local authorities made use of it and sometimes even withdrew concessions after accidents. The businesses concerned argued that they were at a competitive disadvantage and complaints by the businesses concerned to the central authorities in Berlin usually resulted in the reversal of the withdrawal of the concession. However, the entrepreneurs wanted legal certainty and interestingly they therefore pressed for a general duty to obtain a concession, one of the key arguments being that in other countries such concessions were easily obtained.[97] By *Kabinetsorder* (Cabinet Order) of January 1,

[97] *Ibid.* p. 36.

1831[98] such a general duty was introduced. In a regulation of implementation (Instruktion)[99] the requirements for obtaining such a concession were laid out, mostly certain technical specifications. There was a widespread belief that the machinery was controllable if constructed and operated properly. The Instruktion, however, did not require a regular reinspection of the machinery. As the technological development progressed, the rules had to be adjusted accordingly.[100] The Trade, Commerce and Industry Regulation (Gewerbeordnung) (GewO) Act of 1845[101] extended the duty to obtain a concession for all 'dangerous' machinery (§26), under which steam boilers (§27) were expressly included. Still, there were no regular check-ups as to whether the machinery still complied with the conditions under which the concession was originally obtained, and even further amendments due to technological change[102] did not establish such inspections.

Throughout this period there were sanctions (fine or imprisonment) attached to a breach of the concession, which of course also applied to later changes of machinery. As there were no controls, these sanctions in practice were only applied after an accident.[103]

Not until 1856 were regular inspections of steam boilers established in the Act on Steam Boilers (Dampfkesselgesetz).[104] While in the beginning the inspections were carried out by civil servants only, gradually private organisations began to take over the duties, thus establishing the beginning of a system of private self-regulation under state supervision that is still in place in Germany today: the technical supervision associations (Technische Überwachungsvereine).[105]

[98] Allerhöchste Kabinetsorder vom 1.1.1831, die Anlagen und den Gebrauch der Dampfmaschinen betreffend, Preußische Gesetzes-Sammlung 1831, 243.

[99] Instruktion vom 13.10.1831 zur Vollziehung der Allerhöchsten Kabinetsorder vom 1.1.1831, die Anlagen und den Gebrauch der Dampfmaschinen betreffend, Preußische Gesetzes-Sammlung 1831, 244–7. A further Instruktion gave a right of continuance to existing steam boilers, Fernerweite Instruktion vom 21.5.1935 zur Vollziehung der Allerhöchsten Kabinetsorder vom 1.1.1831, die Anlagen und den Gebrauch der Dampfmaschinen betreffend, Preußische Gesetzes-Sammlung 1835, 94–6.

[100] Cf. Regulativ vom 6.5.1838, die Anlage und den Gebrauch von Dampfkesseln und Dampfentwicklern betreffend, Preußische Gesetzes-Sammlung 1838, 262–72.

[101] Allgemeine Gewerbeordnung vom 17.1.1845, Preußische Gesetzes-Sammlung 1845, 41–78.

[102] Regulativ vom 6.9.1848, die Anlage der Dampfkessel betreffend, Preußische Gesetzes-Sammlung 1848, 321–35.

[103] vom Feld, above note 67, at p. 36.

[104] Gesetz, den Betrieb der Dampfkessel betreffend vom 7.5.1856, Preußische Gesetzes-Sammlung 1856, 295 and Regulativ vom 23.8.1856 zur Ausführung des Gesetzes, den Betrieb der Dampfkessel betreffend, Ministerialblatt für die gesamte innere Verwaltung in den Königlich Preußischen Staaten 1856, 210.

[105] For a detailed account of the historical development see vom Feld, above note 67. There apparently is evidence that these private organisations (*Dampfkesselüberwachungsvereine*)

The legal rules on concessions, however, did not concern liability for personal damages from exploding steam boilers. Unlike railways, there were no special regulations so liability required fault,[106] which was very difficult to prove. In any event, the procedural position of the claimant did little to encourage such claims (see above p. 147 *et seq.*). In principle, Part I, Title 6, §26 of the Allgemeines Landrecht für die Preußischen Staaten (ALR) 1794 provided for liability irrespective of fault if the damage would have been avoided had certain police laws (*Polizeigesetze*)[107] that were designed to prevent such damage been adhered to,[108] or, simply put, a breach of these laws had to have resulted in injury. However, no such claim seems to have been brought to the courts. Presumably, the accidents were deemed to have been caused by the fault of boiler attendants or stokers. As until 1903, no legal rules of conduct and supervision for these persons existed, there accordingly could not be a breach of a relevant *Polizeigesetz.*[109] Overall, there seemed to be a general tendency of leniency when the defendants were involved in industrial activity.[110]

All in all, this in essence meant that there was virtually no risk of being liable for personal damages resulting from steam boiler explosions.[111] This changed, at least gradually, with the introduction of the Reichshaftpflichtgesetz in 1871.

5. Reichshaftpflichtgesetz of 1871[112]

The practical problems for claimants in successfully suing for damages caused much debate in Germany in the 1860s, particularly where industrial accidents and especially railway accidents were concerned.[113]

overemphasised the liability risks for exploding steam boilers in their own interest, *ibid.* p. 141. The decisions that were referred to in the publications by these organisations (e.g. RGZ 8, 30–5, no. 7 (3 November 1882), described at n. 155; RGZ 12, 45–50, no. 11 (1 July 1884): liability not precluded by recent inspection by civil servant) often did not concern accidents caused by exploding boilers specifically but to liability under §2 RHPflG in general.

106 With the exception of damage to neighbouring property, see above p. 136 *et seq.*

107 Such 'police laws' were statutes enacted for a very specific purpose.

108 Cf. Reichs-Oberhandelsgericht Entscheidungen des Reichs-Oberhandelsgerichts, vol. 23, 46, no. 15 (11 October 1877) and G. Eger, *Das Reichshaftpflichtgesetz* (6th edn, Hannover: Helwing, 1906), pp. 278 and 284.

109 Cf. vom Feld, above note 67, at p. 131.

110 *Ibid.* p. 132; Ogorek, above note 5, at p. 68.

111 For property damage see above p. 136 *et seq.*

112 The following description of the genesis of the RHPflG, etc. follows the excellent account by Ogorek, above note 5, at p. 98.

113 See Lehmann, above note 49, especially at pp. 65, 84 with interesting data comparing Prussia and England; Lehmann, *Nothstand*, above note 65; and the debates and reports

These were very much in the public domain after some widely publicised accidents and the debate soon focused on personal injury.[114] In 1868, a petition by the National-Liberal Party of Leipzig (Nationalliberale Partei zu Leipzig) reached the Reichstag des Norddeutschen Bundes, demanding a reform of the law regarding accidents and liability, and also pointed out that the fact that different legal rules applied in different parts of Germany further increased the general feeling of injustice and uncertainty.[115] The petition demanded new liability rules for certain undertakings, namely railways, mines and factories, a change of the procedural rules, a duty to compensate not only for personal injury as such but also for the damages arising therefrom, and so on. The petitioners argued that it was the economic, social, political and moral duty of the government to remedy the well-known grievances. The responsible commission of the Reichstag supported the petition, even stating that the situation of victims of personal injuries was actually worse than described by the petitioners.[116]

This led the government to prepare a draft Bill which was presented to the first all-German Reichstag (deutschen Gesamtreichstag) in the fourteenth session on 13 April 1871, following consultations with regional governments. Of those, only three opposed the general notion: Hamburg (stating that the industry should not be burdened with disproportionate costs), Sachsen-Weimar (denying the necessity of such legislation), and Mecklenburg-Schwerin (suggesting that this problem should be dealt with in a general reform of the law of obligations).[117]

After some debate (see below, p. 163 *et seq.*), the Reichstag adopted the Liability Act of the Reich (Reichshaftpflichtgesetz) (RHPflG) on 7 June 1871 with a large majority of the votes.[118]

of the Deutsche Juristentage (e.g. *Verhandlungen des 4. deutschen Juristentages*, vol. II, p. 278; report by Schaffrath, in *Verhandlungen des 6. deutschen Juristentages*, vol. III, p. 252; report by T.E. Karsten, in *Verhandlungen des 9. deutschen Juristentages*, vol. I, p. 390); Benöhr, above note 4, at p. 696; Lohsse, above note 6.

[114] For some railway-related accidents see Lehmann, above note 49, at pp. 22, 28, 76 and the Table on p. 85 and further data on p. 86 *et seq.*, including a comparison of Prussian with English data, pp. 65 and 84; Table 5, in Eckardt, *Technischer Wandel*, above note 43, at p. 406.

[115] Drucksachen des Reichstages, 1868, no. 56. See also Lehmann, above note 49, at p. 109 and Benöhr, above note 4, at p. 695.

[116] Cf. Lehmann, above note 49, at p. 110 and Eckardt, 'Evolution', above note 43, at p. 105.

[117] Ogorek, above note 5, at p. 101, n. 16.

[118] Reichs-Gesetzblatt 1871, 207–9, no. 652. An English translation of the Act can be found in Lawson, above note 40, at p. 195.

(a) *Provisions of the RHPflG*[119]

§1 RHPflG: the liability of railways[120] §1 RHPflG established strict liability for railways,[121] very much along the lines of §25 PGE, with two notable exceptions. First, whilst §25 PGE required the damage to occur 'during train transport' (*bei der Beförderung auf der Bahn*), §1 RHPflG stipulated 'in the conduct of a train' (*bei dem Betriebe einer Eisenbahn*), thus allowing for a greater scope of application. The intention was to include activities like loading.[122] Secondly, unlike §25 PGE, the new §1 RHPflG did not include damage to goods. This can be explained by the focus of the criticism (and the media) on personal injury cases[123] and the practice of almost automatic exclusion clauses mentioned above at pp. 149–50, particularly in employment contracts. Transported goods, however, were covered by the rules in §§422 *et seq.* of the General German Commercial Code (Allgemeines Deutsches Handelsgesetzbuch) of 1861, which was deemed satisfactory.[124] Damage to other goods, e.g. those caused by escaping sparks, now had to be dealt with under the general rules (see above p. 136 *et seq.* and below p. 170 *et seq.*) .

§2 RHPflG: liability of mines, quarries, pits and factories[125] Operators of mines, quarries, pits and factories were subject to a much more leni-

[119] For a detailed commentary on the Act see Eger, *Reichs-Haftpflicht-Gesetz*, above note 108; A. Heucke, *Haftpflichtgesetz* (2nd edn, Berlin: Heymann, 1937); F. Seligsohn, *Haftpflichtgesetz* (2nd edn, Berlin and Leipzig: Springer, 1931); O. von Mühlenfels, *Die Haftpflicht der Eisenbahnen und die Unfallversicherung* (Berlin: Reimer, 1884); H.-J. Finger, *Eisenbahngesetze* (6th edn, Munich: Beck, 1970), p. 26.

[120] Translation by Lawson, above note 40, at p. 195: '§1 If in the conduct of a railway a human being is killed or suffers bodily injury, the entrepreneur (the undertaking) is liable for the damage arising therefrom except in so far as he proves that the accident was caused by *force majeure*, or by the fault of the deceased or injured person.'

[121] In 1871 there were some 30,000 km of railways in Germany, cf. Table 1 in Eckardt, *Technischer Wandel*, above note 43, at p. 402.

[122] Preußisches Obertribunal Entscheidungen des königlichen Ober-Tribunals, vol. 28 (1854), 270, no. 33 (24 April 1854), for example, had interpreted 'during train transport' to mean that the train had to be in motion (at p. 273), thus excluding liability for loading/unloading under §25 PGE. Therefore it was felt that the wording had to be changed to include such activities, see *Stenographische Berichte über die Verhandlungen des Deutschen Reichstags 1871*, vol. I, p. 444 (Dr Schwarze). For details see Eger, above note 108, at pp. 3–42; Heucke, above note 119, pp. 10–23; Seligsohn, above note 128, at pp. 24–46.

[123] Table 5 in Eckardt, *Technischer Wandel*, above note 43, at p. 406, shows data of injuries and deaths caused by railways from 1854–1870.

[124] See Eger, above note 41, at p. 181.

[125] Translation by Lawson, above note 40, at p. 196: '§2 Anyone who carries on a mine, quarry, pit, or factory is, if an authorized agent or representative or anyone employed in the direction or supervision of the undertaking or the workmen causes, by a fault in the

ent liability regime (for debate and criticism see below). §2 RHPflG only established liability of the operator for death or injury of a person where an authorised agent or representative or anyone[126] employed in the direction or supervision of the undertaking or the workmen through his fault had caused the damage in the execution of his contractual duty / in the carrying out of the service arrangements (*in Ausführung seiner Dienstverrichtung*).[127] Among other things, injuries by exploding steam boilers in factories, etc. were covered by this provision.[128]

§3 RHPflG: damages[129] According to §3 RHPflG, the damages to be paid where liability was established under §1 or §2 covered not only damages for personal injury but also the damage resulting therefrom, i.e. loss of income, etc., including maintenance for the dependants in case of death.

§5 RHPflG: prohibition of contracting out of liability[130] In §5 RHPflG, the legislator took into account the existing practice of railway operators and other industrial undertakings to include clauses in their contracts that limited or excluded their liability. The provision prohibited such exclusion clauses and in its second sentence declared existing clauses in such terms to be without legal effect.

§6 RHPflG: assessment of evidence As had been demanded by practitioners, §§6 and 7 RHPflG changed the rules on evidence, giving the judges greater discretion when weighing the evidence.

(b) *Parliamentary debate on the RHPflG*

While the RHPflG enhanced the position of victims of railway and industrial accidents in principle, it nevertheless attracted great criticism both before and after it had been adopted. Many of the Members

carrying out of the service arrangements, death or bodily injury to a human being, liable for the damage arising therefrom.'

126 As is made clear in a decision by the Reichsgericht in 1882: RGZ 8, 30–5, no. 7 (3 November 1882), this comprises any worker, even 'ordinary factory workers', as long as they have a supervisory or instructing function at that moment. It is not necessary that they were hired to have such a function and sufficient that such a function at that point in time was delegated to them. See description of the case below at text to note 156.

127 For details see Eger, above note 108, at pp. 204–317, esp. at p. 298 *et seq*; Seligsohn, above note 119, at p. 185.

128 Steam engines on trains, of course, were covered by §1.

129 English translation in Lawson, above note 40, at p. 196.

130 English translation in *ibid.* p. 197.

of Parliament felt that instead of developing a general solution for this problem area, the Act was a short-sighted and inadequate reaction to public pressure.[131]

The main criticism in Parliament, however, was targeted at the distinction made between railways in §1 and mines, quarries, pits and factories[132] in §2, and the fact that all other industries were left out.[133] Statistical data, it was argued, did not support this distinction, as railway accidents evidently did not occur more frequently than other industrial accidents.[134] Therefore, many MPs supported the view that the rule adopted for railways, strict liability, should extend to all undertakings covered by the RHPflG.[135] Others suggested the opposite and wanted the more lenient rules to apply to railways as well, so that this particular business would not be burdened with excessive costs.[136] In the end, the draft Bill was accepted, including by those who advocated an extension of strict liability, mainly because further resistance might have endangered the entire Bill, leaving most accident victims without any effective remedy at all.[137]

Interestingly, even among those who wholeheartedly supported the Bill, there was widespread agreement that there was a need for a general reform of the law of obligations and the Act was only meant to cover an area where the need was too great to await a general reform,[138] which might include a broader departure from the fault principle. The fact that strict liability for

[131] See, e.g., MP Schulze, calling the RHPflG a '*Notstandsgesetz*' (emergency Act), owing its existence to recent industrial accidents that could not possibly be the last word on this matter, *Stenographische Berichte über die Verhandlungen des Reichstages 1871*, p. 446 Cf. also Zimmermann, above note 6, at p. 1133; Jansen, above note 32, at p. 370.

[132] The use of the word 'factory' was also criticised heavily, as it was unclear exactly which undertakings were meant to be included. *Cf.* Ogorek, above note 5, at p. 113. For a discussion of the cases ensuing from this see Eger, above note 108, at pp. 221–34; Heucke, above note 119, at p. 70; Seligsohn, above note 119, at pp. 160–3.

[133] See especially the MPs von Unruh (p. 201), Becker (p. 210), Bamberger (p. 212), Bethusy-Huc (p. 213), Duncker (p. 214), Ulrich (p. 441) and Schulze (p. 446), all in *Stenographische Berichte über die Verhandlungen des Reichstages 1871.*

[134] von Unruh, above note 142, p. 202 and Duncker, above note 142, p. 214. Cf. also von Mühlenfels, above note 119, at p. 56. But see Eckardt, *Technischer Wandel*, above note 43, at p. 311 (especially at p. 335: the highest number of accidents actually seemed to occur in railway operations) and Table no. 5 at p. 405.

[135] Namely MPs Becker, Bamberger, Bethusy-Huc, Duncker, Ulrich and Schulze, above note 133.

[136] von Unruh, above note 133.

[137] See Russel, in *Stenographische Berichte über die Verhandlungen des Reichstages 1871*, p. 456; Bamberger, *ibid.* p. 212; Becker, *ibid.* p. 211.

[138] See Ogorek, above note 5, at p. 103.

railways was included was due mainly to the fact that, given the existence of §25 PGE, this could be justified as already existing, 'historic' law.[139]

Predictably, the Bill was met with resistance by the industry. It was argued that the Bill, should it be accepted, would worsen the conflict between employers and employees and would in general not be in the workers' interest. Critics predicted lengthy litigation regarding the question of fault under §2 and, should the Bill be accepted, the establishment of special tribunals or courts of arbitration for these disputes.[140] The Association of Private Railway Operators (Verein der Privat-Eisenbahnen) unsurprisingly opposed the strict liability rule of the draft for railway operations.[141]

Apart from that, however, there was surprisingly little criticism, particularly in Parliament, and widespread support for the Bill. This can be explained by the position of the major interest groups: while those advocating stronger liability rules for industrial accidents and a general reform of the law of obligations saw the Bill as a first stepping stone, others supported the Bill as a relief for the complaints of the working class and a preventative measure against the 'spectre' of socialism[142] – an argument that also triggered the next reform of this area of law (see below p. 168 *et seq.*).

(c) *Further criticism and the effect of the RHPflG in practice*

The adopted RHPflG was criticised heavily by academics and politicians. It was found by some to be incomplete and insufficient, particularly since it did not cover property damage as well.[143] This remained so until 1940, some fifty years later, when the Act on Railway Liability for Property Damages (Eisenbahnsachschädengesetz)[144] extended the liability for railways established under the RHPflG to property damage.[145] The objections to the arbitrary distinction between railways and

[139] Achenbach, in *Stenographische Berichte über die Verhandlungen des Reichstages 1871*, p. 449; Ogorek, above note 5, at p. 104, n. 26.

[140] *Stenographische Berichte über die Verhandlungen des Reichstages 1871*, pp. 439, 467.

[141] *Ibid.* p. 439. For a debate on whether a duty to pay damages can actually be seen as an infringement of freedom see Jansen, above note 32, at p. 60.

[142] E.g. MPs von Unruh and Bethusy-Huc, above note 133, p. 213.

[143] J. Unger, *Handeln auf eigene Gefahr – Ein Beitrag zur Lehre vom Schadenersatz* (2nd edn, Jena, 1893), p. 94.

[144] RGBl. 1940 I, 691. For details see Finger, above note 119, at p. 65.

[145] Cf. F. Schack, 'Funkenschäden im Eisenbahnbetrieb' *NJW* 1965, 1702–3.

other undertakings continued.[146] The discussion was soon dominated by arguments regarding the social situation of workers and the underlying legal question took a back seat.[147] Some suggested that the protection of employees now went too far: when someone knowingly accepted a dangerous job, it was argued, the payment already included a premium for that danger; therefore it was inappropriate to burden the employer with further costs.[148] Others saw these provisions as part of labour law and continued to advocate an extension of the rules to all employees, as the selection of certain 'dangerous' industries was rather arbitrary; finally, some suggested that under certain conditions there should be a form of strict liability for all accidents at the workplace and hinted at a development that was soon to come: compulsory insurance.[149]

Furthermore, the RHPflG did not live up to expectations in practice,[150] particularly since the courts, and particularly the Reichs-Oberhandelsgericht, interpreted the Act rather narrowly, specifically as to the meaning of 'factory' (*Fabrik*) and 'in the execution of his contractual duty / in the carrying out of the service arrangements' (*in Ausführung seiner Dienstverrichtung*). This, the courts argued, was a necessity as the RHPflG was an exception to the general rules on liability.[151] In 1877, the Reichs-Oberhandelsgericht held that despite the difference in the wording between §1 ('in the conduct of a train' – *bei dem Betriebe einer Eisenbahn*) and §2 RHPflG ('in the execution of his contractual duty' – *in Ausführung seiner Dienstverrichtung*) this was meant to be the same[152] and that therefore an accident outside of the area of the factory (here: transport of goods from the factory to the station)

[146] See Dreyer, in *Verhandlungen des 18. deutschen Juristentages*, vol. I, p. 280. Cf. also the further criticism by Endemann, 'Das Reichsgesetz über die Haftpflicht der Eisenbahnen, Bergwerke usw.' (1872) 36 *Volkswirtschaftliche Vierteljahresschrift* 33 as cited in Ogorek, above note 5, at p. 108.

[147] Cf. Benöhr, above note 4, at p. 696.

[148] Endemann, above note 146, at p. 59 as cited in Ogorek, above note 5, at p. 109, especially n. 17: this was seen to be the English legal position. Cf. also Eckardt, *Technischer Wandel*, above note 43, at p. 249.

[149] See Ogorek, above note 5, at p. 110 and further references therein.

[150] It seemed, however, to have the effect of making employers aware of the potential liability, leading to increased safety standards, cf. vom Feld, above note 67, at p. 131.

[151] Cf. Reichs-Oberhandelsgericht, Entscheidungen des Reichs-Oberhandelsgerichts (ROGE) 21, 276, no. 91 (4 January 1877) at 281. Here, that court stated explicitly that while it constituted an inequality that some industries were covered by the RHPflG and others were not, this was a necessary consequence of the fact of the exception being made in a special Act (*Spezialgesetz*).

[152] Also Seligsohn, above note 119, at p. 163, has 'no doubt' (*kein Zweifel*) that this is the case.

was not covered.[153] In other cases, claims were denied because the accident happened outside the line of the work described in §§1 and 2 RHPflG, e.g. in the case of a factory worker who died when repairing the roof of the factory as ordered, or of a miner who died digging a railway tunnel as this had happened neither during the operation of a railway nor was it the result of a mining accident as envisaged by §2 RHPflG.[154]

There were also attempts to water down the liability, for example by holding that while fault was established for the relevant person under §2 RHPflG, the claimant was also at fault and hence any liability would be precluded. In a decision in 1882,[155] the Reichsgericht in principle confirmed that fault of the claimant would indeed preclude liability, but to mitigate the impact of this raised the threshold for that significantly and at the same time limited the impact of the 'fellow servant rule'. In that case, a factory smith had been instructed to conduct some repairs. In turn, he requested the assistance of another worker and while they went about completing the repairs in a manner that can only be described as grossly negligent, the worker was injured. The courts of first and second instance (Landgericht Stade and Oberlandesgericht Celle) had found that since the worker had been negligent himself, this precluded liability altogether. The Landgericht had further excluded liability because the factory smith did not qualify as a relevant person under §2 RHPflG because he had not been hired for a supervisory and instructing function. The Oberlandesgericht disagreed on this point and stated that someone in a temporary function of this sort would also fall under §2 RHPflG. Nevertheless, in this case, the Oberlandesgericht held that since the smith and the worker had been conducting the repairs together and both 'worked with their own hands'[156] there was no such function. The Reichsgericht allowed the claim, arguing that while the worker must or should have been aware of the risk, it had been the duty of the factory smith to apply appropriate safety measures. The 'ordinary worker' could not be seen to be at fault just because he did not refuse to do the work or openly disagreed with the instructions.[157]

153 Reichs-Oberhandelsgericht, ROGE 22, 311, no. 70 (13 June 1877); but cf. a later decision by the same court and senate, Reichs-Oberhandelsgericht, ROGE 25, 184, no. 44 (10 May 1879): laying of waterpipes outside the premises is covered if done 'industrially' (*fabrikmäßig*).

154 See Ogorek, above note 5, at p. 114 and references therein n. 4.

155 RGZ 8, 30–5, no. 7 (3 November 1882). The arguments of the lower courts are reported here as well.

156 Apparently it was felt that working with one's hands automatically precluded one from having a leading function!

157 RGZ 8, 34. Interestingly, the case came to the Reichsgericht again later, the damages for pain and suffering then being disputed on a point of law. Again the claimant prevailed: RGZ 11, 61–3, no. 15 (28 March 1884).

The problems of actually obtaining damages in practice eventually led to call for reform of the RHPflG, criticising its arbitrariness, incalculable application and ineffectiveness, especially of §2 where the claimant still had to prove fault, which was rarely successful:[158] often the person at fault had died or the evidence had been destroyed in the accident.

There were reports that fellow workers had been pressurised into not giving evidence and some apparently had been dismissed when they nevertheless did.[159] Apparently, only 20 per cent of the actions brought were successful.[160] Therefore, a proposal was advanced in Parliament to reverse the burden of proof, i.e. the employer would be obliged to show that he acted with due diligence; this, it was argued, would not only provide better social security for workers but also be an incentive for employers to implement precautionary measures and thus reduce the overall number of accidents.[161] A debate ensued,[162] in which the only agreement seemed to be the need for reform. The social democrats suggested extending §1 to all dangerous undertakings.[163] The conservatives proposed to abandon private law liability altogether and to implement a mandatory general insurance for all businesses, modelled on the social miners' and mine-employees' insurance (*Knappschaft*).[164] In the end the question was referred to a commission. The final report of this commission led to the Accident Insurance Act (Unfallversicherungsgesetz) of 1884.

6. Unfallversicherungsgesetz of 1884

The Accident Insurance Act (Unfallversicherungsgesetz) (UVG) of 1884 was quite explicitly an Act to appease the social democrat and socialist

See also RGZ 1, 48, no. 24 (23 January 1880): not to be held own fault under §1 RHPflG where juvenile employee acted contrary to general instructions with (tacit) approval of his supervisor.

158 See the detailed account by vom Feld, above note 67, at p. 136. Cf. also Eckardt, *Technischer Wandel*, above note 43, at p. 351 and above p. 147 *et seq.*

159 Cf. vom Feld, above note 67, at p. 136.

160 Eckardt, *Technischer Wandel*, above note 43, at p. 352. See also vom Feld, above note 67, at p. 136.

161 MP Hirsch, in *Stenographische Berichte über die Verhandlungen des Reichstages 1878*, vol. II, p. 838.

162 For a detailed description of the debates, cf. Ogorek, above note 5, at p. 115; Eckardt, *Technischer Wandel*, above note 43, at p. 362 (here particularly the table on p. 370).

163 MPs Kapell and Hasenklever and others, *Stenographische Berichte über die Verhandlungen des Reichstages 1878*, vol. III, no. 128 (Appendices).

164 MP Stumm, *Stenographische Berichte über die Verhandlungen des Reichstages 1878*, vol. II, p. 844; cf. Eckardt, *Technischer Wandel*, above note 43, at p. 367.

groups in Germany, who were seen as 'dangerous to the public safety'.[165] These groups, it was said, needed to be combated not only by repressive means but also by positive steps that improved the social situation of the workers. This was not only a 'humanitarian and Christian duty', but would also serve 'the interest of the state'.[166]

The Bill presented explicitly rejected suggestions that private law liability should be widened (e.g. that the strict liability of §1 RHPflG should apply to all industrial undertakings), and preferred a public law solution. This, it was said by the member of government and Minister of State, Hofmann, would forge a social bond between the employers and workers, unlike private law liability which because of its very structure would create animosity between the parties and worsen the conflict.[167] In addition, strict liability for the industry was untenable as this would impede their ability to compete in the international market.[168] An extension of §2 RHPflG to all undertakings was also not seen as feasible; the practical problems in proving fault had rendered the provision almost unworkable, and also reversals of the burden of proof would not solve the problem, merely shift it: then the debate would ensue when undertakings tried to prove that they had not committed any fault.[169] This would create immense litigation, as the worker had nothing to lose and had no risk of costs as he would be suing under the '*Armenrecht*', whereas the business would have to litigate as insurance companies would only pay when a legal title was presented to them. Therefore, the parties were almost forced to litigate, although the result would never be fair: if the worker won, then he would receive more than what was fair and adequate from a welfare point of view; if the worker lost, then he and his family would fall into poverty and the state would have to pick up the bill.[170] Hence there needed to be a public law solution which because of its nature would cover all accidents at the workplace, regardless of fault, but not pay full damages, but only

[165] For the rising number of votes and seats in Parliament of the social democrats see the table in Eckardt, *ibid.* p. 344.

[166] Cf. *Stenographische Berichte über die Verhandlungen des Reichstages 1881*, vol. III, Anlagen, no. 41, p. 222; Ogorek, above note 5, at p. 116; Kötz, above note 32, at p. 6.

[167] *Stenographische Berichte über die Verhandlungen des Reichstages 1879*, vol. I, p. 141; cf. Eckardt, *Technischer Wandel*, above note 43, at p. 352.

[168] MP Hirsch, in *Stenographische Berichte über die Verhandlungen des Reichstages 1878*, vol. II, p. 841. As Eckardt, *Technischer Wandel*, above note 43, at p. 347, notes these are the same arguments used today in similar debates.

[169] *Stenographische Berichte über die Verhandlungen des Reichstages 1881*, vol. III, Anlagen, no. 41, p. 229.

[170] *Ibid.* p. 229. See also Eckardt, *Technischer Wandel*, above note 43, at p. 352.

ensure that the victim received fair compensation that covered his future needs.[171]

Despite continuing objections by liberal and social democrat MPs,[172] the third draft of the Bill was adopted on 27 June 1884.[173] The following years saw a gradual extension of the public law model[174] which essentially is still in place in Germany today.

In brief, under the UVG[175] damages for personal injury[176] were awarded even if the worker himself had been negligent. These damages however, were limited to compensation for the damages actually suffered, especially loss of income and medical costs, as well as a pension in case of invalidity or maintenance for dependants in case of death (§§5, 6). No other claims for damages for personal injury were possible.[177] Disputes were primarily to be dealt with by arbitration tribunals (§§46–50), and appeals to the Reichsversicherungsamt (an administrative body responsible for insurance, §§87–93).

The paradoxical aspect of the UVG is, as Ogorek succinctly puts it, that the most important Acts for the social protection of workers were in fact created by people that were openly hostile towards the working class, while those opposing them most strongly were representatives of the social democrats.[178] Still, it cannot be denied that the situation of workers was improved considerably through this legislation. In any event, the questions of liability were to a very large extent removed from the private law sphere and for years to come debated almost entirely in a public law context. Liability was replaced by public insurance, a development that was here to stay (see below p. 174 *et seq.*).

7. Claims for damages not covered by the RHPflG or the UVG

Cases of personal injury not covered by the RHPflG or the UVG, particularly accidents not outside the workplace, were still covered by the general

[171] Cf. the explanatory statements to the third and final draft of the Act, *Stenographische Berichte über die Verhandlungen des Reichstages 1884*, vol. III, Anlagen, no. 4, p. 66.

[172] Cf. e.g. *Stenographische Berichte über die Verhandlungen des Reichstages 1882*, vol. I, p. 200.

[173] *Stenographische Berichte über die Verhandlungen des Reichstages 1884*, vol. II, p. 1129.

[174] See Ogorek, above note 5, at p. 119 (beginning on 28 May 1885 with the Gesetz über die Ausdehnung der Unfall- und Krankenversicherung, RGBl. 1885, 159, and eventually leading to the *Reichsversicherungsordnung* of 30 May 1911, RGBl. 1911, 509.

[175] See Table 7 in Eckardt, *Technischer Wandel*, above note 43, at p. 408 for data regarding the number of persons insured in 1885–1900, and Table 8, *ibid.* p. 409 for data on the number of insured persons that were injured in 1885–1900.

[176] As under the RHPflG damages to property were not covered: see below p. 171 *et seq.*

[177] Eckardt, *Technischer Wandel*, above note 43, at p. 354.

[178] Ogorek, above note 5, at p. 120.

and special (e.g. PGE)[179] rules in place. As outlined above, claims were mostly unsuccessful, particularly because of the difficulties in proving fault.

The situation regarding property damage, however, was quite different. Some interesting cases here concern damage to property through fires cause by escaping sparks, which were handled very much as before the enactment of the RHPflG.[180] In a landmark decision in 1886, the Reichsgericht[181] awarded damages under the *actio negatoria* despite explicitly finding that there was no fault. A duty to pay damages was seen as a corollary to obtaining a concession to undertake an inevitably dangerous enterprise. Interestingly, the court also expressly rejected the earlier statement by the Oberappellationsgericht München in 1861[182] that the operation of a locomotive as such was to be seen as necessarily constituting a culpable act, as the escape of sparks was foreseeable and inevitable. This could not be the case as the railway was operating under a concession, which excluded fault. Still, the operator was held liable because this could be justified under other grounds, namely because the operation of the railway by its very nature would cause damage. Fault, the court stated, was only one of the grounds on which a claim could be based. Since the claimant was barred from obtaining an injunction as the railway operated with a concession, not allowing him to claim damages would amount to an unacceptable infringement of his property rights. The operator would not be liable for unforeseeable damage, but escaping sparks and their consequences did not fall into that category.[183] The court referred to its own decision made four years earlier;[184] in this case, a piece of land had decreased in value as building permission could not be obtained because of the danger of fire because of sparks escaping from a nearby railway;[185] the court had awarded damages for the loss suffered.

8. Liability under the Bürgerliches Gesetzbuch

While criminal legal theory in Germany distinguished between criminal wrongs that required fault and civil wrongs that did not,[186] this was largely

179 Eger, above note 41, at p. 131.

180 For an overview see Schack, above note 145.

181 RGZ 17, 103, no. 24 (7 December 1886).

182 OAG-München, Seufferts Archiv, vol. 14, 354, no. 208 (16 April 1861), see Ogorek, above note 5, at p. 67; Jansen, above note 32, at pp. 1, 309, esp. n. 275; and above p. 144 *et seq.*

183 See also Ogorek, above note 17, at p. 71.

184 RGZ 7, 265, no. 74 (20 September 1882).

185 The Erlaß des Ministers des Inneren und der öffentlichen Arbeiten, dated 23 July 1892 (EVBl. 93, 152) stipulated certain minimum distances between railways and buildings, cf. Eger, above note 41, at p. 107.

186 See, e.g., K. Binding, *Die Normen und ihre Übertretung* (Leipzig, 1872/1878), vol. I, p. 233: 'Die Strafe wird verhängt, um der Schuld des Verbrechers willen, Ersatz wird geleistet, um

ignored by civil legal theory. Despite some debate,[187] the first draft of the Civil Code (Bürgerliches Gesetzbuch) (BGB)[188] makes clear that it will adhere to the established principle that liability requires fault.[189] Strict liability, it was felt, was a concept alien to the civil law, and should remain confined to special acts where required; this approach still prevails today in German law.[190]

With the exception of liability for (certain) animals,[191] the BGB therefore requires fault of the defendant himself for a successful claim in tort law (§§823 *et seq.* BGB), and this in principle also includes vicarious liability (§831); here, however, the claimant is privileged as the defendant needs to prove that he was not at fault regarding the selection and supervision of his vicarious agent.[192] If, however, the claim is based on contract the situation of the claimant is much better. Then §278 BGB applies and the defendant is liable for a fault of a person whom he employs to perform his contractual obligation to the same extent as for his own fault.[193]

However, in essence the situation had not changed. Tortious liability in principle required fault, with exceptions to be found in special Acts

des Schadens und des Geschädigten willen' and p. 226: 'Strafe und Schadensersatz verfolgen verschiedene Zwecke'; see also, A. Thon, *Rechtsnorm und subjectives Recht* (Weimar, 1878), and Bienenfeld, above note 14, at p. 36; Ogorek, above note 5, at pp. 124 and 84.

187 Ogorek, *ibid.* p. 127, especially the references on p. 134 in n. 134.

188 The final draft was adopted on 18 August 1896 and came into force on 1 January 1900.

189 *Motive zu dem Entwurfe eine Bürgerlichen Gesetzbuches für das Deutsche Reich*, vol. II, p. 744; Zweigert and Kötz, above note 2, at p. 648; cf. also Benöhr, above note 4, at p. 709. The only real exception was the liability for animals, §833 (see above, esp. note 6). An interesting decision in that respect is Reichsgericht Eisenbahn- und Verkehrsrechtliche Entscheidungen und Abhandlungen (RG EE), vol. 27, 66, no. 25 (17 October 1910), where it was held that the failure to implement immediately the most modern equipment to prevent damages could not as such constitute fault.

190 Cf. Jansen, above note 32, at p. 370. A number of special Acts today establish strict liability under certain circumstances or for certain risks, e.g. the successor of the RHPflG, the Liability Act (Haftpflichtgesetz, BGBl. 1978 I, 145), Road Traffic Act (Straßenverkehrsgesetz, BGBl. 2003 I, p. 310, 919), Air Traffic Act (Luftverkehrsgesetz, BGBl. 1999 I, 550), Water Balance Act (Wasserhaushaltsgesetz, BGBl. 2002 I, 3245), Nuclear Power Act (Atomgesetz, BGBl. 1985 I, 1565), Act on Genetic Engineering (Gentechnikgesetz, BGBl. 1993 I, 2066), Act on Liability for Environmental Damages (Umwelthaftungsgesetz, BGBl. 1990 I, 2634), etc. (all citations of the BGBl. refer to the latest re-enactment). For an account of the development of railway liability after the BGB came into force see also Lohsse, above note 6.

191 See Zimmermann, above note 6, at p. 1116, and Scherpe, above note 6.

192 For a brief account how this affects railways' liability, see Eger, above note 41, at p. 131.

193 For the genesis of this provision see the account by Ogorek, above note 5, at p. 129, and the references therein.

like the RHPflG, which remained in force.[194] The Introductory Act of the BGB (EGBGB) in its articles 105 and 218 even allowed the German states (*Länder*) to pass acts that constituted a sterner liability for railway operation and other undertakings than the BGB provided. Article 105 further stated that acts by the German states stipulating stricter liability than the BGB would remain in force; this meant that while §25 PGE was replaced by the RHPflG in respect to personal injury, the provision remained applicable to property damage.[195] In addition, article 125 EGBGB allowed the states to extend the rules laid down in §26 GewO to railway operations.[196]

So, while in the German law of torts the maxim laid down by Rudolph von Jhering, 'Nicht der Schaden verpflichtet zum Schadensersatz, sondern die Schuld' ('It is not the occurrence of the harm which obliges one to make compensation, but fault')[197] prevailed, the courts stuck to the path they had taken before where damage to property was concerned. They protected the owners of property without the need for fault, in effect ignoring the intention of the legislator and to some extent even reversing it.[198]

Following its decisions from 1882[199] and 1886,[200] the Reichsgericht in two decisions in May[201] and October[202] 1904 at first stated that §§904, 1004 or §26 GewO[203] could not be extended to cover damages – this would be contrary to law.[204] It nevertheless awarded pecuniary compensation, referring to

[194] With some modifications brought about by art. 42 EGBGB, but §§1 and 2 remained unchanged.

[195] Cf. Prinz, above note 38, at p. 480; Eger, above note 41, at p. 131. In Bavaria, Brunswick (Braunschweig), Anhalt, Hessia, Lübeck, the Saxonian duchies and Würtemberg, Acts regarding property damage also were in place (*ibid.* p. 133, and E. Rundnagel, *Die Haftung der Eisenbahnen für Verlust, Beschädigung und Lieferfristüberschreitung nach deutschem Eisenbahnrecht*, (3rd/4th edn edited by K. Fritsch and A. Sperber, Leipzig: Weicher, 1924), p. 11, n. 1).

[196] For an overview of which states made use of either art. 105, art. 125 or both see Prinz, above note 38, at p. 480; but cf. Ogorek, above note 5, at p. 133, n. 44. See generally above p. 136 *et seq.*

[197] von Jhering, above note 11, at p. 199: translation by Tony Weir in Zweigert and Kötz, above note 2, at p. 630.

[198] Cf. Ogorek, above note 17, at p. 73 *et seq*, esp. p. 77. For a detailed description of the way claims arising from damage caused from sparks were regulated under the BGB see Prinz, above note 38, at p. 473.

[199] RGZ 7, 265, no. 74 (20 September 1882), see above pp. 170–1.

[200] RGZ 17, 103, no. 24 (7 December 1886), see above pp. 170–1.

[201] RGZ 58, 130, no. 32 (11 May 1904) (a case concerning fires caused by sparks escaping from railways).

[202] RGZ 59, 70, no. 21 (12 October 1904) (damages claimed because of the noise of a nearby railway).

[203] See above p. 136 *et seq*. On the claims of owners of neighbouring property generally see Gordley, above note 19.

[204] RGZ 58, 131.

§906 BGB, regardless of whether the defendant was at fault or not.[205] While the BGB did not expressly provide for such a remedy, the court held that it was nevertheless quite clear from the other provisions that such a remedy was a necessity. Under §1004 BGB, the claimant could obtain an injunction without having to prove fault. If because of a concession this path was not open to him, not awarding him damages would in effect deprive him of his property rights completely and leave him without any protection.[206] This legal principle, as enshrined in the §904 BGB, §25 PGE and §26 GewO (all of which were not applicable in the cases), also found its legal basis in article 9 of the Prussian Constitutional Charter: the right to one's property was inviolable and could only be confined in the public interest.[207] Since in these cases the property rights of the claimants were infringed considerably and because of special provisions or concessions they were barred from bringing actions against the defendant to stop those infringements, they had to be awarded compensation. Any other decision would in effect amount to an expropriation.[208]

This train of thought was followed in a number of decisions.[209] The courts thus circumvented the intentions of the BGB legislator in a remarkable way, to ensure what they perceived to be the fair outcome. As the legislator did not choose to enact legislation preventing this circumvention, this in principle is still the law today.

9. Social security law

The Accident Insurance Act (Unfallversicherungsgesetz) (UVG) of 1884 was the starting point for the transfer of (private) liability to the public

[205] RGZ 59, 74; cf. W. Janzer, 'Haftung der Eisenbahnverwaltung für die durch Funkenflug verursachten Brandschäden außerhalb eines bestehenden Schuldverhältnisses', *Zeitung des Vereins Deutscher Eisenbahnverwaltungen* 1905, 681–3.

[206] RGZ 58, 134. [207] *Ibid.* [208] RGZ 59, 75.

[209] Cf. the references in Ogorek, above note 17, at p. 75, n. 110; the 1932 decision of the Reichsgericht, RGZ 139, 29, no. 7 (26 November 1932) is particularly enlightening: here the court discussed the question and prerequisites of liability in detail, particularly the fault requirement, and expressly held that fault was not required where the owner of a piece of land had to suffer infringements against which he could not obtain an injunction (e.g because of a concession, §26 GewO, etc.). See also Prinz, above note 38, at p. 473 and Janzer, above note 205. For post-Second World War decisions concerning fires caused by escaping sparks, see, e.g., OLG Dresden, NJW 1949, 627 (28 May 1948); Kammergericht Berlin, NJW 1949, 908 (15 July 1949); OLG Schleswig, NJW 1953, 106 (24 April 1952), liability under the Act on Railway Liability for Property Damages (Eisenbahnsachschädengesetz) from 1940, see text to note 144 and Finger, above note 119, at p. 76); Bundesgerichtshof, VersR 1954, 297 (28 April 1954).

insurance model, which is still in place in Germany today but has grown in complexity.[210] The most important branches of the social security system are the national health insurance (*Gesetzliche Krankenversicherung*) (GKV);[211] national accident insurance (*Gesetzliche Unfallversicherung*) (GUV);[212] national pension insurance (*Gesetzliche Rentenversicherung*) (GRV);[213] and *Soziale Pflegeversicherung* (a special insurance covering the costs for those in need of long-term special care).[214] These compulsory[215] public insurance schemes now cover most of the population,[216] not only those in employment, but schoolchildren, students, pensioners, unemployed people and people dependent on public welfare, people in special institutions, and so on.[217] As a result, particularly cases concerning injury caused by asbestos (see below) are dealt with almost entirely under social security law and in special social courts (Sozialgerichte). Because of the nature of the scheme, most of the questions that could be problematic in a tort law context (e.g. fault and causation) are in practice irrelevant.

In brief, anyone covered by the social insurances has a right (i) to receive treatment to protect, preserve, ameliorate and restore his health and capacity, and (ii) to economic security in case of illness, maternity, loss of earning capacity and old age for himself/herself and surviving members of the family.[218] Thus, both occupational and other illnesses are covered, as are vocational disability (*Berufsunfähigkeit*) and complete loss of earning capacity.

To a large extent, social security law has taken over one function of tort law in the areas covered by it. However, given that the general purpose of the (public) social security law is to provide basic social security and thus is in essence restricted to that, tort law in theory could remain important for claims that go beyond those limits, namely in two areas:

[210] Space precludes a detailed description; for a more detailed account of the structure of the social security system, etc. see U. Magnus and J. Fedtke, 'Germany' in U. Magnus (ed.), *The Impact of Social Security Law on Tort Law* (Vienna and New York: Springer, 2003), pp. 86–115, including a brief account on how the system is financed at p. 95 *et seq.*

[211] Social Security Code, Part V (Sozialgesetzbuch V).

[212] Sozialgesetzbuch VII.

[213] Sozialgesetzbuch VI.

[214] Sozialgesetzbuch IX and Nursing Care Insurance Act (Pflegeversicherungsgesetz), BGBl. 1994 I, 1014.

[215] It is possible to opt out of that scheme under certain conditions and replace it with private insurance, and indeed in some cases this is necessary (e.g. those having the status of a civil servant, or who are self-employed). For details see Magnus and Fedtke, above note 210, at p. 86 *et seq.*, especially pp. 92–4.

[216] Cf. the data cited by Magnus and Fedtke, *ibid.* p. 92 *et seq.*

[217] *Ibid.*

[218] §4 Sozialgesetzbuch I.

(1) recourse against tortfeasors (either by the social insurance institution, the state or private insurance companies);[219] and
(2) compensation for pain and suffering and other non-pecuniary loss.

Yet, further provisions limit the scope for regress to general tort law in both areas mentioned.

As for recourse, §116 of the Social Security Code, Part X (Sozialgesetzbuch X)[220] stipulates an automatic *cessio legis* when damages, etc. have been paid under the social security laws, but the conditions for this are surprisingly narrow: in case of accidents caused on the job by co-workers and employers, the *cessio legis* §116 of Sozialgesetzbuch X and equivalents do not apply,[221] unless caused by intent or gross negligence.[222] This leaves almost no room for recourse for injury caused at the workplace (or equivalent). The Bundesgerichtshof in a decision concerning a fight between two schoolboys held in 1979[223] that the intent or gross negligence must relate to causing the damage as such, not simply the wrongful behaviour; this, it was held, is because it is the very nature of the social security system to indemnify against certain claims, particularly when there is some form of wrongful act, in order to keep the workplace (or in this case school) free from additional conflict. Only when the relevant action (or omission) was intentional or grossly negligent should recourse be possible, instead of burdening the collective of the insured with the costs.[224]

Remarkably, claims for pain and suffering and other non-pecuniary losses (e.g. under §847 BGB) are barred under §104 *et seq.* Sozialgesetzbuch VII if the injury in principle falls into the area of the social security system – unless the injury was caused with intent.[225] This bar has been challenged on constitutional grounds, but without success; the Bundesverfassungsgericht[226] held that barring claims for pain and suffering, etc. did not constitute an unjustifiable discrimination under Article 3 of the German Basic Law (Grundgesetz, the German

[219] For details on the recourse see Kötz and Wagner, above note 2, at p. 294; see also H.-L. Weyers, *Unfallschäden* (Frankfurt, 1971), p. 401.

[220] Similar provisions exist for the equivalents of the social insurance, cf. §67 of the Insurance Contract Act (Versicherungsvertragsgesetz); §87a of the Federal Civil Servants Act (Bundesbeamtengesetz); §52 of the General Civil Servants Act (Beamtenrechtsrahmengesetz).

[221] §§104, 105 Sozialgesetzbuch VII.

[222] §§110, 111 Sozialgesetzbuch VII; this is seen as an incentive for preventative measures, cf. Bundesgerichtshof, Entscheidungen des Bundesgerichtshofes in Zivilsachen (BGHZ) 75, 328 (20 November 1979) at 331 and Magnus and Fedtke, above note 210, at pp. 98 and 108.

[223] Bundesgerichtshof, BGHZ 75, 328 (20 November 1979). [224] *Ibid.* 330–3.

[225] §§104, 105 Sozialgesetzbuch VII. [226] BVerfG, NJW 1973, 502 (7 November 1972).

Constitution). The petitioner alleged that there was no justification for the different liability systems available: on the one hand, the BGB, which would allow for claims for pain and suffering; and the social security system on the other hand, which denied such claims (as by its very nature it focuses on need rather than full compensation and especially compensation for pain and suffering); and that redress to the BGB provisions should be allowed, as failure to do so would unjustifiably penalise the petitioner when compared to those who could claim under general tort law principles. Yet the Bundesverfassungsgericht held that the differential treatment was justifiable because of the privileges obtained at the same time: the claim under the state social security scheme was completely independent of fault, so there was no risk of litigation involved in that respect, including possible contributory fault and/or negligence (which under ordinary tort law principles could lead to a reduction of the claim); further, there was no risk of insolvency of the debtor which in itself was a considerable privilege. The bar was also meant to protect the employers, who alone cover the entire costs for the GUV; in particular, it was meant to make it possible for them to calculate their liability risk; this protection would be removed were claims for pain and suffering allowed. A further argument was one already known from Bismarck's time:[227] not having to sue your employer was said to prevent animosity between the parties and to avoid further conflict.[228] The latter line of argument was also used to justify the privilege of co-workers.[229] Finally the injured person would also receive further medical treatment, especially for occupational rehabilitation which was paid for by the scheme. All in all, therefore, the bar was seen as justifiable and hence constitutional.[230] A further argument put forward in the case was also rejected. Allowing claims for pain and suffering in case of intentional causing of injury was not seen as an unjustifiable breach of the system; on the contrary, by intentionally causing harm the tortfeasor himself had left the foundations on which the social security system rests and thus allowing claims for pain and suffering was consistent with the general approach taken in such matters.[231]

Thus, while in theory tort law (especially §823 *et seq.* BGB) remains applicable in cases of occupational accidents and also accidents at school or university, etc. (which includes most asbestos related claims), virtually all

[227] Used, e.g., by the responsible minister Hofmann in 1879, cf. *Stenographische Berichte über die Verhandlungen des Reichstages 1879*, vol. I, p. 141. Cf. also Eckardt, *Technischer Wandel*, above note 43, at p. 352 and above p. 171 *et seq.*

[228] See BVerfG, NJW 1973, 502 at 503 (7 November 1972). [229] *Ibid.* 504. [230] *Ibid.* 503.

[231] *Ibid.* 504.

claims based on general tort law are barred by §104 *et seq*. Sozialgesetzbuch VII.[232] This means that no claims can be brought against employers for injuries arising from the employment unless there was intent. Claimants are effectively limited to claims under the social security system; if they cannot bring a claim under this system, they cannot bring a claim at all.[233]

In order to succeed in a claim under the GUV it is generally sufficient to prove that one suffers from one of the diseases recognised as work-related. Therefore, questions of causation are of almost no relevance in this context,[234] especially as benefits under the GKV, which covers the medical costs, are granted without any inquiry into the cause of the illness. Even where proof of causation is required, fault is irrelevant, including contributory negligence.[235]

This might explain why the cases concerning asbestos-related disease seem to be restricted to the Sozialgerichte and mostly focus entirely on whether the disease contracted was one of those recognised as a work-related disease.[236]

10. Asbestos

Asbestos was used widely in Germany until it was banned in 1993. Just as in other countries, a large number of injuries were caused. Yet, this did not have a large impact on private law as most of the claims were brought under the social security systems and hence were governed by public law. Indeed, as stated in one of the leading German textbooks on tort law (see above p. 174 *et seq.*), claims for damages based on asbestos-related diseases are unheard of in the German private law arena, and no relevant decisions of private law courts are to be found – they are simply absorbed by the state social insurance scheme.[237] Nevertheless, injuries caused by asbestos have

[232] Cf. Magnus and Fedtke, above note 210, at p. 97. [233] *Ibid.* 102.

[234] *Ibid.* 99 and references there.

[235] E. Deutsch, *Allgemeines Haftungsrecht* (2nd edn, Cologne: Heymann, 1996), p. 474; Magnus and Fedtke, above note 210, at p. 99. This extends even to the recourse in most such cases (for exceptions see *ibid.* p. 100.).

[236] Cf., e.g., Bundessozialgericht, HV-INFO 1993, 588 (29 September 1992); Bundessozialgericht, HV-INFO 1989, 1177 (6 April 1989); Bundessozialgericht Entscheidungen des Bundessozialgerichts 79, 250 (14 November 1996); Landessozialgericht Rheinland-Pfalz, HV-INFO 1989, 314 (26 October 1988); Landessozialgericht Rheinland-Pfalz, HV-INFO 1992, 2595 (29 July 1992); Landessozialgericht Rheinland-Pfalz, HV-INFO 1992, 2783 (9 October 1991).

[237] Kötz and Wagner, above note 2, at p. 222. Consequently, asbestos cases are not much discussed in German legal textbooks – and if they are, then only by reference to English or American cases (cf. *ibid.* pp. 222, 242).

been a major topic in German health politics since the 1970s. The number of deaths caused by asbestos-related diseases has risen sharply over the years, from 74 in 1980 to 957 in 2000.[238]

(a) *Use of asbestos in Germany*

In the Federal Republic of Germany, the usage of asbestos rose steadily from 1948 to 1965, reaching 170,000 tons used per year. Until 1980, the usage remained at this level. After 1980, the German industry began using alternative products, which led to a decrease in the usage of asbestos to 50,000 tons per year in 1989. In the former German Democratic Republic, asbestos was used from about 1960, with a maximum usage of 74,000 tons per year in 1980; in 1987, 53,000 tons of asbestos were used.

Gradually, measures were taken for the protection of workers. In 1940, the Employer's Liability Insurance Association issued guidelines for the protection of workers from asbestos dust, requiring proper ventilation and exhaust air installations. In 1979, the spraying on of asbestos was banned by law; in 1982, the use of asbestos for certain products was prohibited. In 1990, the use of asbestos was classified as 'high risk' for the purposes of the Gefahrstoff-Verordnung (listing carcinogenic substances and restricting the usage of such substances); and, finally, in 1993 the use of asbestos was made illegal altogether through an amendment of the Gefahrstoff-Verordnung.[239] By this time, of course, a large number of workers had been exposed to asbestos. As asbestos-related diseases can have an extremely long latency period,[240] the German social security system has been facing a dramatic increase in the number of claims made and the peak of these claims is expected between 2010 and 2015.[241]

(b) *Asbestos-related diseases as occupational diseases*

The possible contraction of dust-related diseases[242] at the workplace has been known in Germany for a long time – the first reports actually date

[238] Hauptverband der gewerblichen Berufsgenossenschaften (HVBG) with C. Drechsel-Schlund, M. Butz, G. Drexel, W. Plinske and H.-P. Francks, *Asbestverursachte Berufskrankheiten in Deutschland – Entstehung und Prognose* (Sankt Augustin, 2003), p. 11.

[239] HVBG, above note 238, at pp. 13 and 26.

[240] But the latency period can be very short; in one such case a 36-year-old died of mesothelioma very shortly after having been exposed to asbestos dust, see *ibid.* p. 21.

[241] *Ibid.* pp. 73, 76.

[242] Including asbestos and other dust.

back to 1556.[243] Nevertheless, they were not included as 'occupational diseases' in the original version of the Unfallversicherungsgesetz of 1884.[244] It was only in 1925 that the insurance cover was extended to cover what are today known as 'occupational diseases' and eleven such diseases were recognised.[245] However, these first recognised occupational diseases did not include any caused by asbestos or silicone dust. Some four years later, the Zweite Berufskrankheiten-Verordnung (BKV) in 1929[246] added '*schwere Staublungenerkrankung (Silikose)*' (severe silicosis) to the list of recognised occupational diseases. Another seven years later '*Schwere Asbeststaublungenerkrankung (Asbestose)*' (severe asbestosis) was included as well.[247] The requirement 'severe' meant that the earning capacity of the person must have been reduced through the disease by 50 per cent or more, which caused some practical difficulty.[248] This might explain why, for example, in 1950 only seventeen cases of asbestosis led to claims made under the public insurance scheme (UVG).[249] In 1952, '*schwere*' (severe) was deleted as a requirement for claims for asbestosis-related damages[250] which led to an increased number of claims; for example, in 1960 thirty-five cases of asbestosis were notified.

Lung cancer or cancer of the throat caused in connection with asbestosis as a separate occupational disease was recognised in 1943 for the purposes of the public insurance scheme.[251] However, because of the long latency period, not many claims were made before the 1980s. Indeed, in 1960 and 1965 only one case was notified, in 1970 two cases were notified, and fifteen in 1975.[252]

[243] HVBG, above note 238, at p. 15, referring to works by Agricola on lung diseases causing cancer from 1556, Paracelsus' *Von der Bergsucht oder Bergkrankheiten* from 1567 and Pansa's *Beschwerliche Berg- und giftige Lungensucht* from 1614.

[244] Above p. 168 *et seq*. This was despite a well-known study by Hirt about such diseases (a study dealing with 'Staubinhalationskrankheiten Pneumoconiosis', see HVBG, above note 238, at p. 15).

[245] See Verordnung über die Ausdehnung der Unfallversicherung auf gewerbliche Berufskrankheiten of 12 May 1925, RGBl. 1925 I, 69.

[246] 11 February 1929, RGBl. 1929 I, 27.

[247] Dritte Verordnung über die Ausdehnung der Unfallversicherung auf Berufskrankheiten, 16 December 1936, RGBl. 1936 I, 1117.

[248] HVBG, above note 238, at p. 15. [249] *Ibid*. p. 18.

[250] This essentially meant that now a reduction of the earning capacity of 20 per cent sufficed for such claims, cf. HVBG, above note 238, at p. 18.

[251] Vierte Verordnung über Ausdehnung der Unfallversicherung auf Berufskrankheiten, 29 January 1943, RGBl. 1943 I, 85.

[252] HVBG, above note 238, at p. 20.

It was only in 1976 that asbestos-related mesothelioma was recognised as an occupational disease.[253] The number of people suffering from this rose very quickly in the 1980s: in 1980, forty-eight cases were notified, in 1984 the number had risen to 160.

Today, the Berufskrankheiten-Verordnung[254] lists the following diseases related to asbestos:

(i) under no. 4103: asbestosis or asbestos pleural disease;
(ii) under no. 4104: lung cancer or cancer of the throat caused in connection with asbestosis or asbestos pleural disease or in connection with proof of a certain cumulative exposure to asbestos dust;
(iii) under no. 4105: asbestos-related mesothelioma.

Procedurally these diseases are classified as 'workplace accidents', which might make the day of the contraction of the disease relevant and a potential cause of dispute. However, as early as 1936 it was made clear for silicosis that the last day where the insured person was working in an environment which because of its nature could have caused or contributed to the disease was seen as 'the day of the accident'.[255] This was particularly relevant for the calculation of the pension, etc. as such pensions are usually based on the income at the time of the accident. This rule was in 1963 extended to all occupational diseases.[256] Further, it was regulated that in case of death of an insured person there was a presumption that the asbestos-related disease caused the death if there was a recognised reduction of earning capacity of 50 per cent or more because of such a disease, unless it was 'evident'[257] that the death was not caused by the disease. Importantly, a *post mortem* could not be requested to rebut the presumption.[258]

In 1973, the *Zentrale Erfassungsstelle für asbeststaubgefährdete Arbeitnehmer* was established. Since then, this central agency has been in charge of the medical care of people who were exposed to asbestos and

[253] Berufskrankheitenverordnung of 8 December 1976, BGBl. 1976 I, 3329 (at first limited to mesothelioma of pleura and peritoneum, in 1992 extended to include pericardium).

[254] 31 October 1997, BGBl. 1997 I, 2623.

[255] §3(3) of the Berufskrankheitenverordnung of 16 December 1936.

[256] Gesetz zur Neuregelung des Rechts der gesetzlichen Unfallversicherung of 30 April 1963, BGBl. 1963 I, 241.

[257] According to the Bundessozialgericht, 'evident' in this context means 'beyond reasonable doubt having caused or contributed to the death or beyond reasonable doubt having accelerated the death by at least one year': Entscheidungen des Bundessozialgerichts 28, 38 (14 March 1968) and HVBG-INFO 2006, no. 6, 714–21 (7 February 2006).

[258] See §63 Sozialgesetzbuch VII.

are thus likely to contract asbestos-related diseases. This also includes regular medical check-ups. In 2003, 254,000 persons were registered with the agency, plus another 122,000 from the former German Democratic Republic.[259]

The impact of asbestos-related diseases on the German social security system is significant and will continue to be so for some time. However, the number of claims for asbestosis and asbestos-related lung cancer and throat cancer has fallen by 10 per cent from 1998 to 2000; mesothelioma cases, on the other hand, still increase by 6–10 per cent annually.[260] From 1980–2000, 18,250 cases of asbestosis were recognised as occupational disease; so were 6,797 cases of asbestos-caused lung or throat cancer and 6,663 cases of mesothelioma.[261] All in all, therefore, a total number of 31,710 cases can be attributed to asbestos-related diseases.

In 1999, just over half of the 1,800 deaths in the occupational context 903 cases or 50.2 per cent) were attributed to asbestos-related diseases.[262] The number of deaths has been rising steadily since 1980, with the figure doubling from 1991 to 1996.[263] Since then, there has been a gradual decrease in the numbers.

The vast majority of people suffering from asbestos-related diseases have been exposed to asbestos between 1950 and 1970; the high risk groups identified are workers in the chemical industry and people whose job comprised insulation and sealing of buildings.[264]

Projections as to future numbers vary, but it is assumed that the peak of claims for injuries suffered from asbestos-related diseases will be reached between 2010 and 2015.[265]

(c) Asbestos in a non-occupational context

As outlined above, most cases concerning asbestos are dealt with by the Sozialgerichte as they occur in an occupational (or quasi-occupational) context. Outside of this context, the general principles of the BGB apply (see above p. 171 *et seq.*). Yet, as has been stated, personal injury claims concerning asbestos are unheard of in German civil law courts.[266]

259 HVBG, above note 238, at p. 31. For more detail on asbestos-related diseases in the GDR, see *ibid.* pp. 34–40.

260 *Ibid.* p. 44.

261 HVBG, above note 238, at pp. 50–2.

262 2.7 per cent were attributed to asbestosis, 26.9 per cent to asbestos-caused lung or throat cancer and 20.6 per cent to mesothelioma. Silicosis was the cause for 28.2 per cent of the deaths: *ibid.* p. 55.

263 *Ibid.* p. 56.

264 *Ibid.* p. 87.

265 *Ibid.* pp. 73, 76.

266 Kötz and Wagner, above note 2, at p. 222, and also see above note 237.

In the civil law courts, asbestos mostly features in the context of rent disputes, i.e. where the rented dwelling is found to be contaminated with asbestos (mostly from off-peak storage heaters) and either reduction of the rent or removal of the contamination is claimed.[267] In any event, liability here is independent of fault: if asbestos is found in the rented dwelling (including fittings, etc.) then this is deemed a breach of contract and there is liability.

Another area where asbestos has been dealt with is where in the course of building works (particularly those for asbestos decontamination) other property has been contaminated; here the general principles of tort[268] and contract law[269] apply. These cases, however, seem to be restricted to property damage, based on the expertise guaranteed by the company, and little is to be gained from these cases in way of general principle.

In some labour law conflicts, employers have been sued to provide adequate protection where there was a danger of asbestos contamination,[270] and in sales contracts the duty to disclose whether asbestos products have been used has been litigated.[271]

Finally, manufacturers of products containing asbestos have been sued for damages under tort law. In one case, a company produced off-peak storage heaters in 1972/73 containing asbestos. The court held that there was no fault and hence no liability under §823 BGB. A manufacturer could not be considered at fault if his product was at a later point in time found to be harmful. This would only be so if at the time of production the state of knowledge was such that the dangerousness should have been known;

[267] Landgericht Dortmund Wohnungswirtschaft and Mietrecht (WuM) 1996, 141 (16 February 1994); Amtsgericht Bonn, WuM 1991, 543 (2 July 1991); Landgericht Kassel, ZMR 1996, 90 (8 November 1994); Amtsgericht Heidelberg, NJWE-MietR 1996, 267 (8 May 1996); Landgericht Berlin, WuM 1996, 671 (19 July 1996).

[268] Scc, c.g., Oberlandesgericht Hamm, MDR 1992, 558 (5 March 1992). In that case, the company claimed to have expertise in this kind of work but let a non-specialist subcontractor perform the work (there was actually a 'chain' of subcontractors). The fault hence was seen in hiring this subcontractor and in the fact that the nature of the work was known to be dangerous and likely to contaminate; failure to employ properly trained personnel therefore constituted fault.

[269] See, e.g., Oberlandesgericht Köln, Az. 13 U 240/92, unpublished (11 November 2002). See also Kötz and Wagner, above note 2, at p. 242.

[270] Landesarbeitsgericht Köln, AiB 1993, 393 (22 January 1993); Bundesarbeitsgericht Entscheidungen des Bundesarbeitsgerichtes 75, 332 (2 February 1994); Bundesarbeitsgericht Entscheidungen des Bundesarbeitsgerichtes 85, 155 (19 February 1997).

[271] OLG Celle, Oberlandesgerichtrechtsprechung (OLGR) Celle 1996, 51 (8 December 1995); Landgericht Hannover, MDR 1998, 1474 (25 May 1998); Oberlandesgericht Koblenz, ZMR 2003, 185 (27 June 2002).

unless he should have known, he was not at fault and thus there was no liability.[272] Ultimately, this is nothing but the application of the general principle of the BGB, as stated by Rudolph von Jhering in 1879: 'Nicht der Schaden verpflichtet zum Schadensersatz, sondern die Schuld'.[273]

11. Conclusion

This chapter has tried to show the development of liability for sparks, exploding boilers and asbestos in Germany. As for escaping sparks, the approach taken in Germany in general was a property law one, which can still be found today. Nevertheless, with §25 PGE and Acts in other parts of Germany that followed the Prussian model, a remarkable legislatory precedent for strict liability was set as early as 1838. The RHPflG followed in 1871, not only covering railways but also industrial accidents, including exploding boilers. Those Acts, however, fortified the position of the concept of strict liability as alien to the general law, and up until today rules of strict liability (with the exception of §833 BGB) are confined to special Acts. In any event, the existence of the strict, or at least stricter, liability regimes seems not to have had a negative impact on industrial development.

As the RHPflG and the general law could not deal with the growing industrialisation satisfactorily, and to combat the 'spectre of socialism', in 1884, with the UVG, a compulsory public insurance scheme was introduced by Bismarck. This scheme has developed into today's social security laws, which cover almost all cases concerning asbestos-related diseases. In Germany, tort law and its function have thus in the occupational (and quasi-occupational) context largely been replaced by public insurance and social security laws.

[272] Oberlandesgericht Koblenz, Az. 5 U 1416/93, unpublished (1 June 1994).

[273] von Jhering, above note 11, at p. 199 (translation by Tony Weir in Zweigert and Kötz, above note 2, at p. 630: 'It is not the occurrence of the harm which obliges one to make compensation, but fault').

5

Technological Change and the Development of Liability for Fault in Italy

CHIARA FAVILLI

1. Introduction

All the selected topics relate to the harmful implications of technological progress – the escape of sparks from railway engines in the mid- to late nineteenth century, exploding steam boilers in the late nineteenth century and asbestos-related injuries and illnesses in the twentieth century – as telling perspectives on the evolution of liability for fault in the Italian legal system. In fact, it will be noted that the distribution of risks and costs between the victim and the tortfeasor follows social choices in balancing rights and values with technological development in each field.

The number of decisions varies considerably for each topic. Decisions about the escape of sparks from railway engines are abundant in the period from 1870 to 1940, above all in relation to the scant number of law reports published during this period, whilst cases involving explosions of steam boilers are rare – just a couple of decisions at the end of the period under consideration. Case law on asbestos is very recent, due to the long latent period of the illness and the late awareness of the link between the disease and exposure to asbestos. Most of the cases were settled in the last decade of the twentieth century. By contrast, asbestos litigation is currently increasing as knowledge of the consequences of asbestos spreads.

2. Escape of sparks from railway engines in the mid- to late nineteenth century

The first railway in Italy was opened in 1839 in the Reign of Sicily. It linked Naples to Portici and it was approximately 7 km in length, under half the length of the Liverpool to Manchester line, opened in 1830. Railway construction in other parts of the country then followed in the Kingdom of Lombardo-Veneto (Milan to Monza in 1840), in the Grand Duchy of Tuscany (Pisa to Livorno in 1844), in the Papal States (Rome to Civitavecchia in 1859) and in the Kingdom of Sardinia (Turin to Moncalieri in 1848). By the mid-nineteenth century, most parts of northern Italy were linked by railway. However, Italian railway connections were of minor importance: in 1861, when the Reign of Italy was proclaimed, there were just 2,521 km of railways, compared to 14,600 km in England, 11,000 km in Germany and 4,000 km in France. From 1861 to 1876, railway connections had increased to 7,780 km in Italy, 22,000 km in France, 28,000 km in Germany and 30,000 km in England.

Because the railways were constructed by several companies, according to the wishes of the local sovereigns, each project followed its own criteria and the distribution of the railways over the country was irregular. The same was true for management. As long as Italy was divided into states and kingdoms, railway connections were run by nearly twenty companies; as a consequence, the carriage of goods and the transport of people were complicated by the increase of fares, schedules and connections.

After unification, the idea of a national rail network was put into practice. Italy started reducing the number of railway licences and unifying the management of the railway connections; in particular, national lines were grouped together in three networks, each granted to a different company on the basis of a convention. Some fundamental regulations in this field were approved at this point: l. 20 March 1865, all. F on Public Works, whose Title V set out major provisions concerning the operation of the railways; Reale Decreto (R.D.) of 31 October 1873, no. 1687, regulated policy, safety and regularity in the use of the railways. Then, with the Law of 22 April 1905, no. 137, most of the networks and smaller lines were taken over by the state itself. The Azienda Autonoma delle Ferrovie dello Stato (formerly the FFSS and today the FS) was created in order to manage the 75 per cent of Italian railways owned by the state. The remaining 25 per cent, owned by private companies, were mainly concentrated in the south-east, where they still operate today.

In the mid- to late nineteenth century, steam, oil and electricity were the most frequently used energy sources. In Italy, the use of steam was short-lived. The prototypes of electric trains were introduced in 1870 and new experiments were attempted in 1890. Nonetheless, in spite of the good results of these experiments, the first notable achievement in electric railway lines was as late as the beginning of 1900.

During the Second World War, the system suffered considerable damage and extensive reconstruction was necessary. The electrified network was restored and greatly extended. Damage was so great that the network only reached its original size at the beginning of the 1970s.

After the Second World War, the popularity of the steam locomotive dwindled in favour of diesel and electric traction, which are more efficient. Gradually, the use of steam locomotives was abandoned; in a very short time, Italy became the first producer of electric and diesel engines, and on 22 July 1976 the last steam locomotive was made in Italy.[1]

A decline in production and the use of steam locomotives did not mean the end of accidents.[2] Sparks from engines were still supposed to be a primary cause of forest fires. Precautionary measures were introduced, both at local and national level.

Together with civil actions for damage to goods, cases involving the escape of sparks from railway engines represented the vast majority of civil actions brought against the Italian railways for pecuniary damage; however, claims for personal damage related only to accidents occurring during movement outside or inside the train (an unguarded level crossing or the opening of a door).

(a) Liability issues

In the nineteenth century, when civil codifications inspired by the Napoleonic Code were in force throughout the country, real property was the fundamental institution of the economy, and it was therefore especially protected from every possible attack and interference. By the end of the century, however, the rights of land-owners came into conflict with interests that were becoming more and more relevant as industrialisation developed.[3] As well as enduring expropriation provisions necessary for the construction of railway lines, agrarian property was exposed to damage connected with the increase of traffic on tracks, in particular

1 For more references regarding the use of steam locomotives in Italy, see 'Railway', *Encyclopaedia Britannica* (Chicago/London/Toronto/Geneva: Encyclopaedia Britannica, 1962), vol. XVIII, p. 933. On the development of the organisation of Italian railways, cf. G. Correale, 'Ferrovie dello Stato' in *Enc. Giur.* (Rome: Istituto dell'Enciclopedia, 1989), vol. XIV, p. 1; A. Rabbeno, 'Ferrovie', in *Dig. It.*, XI, p. II, 1, 1926, p. 31; A. Mocci, 'Ferrovie, (I) Ferrovie dello Stato' in *Enc. Dir.* (Milan: Giuffrè, 1968), vol. XVII, p. 218.

2 See, e.g., Trib. Roma, 26 March 1979, *Foro it.* 1982, I, 605; Cass., 3 June 1991, no. 6231, *Arch. circolaz.* 1991, 746.

3 Rabbeno, above note 1, at p. 31.

to the destructive consequences of the frequent escape of sparks from railway engines.[4]

Most decisions on this subject were determined between 1870[5] and 1941. The lack of decisions in the first twenty years under observation is partly due to political reasons: the first reviews of Italian case law begin in 1861, with the unification of the country, replacing the anthologies of decisions published in the local kingdoms.[6] However, at the very beginning these reviews had no place for civil liability, except to provide information about the projects of legislation on liability for railway accidents in foreign countries such as England, Austria and France. On the other hand, the lack of cases is possibly also due to the initial low number of steam trains and their limited movement on tracks in the years immediately before and after unification. As regards the end of the period under consideration, the almost total destruction of the railways at the beginning of the Second World War and the focus of attention of law reports concerning the war and the regime do not permit verification of the approach of case law to this topic under the Civil Code of 1942.[7]

As regards the parties to litigation, claims were usually brought by land-owners, most of whom were members of the rural aristocracy, against railway companies, in order to obtain compensation for damage to their property and related economic activities caused by the escape of sparks. There were very few exceptions to this pattern. On the one hand, the claim could be brought not only by the owner, but also by a person having a contractual right to use and enjoy the damaged property.[8] On the other hand, the action was brought against the counterpart of the

[4] G. Alpa, M. Bessone and A. Fusaro, *Poteri dei privati e statuto della proprietà* (Rome: SEAM, 2001), p. 45.

[5] Cass. Torino, 20 December 1870, *Monit. Trib.* 1872, 1213 seems to be the very first decision in this field as the case note refers to the application of the same rule by Cass. Torino, 18 May 1869, *Monit. Trib.* 1870, 482 in a different case involving damage caused by an overflow of water.

[6] In particular, the law report *Monitore dei Tribunali* ('*Monit. Trib.*') was first published in 1865.

[7] A brief outline of the later approach is given below p. 199 in relation to the recent developments in case law, after the end of the period under consideration.

[8] In App. Palermo, 8 September 1916, *Monit. Trib.* XVIII, 1917, 92, the claim was brought by an employee of the Italian railways. For work reasons he was living in a house owned by the railways; sparks caused by a passing train set fire to the house. The court rejected his claim for compensation for damage, since he could not satisfy the burden of proof. He also tried to ground the liability of the railways on contractual relations. The claimant maintained that he was not obliged to bear the risks connected to the things of which he had custody in the course of his work, but that the railways were obliged to indemnify him against possible damage. The court held that, since the Italian railways were under

contract, attempting to base liability on the lack of precautionary measures connected to the contractual relation.[9]

In this field, the ordinary courts normally had jurisdiction to decide the case, both when the railways were in private management and also when they passed to public authority. According to article 2 of the Law on administrative litigation of 20 March 1865, the administrative courts were to deal with the legality of the activity of the administration, whereas the ordinary courts were to deal with any damage caused by the omission or unlawful exercise of that activity.[10]

Conflicts between railway companies and land-owners were always resolved according to the rules of civil liability. In particular, the courts applied the rules on liability for fault. According to article 1151 of the Civil Code (c.c.) of 1865, 'any human act which causes damage to another obliges the person at fault to compensate the damage' and, according to article 1152 c.c., everyone is liable for damage caused not only by a wilful act, but also by negligence or imprudence.

The case law shows that the land-owners usually claimed for compensation for direct damage to their property caused by the escape of sparks from railway engines. In one case, however, the claimant asked for compensation for indirect or consequential damage to grazing pasture which was permanently destroyed by fire.[11]

Initially, these claims were successful, regardless of the recklessness or negligence of the defendant. All the plaintiff had to prove was the interference with property rights, e.g. that the damage was caused by the escape of sparks from trains.

No civil actions for personal injuries caused by the escape of sparks are to be found in the period under investigation.[12]

public authority, the claim was regulated by administrative law, which did not impose civil liability on the railways for damage to property in the custody of an employee.

[9] In Cass. 30 March 1935, *Monit. Trib.* LXXVI (1935), 691, the civil action was brought by the land-owner against an agister (*soccidario*), since he did not adopt the measures imposed by law to prevent the spread of fire. The court denied liability, considering that no clause in the contract imposed the adoption of such measures on the part of the *soccidario*.

[10] SU, 15 April 1920, *Monit. Trib.* LXI, 1920, 493.

[11] App. Lucca, 12 August 1913, *Monit. Trib.* 1913, 994. On indirect damage, cf. also Cass. Torino, 31 December 1903, *Monit. Trib.* 1904, 428.

[12] On damage to persons caused by railway accidents, such as a rail crash, see Rabbeno, above note 1, at p. 86 and case law quoted in *ibid.* note (1) at p. 87; G. Cazzetta, *Responsabilità aquiliana e frammentazione del diritto comune civilistico (1865–1914)* (Milan: Giuffrè 1991), p. 341, note 43.

In the only case of injunction, the plaintiff asked that the railway administration be ordered to raise a barrier in order to avoid damage to property. The court refused the injunction, since according to the legislation in force (article 534 c.c., article 234 *et seq.* of the Law of 20 March 1865 on Public Works) these decisions fell under the jurisdiction of the public administration only.[13] As was subsequently affirmed, since the railway service is under public law, private owners are not entitled to ask the public administration for the adoption of technical methods or security measures; all they can do is bring an action against the acts which have been illegitimately adopted.[14]

Claimants usually asked for the ending of railway activities as a way of avoiding damage. This possibility was considered in one decision as the only effective remedy, but its adoption would be more harmful to individuals and the economy than continuing with traffic on those rail lines.[15]

(b) Case law

Five introductory matters need to be considered.

(i) Railway companies were not immune from liability for damage caused by the escape of sparks. The existence of such special immunity was expressly denied,[16] quoting the statute of 30 March 1865, paragraph F on Public Works, which introduced preventive measures in order to protect private properties from several kinds of danger: one specific provision, in Title V (on railways), forbade building and planting at a distance of less than 6 metres from the railway (article 235); one general provision, in Title II (on ordinary streets), forbade the planting of a wood at a distance of less than 100 metres from the road (article 71).[17] None of these provisions provided immunity for the railways as a sanction for private subjects. Legislation only gave the railways the right to ask for the demolition, on indemnity, of the factories and (isolated) plantations situated at a shorter distance at the time of the creation of the railways, provided that this was necessary for public security and for free passage (article 240). As for (groups of) plants, bushes and woods situated at less

[13] SU, 25 November 1911, *Monit. Trib.* 1912, 131; SU, 5 October 1911, *Monit. Trib.* 1911, 901.

[14] As stated by SU, 15 November 1920, *Monit. Trib.* LXI, 1920, 493.

[15] App. Bologna, 13 February 1914, *Foro it.* 1914, 1, I, 694.

[16] App. Firenze, 12 March 1877, *Foro it.* 1877, 2, I, 36, commentary by Giuliani.

[17] On this rule and on the concurrent application of art. 579 c.c. 1865, cf. Pret. Codogno, 3 June 1870, *Monit. Trib.* 1870, 678 overruled by Trib. civ. e correz. Lodi, appeal section, *Monit. Trib.* 1871, 792.

than 100 metres from the road, the legislation provided that they must be tolerated, unless they were a hazard. Nevertheless, when they become ripe or withered, they could not be replaced closer than 100 metres from the track (article 73).[18]

(ii) Case law does not show obstacles to recovery caused by periods of limitation, causation hurdles or difficulties of proof. Proof of causation, as will be seen, was 'absorbed' in (i.e. was considered sufficient for) the demonstration of the other prerequisites of the liability for fault, especially at the first stage in the development of this issue.

(iii) The liability of the railways was not to be found in formal sources. Neither the regulations concerning the policing, safety and regularity of the use of railways of 31 October 1873 nor the statute of 30 March 1865 all. F on Public Works made provision regarding the civil liability of the railways. Nevertheless, as one decision[19] makes clear, one theoretical approach tried to infer the strict liability of the railways from the rules contained in the legislation (article 2 of the regulation on railways and article 317 of the statute on Public Works). The argument was that the duty 'to take all the measures and to use all the precautions suggested by practice and science in order to prevent and avoid accidents' could justify the absolute liability of the railways for damage independently from fault, in a way similar to the legislation on industrial accidents.

(iv) In case law, no references are made to insurance law. However, the Commercial Code of 1882 had specific provisions (articles 441–444) on insurance against damage caused by fire.

(v) Under the 1865 Civil Code, the liability of the railways was never based on the hypothesis of strict liability or no-fault liability, as happened once the Civil Code of 1942 became effective.[20] In reality, within the hypothesis of strict liability contained in the former code, there was no

[18] The description follows the approach of Cass. Roma, 13 August 1910, *Giur. It.* 1911, 1226, in order to distinguish between the application of arts. 73 and 235 to 'plants'.

[19] Cass. Torino, 26 June 1913, *Giur. It.* 1914, 313 and *Foro it.* 1915, I, 41.

[20] After the Civil Code of 1942 came into force, Pret. Patti, 15 June 1963, *Giust. Civ.* 1963, I, 2716 applied art. 2051 c.c. (damage caused by things in custody) to the public administration as being the guardian of the escarpment on which the fire broke out. The railway service was considered a dangerous activity by Cass. 27 February 1984, no. 1393, *Foro it.* 1984, I, 1280 and *Foro it.* 1985, I, 1497, commentary by M. Comporti, 'Presunzioni di responsabilità e pubblica amministrazione: verso l'eliminazione di privilegi ingiustificati'. On the other hand, App. Bari, 5 July 1958, *Corti Bari, Lecce, Potenza* 1958, 717 excluded the application of art. 2050 c.c. (dangerous activities) to the cases under consideration. In general, on the civil liability of the railways, cf. M. Bona, 'Responsabilità delle Ferrovie dello Stato ed onere probatorio: verso la strict liability dell'amministrazione ferroviaria?' *Danno e resp.* 1999, 1233.

equivalent of article 2050 on liability for dangerous activities. Article 2050 states that anyone who causes damage in the carrying out of an activity which is dangerous by nature or through the kind of instruments used is liable unless he proves that he adopted all the appropriate measures to avoid the damage. Nevertheless, the escape of sparks could be considered as coming within article 1153, first paragraph, 1865 c.c., which includes, within the hypothesis of vicarious liability, responsibility for the care of things.

Instead, a rule of strict liability was developed by the courts from the rules on liability for fault. As will be seen, different approaches to the fault liability rule in article 1151 c.c. were the basis of opposing outcomes during the evolution of the jurisprudence.

Following this evolution, case law can be divided into two stages, although inevitably the transition is not clear. In particular, between 1909 and 1914, both approaches to the fault liability rule were present, albeit to differing degrees. The first stage was still predominant as the second stage began to appear. For this reason, the year 1915 can be considered as a temporal divide, since after that there were no more decisions that chose the approach of the first phase; however, from 1909 onwards it is possible to find the first applications of the approach which would characterise the second stage of the case law on this topic. This second phase was subsequently articulated independently.

(i) First period decisions (1870–1915)

In the first period, the main argument used in case law was the application of the general rule on fault liability – article 1151 c.c. 1865 – as it contained a hypothesis of strict liability.[21] In the enforcement of this provision the courts held that the fault of the defendant did not have to be proved, as in general it would have to be, but was to be presumed *(culpa in re ipsa)*. The purpose was twofold: to favour the claimant by placing on the defendant the burden of proving the absence of fault, and to guarantee compensation for damage in connection with the simple outcome of the *iniuria* and causation.[22] In practice, according to case

[21] Real and genuine hypotheses of strict liability or the presumption of liability are especially provided by the Civil Code of 1865 in arts. 1153, 1154 and 1155 and by legislation on accidents at work.

[22] Cf. S.G. Pitocchi, 'Ancora sulla irresponsabilità per danni incolpevoli, e sull'art. 73 della legge sulle opere pubbliche' *Giur. It.* 1917, 855. In the commentary to Cass. Firenze, 26 March 1917, he emphasised that, in this decision, fault was considered as the infringement of a legal duty, primarily of the duty of *neminem laedere*. Consequently, the violation of

law, in order to be exempted from liability it was not enough for the railways to prove the absence of fault by showing compliance with the laws and protective measures. Moreover, in some decisions the railways' defence was barred by the judges refusing to admit any relaxation of the burden of proof; and in other cases, the courts constructed the burden of proof in such a way that the defendant was prevented from making use of it. An examination of the case law might be useful in order to clarify this outline.

Sometimes the presumption of fault was asserted without any argument: 'fault of the railways for the damage caused by the escape of sparks is always presumed. Due to this presumption, the proof of fault did not lie with the claimant, but the reversal of the burden of proof lies with the defendant'[23]; or it was simply argued that fault was to be considered in the light of the same harmful fact of the escape of sparks or of the danger of burning materials to the surrounding property.[24] Finally, courts assumed that fault in article 1151 c.c. was demonstrated by causation.[25]

Some decisions, on the other hand, did justify the presumption of fault, connecting it to a presumed negligence. In the Court of Appeal of Florence decision on 12 March 1877, for example 'from the escape of sparks it follows that the adoption of precautionary measures was defective and thence the negligence or imprudence of the railway personnel'[26]; in the opinion of the Court of Cassation of Turin on 20 December 1870, the railways were liable for not having sought to have the factory demolished before the fire occurred, when it was situated at a distance that was less than that prescribed by law. It had failed to do so in order to avoid

the right of the land-owner assumed the fault of the railways, and the proof of the event represented, at the same time, demonstration of the damage and fault.

23 Trib. Roma, 2 June 1882, *Temi rom.* 1882, 560 and *Foro it.* 1882, 1233. Cf. App. Firenze, 12 March 1877, *Foro it.* 1877, I, 36, commentary by Giuliani, which reported that the trend of case law was to find in favour of the liability of the railways, even if they had not violated administrative regulations.

24 See App. Lucca, 12 August 1913, *Foro it.* 1914, I, 48; App. Torino, 18 April 1913, *Giur. It.* 1, II, 437 and *Monit. Trib.* 1913, 972; App. Bologna, 13 February 1914, *Foro it.* 1914, 1, I, 694; Cass. Napoli, 9 September 1911, *Foro it.* 1912, I, 20 and *Giur. It.* I, 1, 1911, 1239. Cf. S. Rodotà, 'Proprietà e industria. Variazione intorno alla responsabilità civile' *Politica del diritto* 1978, 421, who underlined the merely formal consideration of the prerequisite of fault by these courts.

25 This approach is referred to and criticised by Cass. Torino, 26 June 1913, *Giur. It.* 1914, 313 and *Foro it.* 1915, I, 41.

26 App. Firenze, 12 March 1877, *Foro it.* 1877, I, 36, commentary by Giuliani. Against this approach ('reasoning from the unknown to the unknown'), cf. App. Roma, 28 May 1910, *Foro it.* 1911, 1, II, 980.

the payment of an indemnity.[27] In other cases, fault was presumed as a consequence of the (presumed) absence of an act of God: the involvement in the event of the 'hand and act' of man established the accountability, the link of causation between human event and damage, and, as a direct consequence, liability.[28] In other words, human intervention excluded an act of God, making the fault implicit.[29] In other decisions, the presumption of fault was based, alternatively, on the foreseeability of the damage and on the awareness of its inevitability. On the one hand, judging by the precautions taken and the recommendations given to staff, it followed that damage caused by sparks was always foreseeable by the railways.[30] On the other hand, the awareness that the escape of sparks was an inevitable consequence of traffic on the railway excluded an act of God, which is a fortuitous event, independent from any will or prediction.[31] Finally, the fault was simply considered an inevitable consequence of the dangerousness of railway activity.[32]

Among the principles used to justify the presumption of fault, case law makes reference to the principles of 'natural law':[33] *nemo locupletari cum aliena iactura* and *cuius commoda eius et incommoda.*

In the first sense, it underlined the decision of the Court of Cassation of Naples on 9 September 1911, which emphasised that economic utilities obtained by the state through the control of the railways could not result in damage to others; for this reason, damage caused by trains must be compensated according to the principle of *neminem laedere.*[34] On the other hand, this approach was influenced by the debate about the strict liability of the public administration for its lawful acts, too. It follows from principles of 'natural' law, i.e. that nobody has the right to gain at the expense of someone else and that the costs of public services should be allocated according to the criteria of distributive justice.[35]

[27] Cass. Torino, 20 December 1870, *Monit. Trib.* 1872, 1213.

[28] Cass. Napoli, 10 February 1882, *Giur. It.* 1882, I, 1, 460.

[29] Cass. Firenze, 26 March 1917, *Monit. Trib.* 1917, 376.

[30] Cass. Torino, 20 December 1870, *Monit. Trib.* 1872, 1213; App. Lucca, 12 August 1913, *Foro it.* 1914, I, 48.

[31] Cass. Firenze, 18 May 1914, *Giur. It.* 1914, I, 1, 1000.

[32] Cass. Torino, 20 December 1870, *Monit. Trib.* 1872, 1213.

[33] Cf. App. Bologna, 13 February 1914, *Monit. Trib.* 1914, 449.

[34] Cass. Napoli, 9 September 1911, *Giur. It.* I, 1, 1911, 1239, commentary by L. Malpeli, 'Considerazioni sul parziale riconoscimento probatorio delle perizie preventive' *Monit. Trib.* 1912, 70 and *Foro it.* 1912, I, 1531.

[35] Trib. Roma, 27 February 1914, *Giur. It.* 1914, 312, commentary by S.G. Pitocchi, 'Irresponsabilità per danno incolpevole', *ibid.* 315. Cf. Cazzetta, above note 12, at p. 351.

A clear application of the second principle was demonstrated by the Court of Cassation of Turin on 20 December 1870:

> The railway activity by means of locomotives, from which sparks could escape, constitutes in itself such a serious danger to the surrounding property, that it cannot be admitted that for the authorisation of similar exercise and in spite of the observance of the prescribed precautions, if they transpire to be insufficient, the railways are exempted from compensating the damage which is a direct and foreseeable consequence of its behaviour.[36]

Applying the principle *cuius commoda eius et incommoda*, the Court of Cassation of Florence held on 10 February 1921 that 'if damage to private property is an unavoidable consequence of an activity, this sacrifice must be compensated, even if the activity is a public service, except in the case of an act of God'.[37]

To be exempted from liability, as has been seen, it is not sufficient for the defendant to prove the absence of fault. Several decisions, following the model of French doctrine,[38] underline that it is not enough for the defendant to show that it was authorised by the government, or that it attended to the relevant requirements and adopted instruments and criteria according to the best science and experience.[39] Nevertheless, in one decision, this diligent respect for legal requirements led the court to presume that the damage was caused by an element outside the control of the railways, that broke the chain of causation.[40]

In this connection, the statement by the Tribunal of Rome on 2 June 1882, in connection with the presumption of fault, was significant. In order to be exempted from liability, the only defences recognised were force majeure and the exclusive fault of the victim.[41]

36 Cass. Torino, 20 December 1870, *Monit. Trib.* 1872, 1213.

37 Cass. Firenze, 10 February 1921, *Giur. It.* I, 1, 272. Against this reasoning, cf. Cass. Torino, 26 June 1913, *Giur. It.* 1914, 313 and *Foro it.* 1915, I, 41.

38 A list of the French authors is contained in the commentary by Giuliani, 'Studio di diritto ferroviario: Se una società ferroviaria debba ritenersi responsabile degli incendi cagionati dalle scintille che escono dalle locomotive', in *Foro it.* II, 1, 561.

39 Cf. Cass. Torino, 20 December 1870, *Monit. Trib.* 1872, 1213; App. Firenze, 12 March 1877, *Foro it.* 1877, I, 36, commentary by Giuliani. The same statements are contained in Cass. Firenze, 26 March 1917, *Monit. Trib.* 1917, 376 and in *Giur. It.* 1917, 849, commentary by S.G. Pitocchi, 'Ancora sulla irresponsabilita' per danni incolpevoli e sull'art. 73 della legge sulle opere pubbliche' and in App. Lucca, 12 August 1913, *Monit. Trib.* 1913, 994 which is ratified by Cass. Firenze, 18 May 1914, *Giur. It.* I, 1, 1914, 999.

40 App. Catanzaro, *Monit. Trib.* 1908, 370, quoted by S.G. Pitocchi, 'Irresponsabilita per danno incolpevole', above note 35, at 324, without indication of the date.

41 Trib. Roma, 2 June 1882, *Temi rom.* 1882, 560, summarised in *Foro it.* 1882, 1233. By analogy in App. Torino, 18 April 1913, *Monit. Trib.* 1913, 972, *Giur. It.* I, 2, 436 and *Foro it.*

As regards force majeure, usually considered in case law together with acts of God to which it is considered equivalent, the defendant had to prove that the escape of sparks was caused by natural unforeseeable circumstances, such as a furious wind, or that the damage was unavoidable, as current scientific knowledge was not able to suggest suitable measures to avoid such danger.[42] At the same time, the opinion of the courts about the foreseeability of the escape of sparks, in reality made it clear that the defence of act of God could not be substantiated in practice.

With regard to the second aspect, the extent to which the fault of the land-owner was recognised was strongly influenced by the interpretation of article 73 of the Law of 20 March 1865 all. F on Public Works. This laid down an obligation on the part of the railways to tolerate 'plants, hedges and woods' on the property of the claimant provided they were not hazardous, but prohibited their replacement unless the distance prescribed by article 71 for planting woods (100 metres from the track) was observed. Apart from the vegetation specifically mentioned, the case law was divided on what to include in the extension of this rule.[43] Was coppicing (and, in particular, the ramifications arising after the periodic cut) and the plants grown from seed that fell after the construction of the railways subject or not to the obligation to respect the distance prescribed by article 73? According to the interpretation given to this provision, the possibility of ascribing the fault to the damaged party – to exclude or reduce the liability of the railways – changed. Italian courts were divided between a restrictive interpretation, which considered natural propagation as part of the original plantation, so the railways were obliged to tolerate it, even though it was within the legally defined zone,[44] and an extensive interpretation which considered the propagation as a new plant,[45] which

I, 1913, 1480; Trib. Termini Imerese, 13 June 1917, quoted by S.G. Pitocchi, 'Ancora sulla irresponsabilità' *Giur. It.* 1917, 852. As to the manipulation of fault by judges to reach a conclusion they considered fairer, see Cazzetta, above note 12, at p. 343.

42 Trib. di Commercio di Bologna, 31 January 1876, *Foro it.* 1876, I, 1, 822; App. Torino, 18 April 1913, *Foro it.* 1913, I, 1480 and *Giur. It.* 1913, I, 2, 436.

43 Cf. App. Lucca, 18 April 1913, *Giur. It.* 1, II, 441; App. Torino, 18 April 1913, *Foro it.* 1913, I, 1480.

44 Cass. Firenze, 13 June 1912, *Giur. It.* 1, I, 1109, *Monit. Trib.* 1913, 27 and *Foro it.* 1912, I, 1526; Cass. Firenze, 18 May 1914, *Giur. It.* I, 1, 1914, 999; Cass. Firenze, 26 March 1917, *Monit. Trib.* 1917, 376 and *Giur. It.* 1917, 850, commentary by S.G. Pitocchi. The same opinion was expressed by App. Lucca, 12 August 1913, *Foro it.* I, 4, 49.

45 App. Catania, 8 August 1890, *Giur. It.* 1900, I, 1, 346; Cass. Torino, 2 March 1900, *Giur. It.* 1900, 1, I, 346; Cass. Torino, 21 August 1912, *Monit. Trib.* 1913, 306 and *Foro it.* 1912, I, 1531; App. Catania, 8 August 1890, *Giur. It.* 1891, II, 125; App. Napoli, 7 July 1923, *Monit. Trib.* 1924, 97; App. Napoli, 2 June 1926, *Giur. It.* 1926, I, 2, 627; Cass. Milano, 19

the land-owner was obliged to cut down in order to respect the distance of 100 metres prescribed for planting woods, following the opinion of French scholars.[46]

Moreover, the decisions were influenced by the perspective chosen in applying this rule. In the opinion of some courts, the infringement of these limits meant the fault of the claimant; in the opinion of others, liability fell on the defendant who was entitled to ask for their respect. In the beginning, the second approach was prevalent. The railways were held liable, even if they proved that the distance from the tracks of plantations and buildings damaged by the fire was less than that prescribed by law. The argument was that the defendant should have had them cut or demolished on request upon payment of an indemnity. If it did not do that, it must pay damages to the claimant.[47]

The consequence of such a restriction for the relaxing of the burden of proof was that the only possibility for the railways to escape liability would be a rule containing an obligation for the victims to bear the consequences of the damage, i.e. an immunity rule – a rule which, as stated, does not exist. In reality, the same conclusion was reached by a decision about the liability of the railways for damage caused by an overflow of water caused by works built by the same railway company. They were sanctioned to compensate the damage in spite of having proved a respect for the prescriptions of the law on public works and of the conditions and burdens imposed by the administrative authority, which had tested the works. In the opinion of the court, a company would only be freed from liability where it could also demonstrate that a special provision of the same law applied under which the injured person was obliged to bear the loss without any right to compensation.[48]

As well as the argument of *culpa in re ipsa* (which, according to the interpretation given by the case law, is more a presumption of liability than a presumption of fault), a property rule argument appears in the case law, based on the law on expropriation. This argument was used

October 1926, *Monit. Trib.* 1926, 952; Cass. Roma, 13 August 1910, *Giur. It.* 1911, 1226, commentary by S.G. Pitocchi, 'Sulla distanza da osservarsi fra la strada ferrata e i piantamenti di boschi'; App. Torino 18 April 1913, *Foro it.* 1913, I, 1480; Cass. Torino, 18 August 1912, *Giur. It.* 1912, I, 1, 1311, commentary by S.G. Pitocchi.

[46] Cf. French case law, quoted by S.G. Pitocchi, 'Sulla distanza da osservarsi fra la strada ferrata e i piantamenti di boschi', above note 45, at 1230.

[47] Cass. Torino, 20 December 1870, *Monit. Trib.* 1872, 1213. The opposite approach is adopted by Cass. Roma, 13 August 1910, above note 45.

[48] Cass. Torino, 18 May 1869, *Monit. Trib.* 1870, 482.

to guarantee compensation for damage, for the judges failed to find the railways negligent by application of rules on fault liability.

In one case, the court, even admitting that the damage was unavoidable (as science was not in a position to suggest suitable equipment to completely exclude the risk) and that, for this reason, the railways could not be liable, stated that the defendant was nevertheless obliged to compensate the damage caused to the land-owner due to the fact that 'property is sacred and inviolable by people and by the state. Nobody is obliged to renounce it or to suffer any material reductions, even if it is in the public interest (it would be a kind of indirect expropriation), without convenient and just compensation.'[49] This conclusion was reached on the basis of the quoted principles contained in the Civil Code of 1865 (article 438 c.c.) and in the Constitution (article 29 of the Statuto Albertino).

On the other hand, moving from the same consideration ('as the event was unavoidable, the railways are not at fault'), the resultant obligation of the defendant to compensate damage caused by the *iniuria* was also reached following the principle of interpretation by analogy ('general principles of law as found in article 3, paragraph 2, of the preliminary provisions of the civil code of 1865'). In particular, the Court of Appeal of Bologna refused the application of article 1151 c.c., as the court of first instance had done, since the fault of the defendant was not proven. Presuming the lack of a rule for such a case, the court used the general rule mentioned above as a bridge to draw a parallel with the provisions on nuisance (article 574 c.c. 1865). According to the court, had the action been brought against a private individual, the claimant would have had the possibility of choosing between an injunction and an action for damages. However, since the defendant was not a private individual, but the public administration, the right to an injunction was converted to a right to just compensation for the burden imposed on the claimant, i.e. the right to an injunction 'is turned into the right to comparable compensation for the harm that he was obliged to suffer in the interests of society'.[50]

Choosing this approach, it is noted that the principles of fairness and of distributive justice (which impose upon the collectivity the burden of

[49] Cf. App. Firenze, 12 March 1877, *Foro it.* 1877, I, 36. A similar approach can be found in Cass. Napoli, 9 September 1911, *Giur. It.* I, 1, 1911, 1236, commentary by L. Malpeli, 'Considerazioni sul parziale riconoscimento probatorio delle perizie preventive'. The property argument is also quoted by App. Lucca, 12 August 1913, *Monit. Trib.* 1913, 994, confirmed by Cass. Firenze, 18 May 1914, *Giur. It.* I, 1, 1914, 999.

[50] App. Bologna, 13 February 1914, *Monit. Trib.* 1914, 449 and *Foro it.* 1914, 694.

the harm caused for a reason of public utility) were applied in the field of tort liability.[51]

Nevertheless, it must be emphasised that while most of the courts continued to apply the presumption of fault of the railways from the simple fact of the escape of sparks, some courts began to criticise the common approach to the fault liability rule. In particular, the Tribunal and the Court of Appeal of Rome defended the genuine interpretation of article 1151 c.c., which put the burden of proof of fault on the claimant and exempted the defendant from liability if this evidence was not given; as the court stated, in this case the cause of damage was to be considered a fortuitous event. Since a fire had broken out, the protection was clearly not suitable or, even if it had been appropriate, it was unable to prevent the fire. In the first case, the proof of fault could not be replaced by way of presumptions. In the second case, the event must be regarded as a mere chance (act of God), for which nobody was to blame. Thus, the damage would lie with the victim, unless the law or a contract provided otherwise.[52]

(ii) Second period decisions (1915–1940)

Within a few years, the critical opinion of the courts of Rome spread all over Italy. The approach of the case law changed, leaving behind a formalistic application of article 1151 c.c. and establishing a genuine interpretation of the responsibility for fault. Now the courts found civil liability only if the injured person was able to prove a fact which was imputable to the defendant in a subjective sense (fraud or fault), besides the *iniuria* and the link of causation to the damage.[53]

[51] This is the opinion expressed by Cazzetta, above note 12, at p. 350.

[52] Trib. Roma, 7–23 April 1909, *Rev. Dir. Pubb.* 1909, 551, commentary by L. Malpeli, 'Della responsabilità delle ferrovie dello Stato per incendi dipendenti dal transito dei treni'; App. Roma, 28 May 1910, *Giur. It.* 1910, 1, II, 980; Trib. Roma, 27 February 1914, *Giur. It.* 1914, 312, commentary by S.G. Pitocchi, 'Irresponsabilita per danno incolpevole'. The same approach is followed in a railway accident case by Cass. Torino, 26 June 1913, *Giur. It.* 1914, 313 and *Foro it.* 1915, I, 41.

[53] App. Roma, 29 July 1915, *Foro it.* 1916, I, 120; Trib. Palermo, 7 April 1916, *Ferr. Ital.* 1916, 308; App. Palermo, 8 September 1916, *Monit. Trib.* 1917, 91; Cass. Roma, 2 March 1916, *Monit. Trib.* 1916, 585 and *Giur. It.* 1916, I, I, 443; App. Palermo 4 August 1916, *Foro sic.* 1916, I, 413; Trib. Catanzaro, 17 July 1918, *Monit. Trib.* 1919, 762; App. Palermo 4 August 1916, *Foro sic.* 1916, I, 413; Cass. Palermo, 16 March 1918, *Monit. Trib.* 1918, 237 and *Giur. It.* 1918, 534; Cass. Palermo, 26 February 1918, *Monit. Trib.* 1918, 305; Cass. Roma, 23 July 1918, *Giur. It.* 1918, 831; Cass. Palermo, 2 May 1918 and 13 August 1918, *Monit. Trib.* 1919, 135; Cass. Napoli, 23 June 1919, *Monit. Trib.* 1919, 551; Trib. Catania, 28 March 1919, *Monit. Trib.* 1919, 789; Cass. Roma, 22 January 1920, *Monit. Trib.* 1920, 250;

In particular, the fault of the defendant was proved where it did not comply with the protective measures required by law[54] or if it applied the measures in an incorrect way. So liability existed if the fire developed on accumulated dry material on the embankment of the railway line which the railway administration should have removed (according to article 237 of the Law on Public Works of 20 March 1865 all. F); or where there had been no maintenance or poor maintenance, or in case of negligent behaviour on the part of railway personnel (for whom the railway company was responsible under the vicarious liability rule of article 1153 c.c. 1865). If the claimant did not succeed in proving fault by means of presumptions or if the defendant was able to demonstrate scrupulous care and respect for legal requirements, the harm would be an act of God not giving rise to a claim for compensation for loss but falling on the victim.[55]

In subsequent years, an approach developed in the direction of a reduction in the extent that fault was found against the railways and thus of their liability. Demonstration by the railways that they had adopted the protective measures required by law was considered sufficient to exonerate them from responsibility. This was the conclusion, even where scientific evidence suggested that there were more effective measures than those adopted. In particular, the Court of Cassation of Milan, in a decision of 19 April 1929, exempted the railways from having to prove that the legally required preventive measures for avoiding damage were suitable.[56]

In order to justify this approach, case law tended to resort to deference towards autonomy of the administration when assessing the measures, the systems and the potential of preventive measures it had required.[57] Such autonomy was not subject to judicial control, unless the administration itself fettered its own discretionary power by means of regulations and decrees, thereby favouring private individuals. As a consequence, the

Cass. Palermo, 11 January 1921, *Giur. It.* 1921, I, 1, 706 (m.); Cass. Napoli, 22 November 1921, *Monit. Trib.* 1922, 248; App. Milano, 5 June 1923, *Monit. Trib.* 1923, 499; Cass. 4 June 1924, *Monit. Trib.* 1924, 886 and *Giur. It.* 1924, 1048; Cass., II sez., 21 March 1928, *Giur. It.* 1928, 954. Later, the United Sections of the Court of Cassation welcomed this approach: SU, 15 April 1920, *Monit. Trib.* 493; SU, 9 June 1922, *Monit. Trib.* 1922, 616; SU, 19 November 1925, *Giur. It.* 1926, I, 1, 425; SU, 23 January 1941, *Giur. It.* 1941, 1, I, 252.

54 Cf. Cass. Roma, 2 March 1916, *Monit. Trib.* 1916, 585 and *Giur. It.* 1916, I, I, 443 which takes into consideration the failure to use the preventive measure of applying protection caps.

55 App. Palermo, 8 September 1916, *Monit. Trib.* 1917, 92.

56 Cass. Milano, 19 April 1929, *Monit. Trib.* 1929, 577.

57 Cf. Cass. Roma, 17 April 1920, *Giur. It.* 1, I, 423; SU, 15 April 1925, *Giur. It.* 1925, 553 and *Monit. Trib.* 1925, 925; SU, 19 November 1925, *Giur. It.* 1926, I, 1, 425; Cass. Roma, 5 April 1927, *Monit. Trib.* 1927, 530.

railways acquired a right to operate within these limits and could have recourse to judicial authority in order to require the administration to exercise its discretionary power.[58] In particular, the R.D. of 31 October 1873, no. 1687, on the regulation of railway policing, safety and regularity[59] (which was enacted when the railways were privately owned and maintained in force by the state administration) ordered the adoption of all measures and precautions suggested by expert knowledge and practice in order to prevent accidents. Obviously, the regulation did not specify the measures; the choice of measures lay within the discretion of the railways. Courts were not competent to make judgments about the availability of alternative instruments or more effective ones. So, where more modern and effective models of protection were available, the failure of the railways to adopt them was not considered as proof of fault. As a consequence, the railways were subject to a lower level of diligence and to a deference which could lead to a merely formalistic approach to fault. As a result, the approach of case law to the liability of the railways in this second period was effectively limited to the failure to maintain the cleanliness of the tracks.

It must be stressed that, in modern times, this approach has been refuted. As was said, damage caused to land by the escape of sparks from railway engines is less common than in the period under observation, but it still happens.[60] For instance, a decision of the Court of Cassation on 29 March 1952 held the railway liable, even though the train was equipped with a protection cap. Because several fires had broken out in the same place before the events in the instant case, the court concluded that the railway was at fault for not intervening to eliminate the cause in order to avoid further fires. Demonstration of the adoption of the protection cap was not sufficient, since the escape of sparks proved it was unsuitable. As the railway was conscious of the risk, it had to adopt further measures, such as assigning employees to supervise the tracks.[61]

In the most recent decision dealing with this topic, the Court of Cassation pointed out what the railway had to prove in order to escape liability. It stated that, in order to be exempt from liability, the railway

[58] These terms are used in Cass. Roma, 17 April 1920, *Giur. It.* 1, I, 423.

[59] These provisions were completed by R.D. 9 May 1912, no. 1447, art. 69 of which prohibited the construction of buildings less than 20 metres from tracks and forbade the accumulation of inflammable material.

[60] For instance, App. Palermo, 12 October 1951, *Giur. Sic.* 1952, II, 24 confirms the liability of the railways in case of failure to adopt or of poor maintenance of protection caps.

[61] Cass., 29 March 1952, *Resp. Civ. Prev.* 1952, 412.

could not rely on discretionary power in relation to the selection of criteria and instruments for the maintenance of the tracks. In fact, its discretionary power was restricted by the common rules on prudence and diligence imposed by the principle of *neminem laedere*. This principle, codified for the management of the railways in article 2 of the R.D. of 31 October 1873, no. 1687, and now found in article 8, Decreto Presidente della Repubblica (D.P.R.) of 11 July 1980, no. 753, requires the railways to adopt all the measures and precautions suggested by expert knowledge and practice in order to prevent accidents. As a consequence, the ordinary courts are today allowed to assess the suitability of the measures and precautions adopted.[62]

In this second period, case law seemed to make a U-turn and be decided upon fault liability only. Perhaps an extra-legal reason for such a different approach can be found in the technical advances in safety and precautionary measures. In the first period, if courts had decided upon fault liability only and according to the genuine interpretation of article 1151 c.c, the lack of efficient systems for protection would have led regularly to finding the railways not liable and the consequent denial of compensation to the victims of damage. In consequence, courts tended to be satisfied if a wrong (*iniuria*) was demonstrated, which allowed the court to assume the existence of fault or ignore fault as a separate requirement. However, the availability of more efficient instruments during this second period allowed courts simply to apply the fault liability rule.

As a conclusion, we would emphasise that in the third period, i.e. after 1950, courts mainly dealt with the topic of the contributory negligence of the land-owner. The Court of Cassation on 29 March 1952 found contributory negligence for failure to maintain the cleanliness of the tracks[63]; the Tribunal of Rome on 3 August 1957 recognised it for the lack of suitable protective measures between woods and tracks[64]; the Tribunal of Naples on 3 October 1968 found it in a failure to respect the distance between trees and tracks prescribed by law.[65] Finally, the Court of Appeal of Rome pointed out that the burden of proof of the contribution of the land-owner to the cause of the damages lies with the railway; in particular, it has to demonstrate failure to protect the estate with the firewalls known as '*roste*'.[66]

[62] Cass., 3 June 1991, no. 6231, *Arch. Circolaz.* 1991, 746.

[63] Cass., 29 March 1952, *Resp. Civ. Prev.* 1952, 412.

[64] Trib. Roma, 13 August 1957, *Temi rom.* 1958, 186.

[65] Trib. Napoli, 3 October 1968, *Giur. It.* 1969, I, 2, 826.

[66] App. Roma, 26 March 1979, *Foro it.* 1982, I, 605, commentary by Di Paola.

(c) *Concluding remarks*

Constant application over time of article 1151 c.c. has led to outcomes that are, in fact, divergent. Changing their interpretation of fault, at one period the courts used this rule to uphold the interests of the land-owners; at another to uphold the interests related to the development of a modern system of public transport, reflecting the evolution of society.

Thus, case law regarding the escape of sparks from railway engines from the mid-nineteenth century to the mid-twentieth century can be taken as an example of the changing perceptions of the interests within society: from the absolute supremacy of rural property to a later period in which this right was balanced against the interests of industrial development. In the final stage, industrial development prevailed over any other economic interests.

Among the elements that could have influenced these substantial changes in the way courts dealt with cases, the following should be noted:

- the reshaping of the real property rights in the Civil Code of 1942; and
- the evolution in the perception of tort law: from an instrument to defend property rights to a device that implements the needs of solidarity and social justice.

3. Explosions of steam boilers (1850–1900)

(a) *Brief outline of the development of steam engines*

The origin of the steam boiler is contemporaneous with the invention of the steam engine. In its earliest stages, during the eighteenth century, the pressure vessel or boiler was constructed of materials and in forms which were subsequently demonstrated to be entirely inadequate. Cast iron, lead, wood and brickwork, often in the form of flat surfaces, were prevalent. The tremendous potential hazards of such materials and designs were soon recognised and the paramount need for safety in both the construction and operation of boilers was reached through dramatic experiences.[67] The technology for obtaining energy by means of steam engines reached its zenith between the end of the eighteenth century and the First World War.

[67] See 'Boilers', *Encyclopaedia Britannica*, above note 1, vol. III, p. 793; S. Lollini, 'Macchine – Macchine a vapore', in *Dig. It.*, XV, I, 1927, p. 31.

However, in Italy the spread of this technology did not reach the same level as in other European countries. In 1910, only 29 per cent of installed engines used steam power, compared to 91 per cent in the United Kingdom. The great difference in these rates can be ascribed to the high price of coal. Italian mineral resources were poor and the users of steam power were obliged to buy most of the fuel abroad at very high prices. For this reason, many boilers in Italy continued to burn wood; some of them were adapted for the mixed use of various fuels like gas, coal and wood without distinction. Among the other reasons used to explain the backwardness of Italy in this area was the initial difficulty experienced by engineers in managing steam boilers. By 1890, there were approximately 1,000 steam boilers operating in Italy,[68] considerably fewer than in other European countries.[69] Some of them were fixed, used mostly in the industrial field; others were locomotives, used mostly in the agricultural field. The use of steam boilers in the workplace focuses our attention towards industrial injuries.

(b) Liability issues

As a scholar emphasised in an article about technological risks written in 1878, the introduction of steam engines offered great advantages, but it also triggered dangers and caused serious damage 'when energy turned to a useful purpose comes for any reason to depart from the direction which it ought to follow'.[70] In the opinion of this scholar, if the risk of injuries was limited to users, it was questionable whether the legislator should interfere with private relations. But where this danger spreads to the whole community, limitations on the freedom of industrial activity were justified by the aim to reconcile it with freedom of the individual. Moreover, apart from precautionary measures, the full protection of personal freedom required the right to compensation for the harm caused to persons or to things. Pointing out the need for control in this field, he underlined that 'in any case, the adoption of preventive or repressive measures by the

[68] C. Bardini, 'I fattori cumulativi nella diffusione di una tecnologia. Le caldaie a vapore in Italia tra la fine del XIX secolo e il 1911' *Rivista di storia economica* 1997, p. 41.

[69] In his comparative analysis, above note 67, at p. 32, S. Lollini emphasises that, in the period under consideration (1871–1880), there were 3,000 boilers in Belgium, 2,400 in Austria, 2,082 in France and 2,000 in England.

[70] M. Amar, 'Legislazione sugli stabilimenti insalubri, pericolosi e incomodi, sulle macchine e caldaie a vapore, regolamenti d'igiene, edilità e polizia locale' 21 *Archivio giuridico*, 1878, p. 145.

authorities would not prevent a private individual bringing an action for harm suffered'. As can be seen, his remarks were looking to the future. Indeed, at that time no general regulation on protective measures was in force nor had any claim been brought for harm resulting from the explosion of a steam boiler.

Focusing attention on the legislation in force after the unification of Italy, boilers are mentioned twice in the Civil Code of 1865. The first mention concerns their legal status. Article 413 c.c. categorises boilers as real property, an immovable 'by destination' if the land-owner has intended it to the service of his property. The second mention in article 574 c.c. is inspired by the risk connected to the use of steam boilers. The provision is contained in the second book ('On Goods, Ownership and its Modifications'), in the paragraph relating to the distances and the intermediary works required for any construction, excavation and plantation. The rule states that:

> A person who wants to build or locate any machine using a steam engine or any other machine where there is a risk of fire, explosion or the escape of noxious vapours near the property of another, must take all precautions and keep all distances which are required by the relevant regulations in order to prevent any harm to the neighbour. In absence of such regulations, a court has the power to determine the relevant matters.

Indeed, article 574 c.c. was a sort of 'programmatic' rule for a very long time until regulations were actually promulgated to prevent any harm to persons. To explain this regulatory inertia, scholars emphasised the aim of avoiding the risks of discouraging commercial private activity and diminishing industrial freedom.[71] But, by contrast, precautionary measures were contained in the Civil Code relating to merchant shipping and other special measures related to railways.[72] As a consequence, there was

[71] Lollini, above note 67, at p. 32.

[72] As regards railways, the relevant provisions were the articles of the Statute on Public Works (contained in the all. F of the Law of 20 March 1865) in the section regulating the licensing and use of rail traffic. Subsequently, regulations were approved by R.D. 31 October 1873: the first, entitled 'for the policing, safety and regularity of the railways', the second 'for the governmental control of the use of the railways'. Thereafter came two ministerial decrees of 4 April 1874: the first, with special rules for admission to the examinations and qualification as an engine driver or ship engineer; the second with instructions about the testing of steam boilers and the machines, fixed and mobile, assigned to the railway service, the examination and testing of mobile machines and the conditions for their use. Finally came the regulation of 3 January 1892 about steam boilers working in ports and at sea which adapted the regulation of 3 April 1890, no. 6793, to this particular kind of boiler.

no comprehensive regulation of the use of steam machines or boilers by the industry for a very long time.

However, at the beginning of the 1890s, other European countries (in particular, France[73] and Belgium)[74] introduced regulations which provided efficient safety measures. They decreed the obligation of a preventive test and approval as a condition for making the steam boiler operational: in particular, in France, regulations prescribed a test by government engineers and a declaration to the local governmental authority about the state and intended use of steam boilers before their sale and use; introduced rules about construction in order to prevent explosions and consequential damage; prescribed the requirement of professional qualifications for stokers, practical instructions to follow for maintenance and preventive measures to be adopted in order to avoid damage.[75] Moreover, in France, the owners of steam boilers formed an association, with its own statute, to cut costs and better control the selection of engineers for their machinery.

As was stressed,[76] the lack of specific general regulations, or bad application of local ones,[77] together with the poor state of the maintenance and construction of boilers and the lack of control by expert engineers, resulted in a great many explosions of steam boilers[78] and consequent fires, with injuries to people and damage to buildings. In particular, in the period between 1871 and 1880, there were six explosions among the

[73] See the general regulations laid down by the Law of 15 October 1910 and followed by several orders and decrees: cf. France, below.

[74] The general regulation laid down in the Law of 21 April 1810 was analogous to contemporary French legislation. It was followed by the Decree of 6 May 1824, regulating the use of steam boilers, and the Decree of 24 June 1839 concerning surveillance staff and technical aspects. Then all the regulations were codified by the Regulation of 4 August 1890.

[75] As far as the references contained in the other Reports, among the regulations introduced in other European countries, one can highlight that the Austrian legislation (the Order of 11 February 1854, the statute and the Imperial Decree, both of 7 July 1871) strengthened the safety measures in comparison with France, adding the obligation of a periodic annual inspection to the precautions about the manufacture, use and inspection of steam boilers. The inspection was made by a government commissioner or by officers designated by the association to which the owner of the boiler belonged and whose selection was approved by the administrative authority.

[76] Cf. G. Curcio, *Commentario della legge di pubblica sicurezza* (Turín: Unione Tip. Ed., 1891), p. 108.

[77] Before the unification of Italy in 1865, some rules had been adopted at a local level: Police Regulation of 20 June 1853 in Tuscany and an Austrian Order of 11 February 1854 in Lombardo-Veneto.

[78] A reference to the frequency of explosions was made by Cons. Stato, 14 June 1901, *Giur. It.* 1901, 318.

888 steam boilers in activity,[79] a very high number in comparison with other European countries, where there were fewer accidents for a larger number of boilers.[80] Subsequently, according to the statistics published by the government, explosions of steam boilers caused eight deaths and twenty-one injured in 1886, nine deaths and seventy-five injured in 1887, and seventeen deaths with thirty injured in 1889.[81] Because of these statistics, authorities began issuing circulars with the aim of raising the level of safety.[82]

Ten years later the wishes contained in the essay of 1878 were fulfilled. A general regulation on the precautions to take in the employment of the new technology was adopted, doubling the level of surveillance and supervision (by technicians and by local government authorities) found in the French regulations.[83] These rules were not limited to the precautions relating to neighbours, as was the rationale of article 574 c.c. 1865; they had a broader application. First of all, some prescriptions about steam boilers were introduced in the legislation on Public Order (Law of 23 December 1888, no. 5888 *decies*), according to which it was forbidden to bring into operation a boiler, new or restored, without a certificate declaring that the boiler satisfied certain requirements, relating to matters such as safety. Issue of the certificate required a compulsory visit and test, repeated at least every four years. These controls were prescribed by the local authority (*prefetto* or *sotto-prefetto*) and performed by an expert selected from engineers or mechanics (article 27).[84] Moreover, no boiler was allowed to work without the continuous assistance of a specialist, selected according to the rules contained in a regulation to be enacted in the future (article 28). The infringement of these prescriptions led to criminal sanctions (article 29).

[79] F. Gill, *Corso elementare delle macchine a vapore* (Palermo: Reber, 1897), p. 333; A. Lion, *Ingegneria legale* (Milan: Hoepli, 1899), p. 261. One can highlight that in Campania there was one explosion among only seventeen steam boilers in operation in this region.

[80] Only one case in each of the countries referred to above note 69.

[81] C.S. Astengo, *La nuova legge sulla pubblica sicurezza* (Rome: Cecchini, 1889), p. 292.

[82] Circular letter of the Ministry of Agriculture, Industry and Trade of 9 February 1882 addressed to the Chamber of Commerce and to all the business associations (organisation of surveillance on steam boilers); circular letter of the Ministry of Agriculture, Industry and Trade of 10 November 1884 (training courses for personnel); circular letter of the Ministry of Agriculture, Industry and Trade of 21 August 1884 (local authorities were requested to introduce adequate rules of protection in the regulations of local policy).

[83] Cf. above note 73.

[84] The articles are quoted in note 1 to the Regulation of 3 April 1890, published in the Appendix to G. Gautero, *Manuale del macchinista e fuochista* (Milan: Hoepli, 1893), p. 148.

As in article 574 c.c., the development of the law on public order entailed the adoption of a regulation (article 138). Within a few months the Ministry of the Interior adopted Regulation 3 April 1890, no. 6793, for the operation and supervision of steam boilers. Besides the specification of the regime of tests (articles 5–7) and inspections (articles 28–35), the regulation stated the methods for the construction of boilers (articles 8–27) and consecutive periodic inspections.

Following the French model, the Regulation established privileged treatment for the associations of steam boiler users if they fulfilled the prescribed requirements. On the one hand, only technicians of the associations were entitled to perform inspections and tests of their steam boilers (article 36). On the other hand, in order to obtain this preferential treatment, associations had to submit their statutes and regulations to the approval of the Ministry of Agriculture, Industry and Trade and they had to demonstrate that their engineers met the requirements of article 4 to qualify for the role of expert (article 37). The provisions of the statutes of these associations make clear that some of them intended to compensate the harm caused by explosions, as far as possible for their members.[85]

There were also rules about the requirements for the person in charge of the continuous support for the boiler (article 38) and provision regarding a list of steam boiler users to be made by the prefecture (articles 42–43). There were also rules imposing criminal penalties in case of infringement of the regulations (article 44).

Finally, in order to complete and clarify the law on steam boilers, Regulation 27 June 1897, no. 890, was enacted with specific prescriptions for activity and supervision by government authorities. Before this regulation, steam boilers owned by private users were not subject to any kind of control, and even then boilers on ships and trains were excluded from the application of this regulation by article 58. As regards compensation for damage, it must be stressed that this Regulation did not contain any statement about civil liability, neither reaffirming the application of general rules in addition to penal sanctions, as French regulations did,[86] nor introducing any special rule. In the absence of a rule of strict liability for dangerous activities, the reference standard was the general rules on civil liability. In particular, the provisions on fault liability were articles 1151 and 1152 (already examined earlier when

[85] *Ibid.* p. 159.

[86] French Imperial Decree of 25 January 1865 stated that, besides penal consequences for the infringement of preventive measures (introduced by the Law of 21 July 1856), the victim was allowed to claim damages.

considering the escape of sparks from railway engines), and article 1153, paragraph 4 c.c. on the vicarious liability of masters in case of harm caused by servants and clerks in the exercise of the commissions with which they were entrusted.

At the end of the period under observation, special provisions related to civil liability were introduced. The Law of 17 March 1898, no. 80, on industrial accidents, established compulsory insurance for industrial injuries that caused violent death or personal injury at work. This insurance, which was compulsory regardless of the number of employees,[87] exempted the employer from civil liability, except in cases of a criminal offence. In that case, the employer had to pay the difference between the indemnity paid by insurance and the damage suffered (article 22).

Before this special regulation came into force, article 442, paragraph 4 of the Commercial Code of 1882 made harm caused by 'explosions or other similar accidents' (including explosions caused by steam)[88] equivalent to damage caused by fire as regards regulations on insurance. In addition, contractual liability regulated by article 1225 *et seq.* c.c. 1865 could apply to damage caused by manufacturing defects of the steam boiler.

(c) *Case law*

Although accidents did occur, as the official reports show, there are few reported judicial decisions during the period under consideration. In particular, any such cases were decided in the first half of this period. For instance, the author of the above-mentioned article of 1878 did not quote any decision; he merely speculated about a hypothetical claim for damages for the explosion of a steam boiler. In order to judge such a case, the author suggested following the approach of the case law about harm caused by the escape of sparks from railway engines. In particular, he quoted the contemporary decision of the Court of Appeal of Florence of 12 March 1877,[89] which confirmed the civil liability of the railways 'even if they have not infringed the prescriptions given by the administration in the public interest'. At the time of that article, the case law was in its infancy. Probably the suggestion was an interpretation of the provision on fault liability as a rule of strict liability in order to guarantee compensation for damage connected to the risks of technological development.

[87] Cf. Cass., 6 July 1900, *Giur. It.* 1900, II, 312.
[88] Cf. M. Majnoni, *Incendio (assicurazione contro i danni dell')*, in *Dig. It.*, XIII, I, 1927, p. 383.
[89] See above note 49.

Before 1892, the few references to legislation on steam boilers were collected under the general entry 'public safety' in the Massimario. Reports at that time mention 'Steam engines and boilers' as a specific entry. The case law cited is mainly from decisions dealing with operational effectiveness of safety measures. However, case law on civil liability is restricted to a couple of judgments.[90]

Although the first decision available is subsequent to the regulations on steam boilers mentioned above, it makes no reference to those rules, since it refers to an accident that had happened a long time before. One must emphasise that the various instances of this process referred to by the Court of Cassation of Turin on 27 March 1889[91] seem to reflect the same development in the approach to the fault liability rule that one can also trace in the liability of railway companies for the escape of sparks. In that case, the explosion of a boiler caused the death of three workers. The claim of the heirs to recover damages was brought against the company which constructed the boiler. The action was based on both contractual and fault liability. The demands of the plaintiffs were accepted by the court of first instance and by the Court of Appeal on both grounds in spite of the lack of proof of fault.

In fault, the lack of proof was resolved by copying the approach of courts to finding railway companies liable for damage caused by sparks, i.e. recognising liability on the basis of *culpa in re ipsa*. The defences of mistake and ignorance of the defects in construction of the steam boiler put forward by the defendant were dismissed by the court, affirming liability even if science was not able to explain the defects.

On the contractual issue, the court referred to article 1225 c.c., according to which proof that the breach of contract or the poor performance of the contractual obligations was attributable to an unexplained cause lies with the defendant. He failed to give this demonstration. In fact, in this case the reversal of the burden of proof did not constitute favouritism by the court (as was noted regarding the escape of sparks) but was a strict application of the rules on contractual responsibility. This positive approach to liability was an exception. The court justified the imposition of liability for the cause of the damage, even if the harm resulted from an unpredictable and inevitable cause. In fact, the court stated that the

[90] Cass., 9 July 1996, no. 6235, *Giur. It.* 1996, 1365, recently dealt with a claim for damages caused by the explosion of a steam boiler. A footnote emphasises that there are no precedents on the same issue.

[91] Cass. Torino, 27 March 1889, *Giur. It.* 1889, 348.

constructor was liable to compensate all the damage caused by the defect in construction, not only when the causes were negligence, inaccuracy, ignorance or miscalculation, but also when science was unable to identify the defects. The justification was that:

> liability is the only and best way to regulate human life. Imposition [of liability] is not possible only when an act of God has caused the damage; but it cannot be found when a positive or negative human act has contributed to the cause of the damage. Besides, the mistake does not represent an excuse for liability, since it implies negligence, lack of skill and ignorance.

This was practically strict liability, since it applied even for unforeseeable and unavoidable damage.

However, in overruling the lower courts, the Court of Cassation excluded contractual liability, since the claimant heirs were not parties to the contract. As regards fault liability, the court also held that this depended on proof of negligence according to article 1151 c.c. 1865. Nevertheless, in a case like that in point, where the court was dealing with an inescapable error or an event that could not be avoided, even with the best human diligence and prevention, the defendant was to be considered exempt from liability. In addition, it denied that the contribution of a human act excluded an act of God. In the opinion of the court, it was not true that the mistake regarding the defect always implied negligence. When an error was inescapable or when it was connected to the level of scientific knowledge at the time, these errors should not be equated to fault.

The only decision available after the legislation came into force was given by the Court of Cassation, Penal Section, on 21 April 1897. It concerned the explosion of a steam boiler which caused the death of five people and injury to another thirteen.[92] The machine operator was held liable due to his failure to comply with his duty as a 'selected expert' under the regulation of 1890 (article 28 of the Law on Public Order). His liability was based on fault. The operator breached his obligation to supervise the boiler in order to check its condition. The continuous presence required by the regulation means that this person must supervise the machine in order to avoid disasters. The supervision of the machine was delegated by the operator to a subordinate who did not have the requisite knowledge and did not notice that the internal part of the boiler was corroded.[93]

[92] Cass. Pen., 21 April 1897, *Giur. It.* 1897, II, 304 and *Cass. Unica*, 1897, 851.

[93] On the application of art. 28 of the legislation on public order see Cass. Roma, 27 January 1900, *Foro it.* 1900, II, 138; App. Genova, 22 December 1899, *Cass. Unica* 1900, 634.

(d) *Concluding remarks*

The research on the topic of liability had two strands.

First, the various legislative regulations prescribed, in minute detail, the precautions and safety measures for the use of steam boilers to strengthen the preventive action against the technological risks.

Secondly, under the Civil Code of 1865, compensation for damage caused by industrial injuries was limited by the difficulty of proving the fault of the employer under article 1151 c.c. Statistics show that the employer was found liable in only 20 per cent of claims for damages.[94] On the one hand, the difficulty of meeting the burden of proof explains the temptation for the 'manipulation' of the fault liability rule by introducing a presumption, also observed in relation to sparks from railway engines.[95] On the other hand, a strategy developed by the academics and applied by case law (found in the first decision mentioned above)[96] was to frame the issue as one of contract, the *locatio operarum*, in order to categorise the accident as a breach of contract – in particular a breach of the obligations of supervision, care and protection towards the employee. In the contractual field, the claimant was excused from shouldering the burden of proof of the fault of the employer. The last step was the introduction of compulsory insurance for industrial injuries, which guaranteed the indemnification of the employee, but who is entitled to recover the full delictual measure of compensation only in cases of crime.[97]

4. Asbestos-related injuries and illnesses in the late twentieth century

(a) *Factual background*

Because of its strength, resistance and durability ('asbestos' in Greek means inextinguishable, indestructible), Italy began exploiting this mineral between 1830 and 1840, with the purpose of protecting things and persons from fire and heat. Its special resistance to heat and chemical agents,

[94] Cf. B. Valori and G. Sossi, *Gli infortuni sul lavoro* (Florence: Goemare, 1911), p. 29.

[95] I. Piva, 'Problemi giuridici e politici della responsabilità alle origini dell'assicurazione infortuni sul lavoro in Italia' *Riv. Inf. Mal. Prof.* 1980, 665.

[96] Cass. Torino, 27 March 1889, *Giur. It.* 1889, 348.

[97] Cf. R. Vivante, 'Responsabilità dei padroni che non presero i provvedimenti necessari per prevenire gli infortuni sul lavoro' *Monit. Trib.* 1891, 441; L. Barassi, 'Contributo alla teoria della responsabilità per fatto non proprio' *Rivistà italiana per le scienze giuridiche* 1897, 191.

its tensile strength, as well as its subsequently discovered ability to be an acoustic insulator, led to an exponential increase of the use of asbestos in the iron and steel industries in the first half of the twentieth century. It was used for several purposes: mainly, for insulating boilers and steam generators, in manufacturing protective clothes and friction articles (brakes and clutches) for transport, and for building materials resistant to compression and to traction (cordage, packing, flooring with 'linoleum'). In particular, in the construction sector, one must stress that an Italian firm, Eternit, created an extremely innovative cement-like material ('Eternit') at the beginning of the twentieth century, to which all the special technological skills of asbestos were applied. It had such an enormous commercial impact that Eternit accounted for over 70 per cent of the asbestos used in industry in Italy. As a consequence, higher percentages of incidence of pathologies connected to asbestos affect the employees in such fields of production.[98]

As well as the various possible fields of employment, it is to be stressed that the rich mines in the Alps made possible extensive use of asbestos in Italy.

In fact, asbestos was first mentioned in protective legislation in the Royal Decree of 14 June 1909, no. 442, concerning unsafe work for women and children.[99] In particular, Table B of article 29 attached to the Royal Decree, drawing up a list of the unsafe and dangerous work in which employment of children and women under age was prohibited or admitted with special measures, at no. 12 forbade them from 'spinning and weaving asbestos ... in places where the speedy extraction of the dust cannot be secured'. Moreover, at no. 21, it mentions the 'asbestos mills', excluding women and children from places where dust can be generated easily unless it can be contained effectively.

In spite of the longstanding awareness of the negative effects connected to exposure to asbestos, the exploitation of this material carried on until the Law of 27 March 1992, no. 257 forbade its use. This statute also introduced measures to help workers who had been exposed to asbestos, such

[98] Cf. Various authors, *Eternit: inquinamento e potere* (Zurich: Veritas, 1983); G. Burragato and A. Morone, 'Amianto e rapporti di lavoro. Profili attinenti alla responsabilità civile del datore di lavoro' in L. Spagnuolo Vigorita (ed.), *Rischio amianto* (Milano: Giuffrè, 1997), p. 79.

[99] In the same period, the Tribunal of Turin gave judgment on a civil action for compensation for libel brought by the manager of a branch of British Asbestos Company Ltd against a newspaper which had denounced the 'unbelievable number of victims' caused by the asbestos industries: Trib. Torino, 22 October 1906, *British Asbestos Company Ltd* v. *Pich e Mariani*, quoted by R. Guariniello, 'I tumori professionali nella giurisprudenza penale' *Foro it.* 1999, II, 236.

as early retirement, revaluation of pension contributions, and creation of a Register. It also ordered the destruction of all goods manufactured using this material.

In fact, this statute has been only partially effective. If the trade in asbestos has completely stopped since 1994, its removal has been proceeding very slowly. According to research by the National Centre of Research (CNR), there are still 8 million tons of asbestos and 2.5 billion square metres of covers made from Eternit in Italy.[100] New cases of asbestosis (the occupational illness caused by exposure to strong doses of asbestos fibres) are uncommon today; the incidence of occupational exposure is exclusively tied to demolition and the removal of substantial quantities of asbestos. However, there are still high risks of developing mesothelioma and other forms of cancer for people living in areas contaminated by asbestos, in particular for people living in houses built with materials containing asbestos or located near refuse areas. This is especially the case for the members of the family of ex-workers who were exposed to asbestos fibres.[101]

Moreover, the National Register of Mesothelioma was only created in 2004. At the time of writing, it has certified 3,670 cases of death, but this is only a partial list, since it does not include the data of some regions, nor deaths that did not occur in hospital.[102] At a statistical level, it is predicted that between 4,000 and 5,000 persons will die every year. The period of latency of the illness is very long: between ten and fifteen years for asbestosis and between twenty and fifty years for pulmonary carcinoma and pleural mesothelioma. So the peak rate of mortality is expected between 2015 and 2025, and, in the opinion of some experts, will last until 2040.[103]

[100] See www.ania.it/rc_generale/attivita/03_03%20introduzione%20AMIANTO.pdf.

[101] For instance, in case law, proving the chain of causation has also been possible in case of no direct exposure to asbestos. In one case, a woman had developed mesothelioma as a consequence of washing the clothes of her husband, who was directly exposed to asbestos in his work: cf. Cass. Pen., 15 May 2003, no. 27975, Cass. Pen. 2005, 429, commentary by E. Di Salvo, 'Tumori da amianto e nesso di causalità'; Cass. Pen., 6 February 2001, no. 5037, *Diritto e Lavoro-Rivista critica del diritto del lavoro* 2001, 519, commentary by A. Culotta, 'Ancora sul nesso di causalità nei processi per malattie da lavoro'; Cass. Pen., 30 March 2000, *Foro it.* 2001, II, 278, commentary by R. Guariniello, 'Dai tumori professionali ai tumori extraprofessionali da amianto' *Rivistà critica del diritto di lavoro* 2001, 519 and *Diritto e pratica del lavoro* 2001, 115.

[102] See the parliamentary report on the Bill on 'Disposizioni a favore dei lavoratori e dei cittadini esposti ed ex esposti all'amianto e dei loro familiari, nonché delega al Governo per l'adozione del testo unico in materia di esposizione all'amianto' of 28 April 2006.

[103] Cf. R. Pisano, 'L'amianto, il suo utilizzo e gli effetti sull'uomo' in A. Di Amato (ed.), *La responsabilità penale da amianto* (Milano: Giuffrè, 2003), p. 7.

(b) *Liability issues*

As a consequence of the relevance of the facts to criminal liability, claims for compensation often form part of the criminal judgment on a charge against the employer for manslaughter (article 589 of the Codice penale (c.p.)) or for personal injuries (article 590 c.p.). Both provisions consider the infringement of the safety rules on industrial accidents to be an aggravating circumstance.[104]

Victims usually claim compensation through an action in contract under articles 2087 c.c. and 1218 c.c., as well as through the general action for direct liability based on article 2043 c.c. As regards the burden of proof, in practice there are no differences between these alternatives.

As a consequence, any judgment on the existence of liability can be broken down into four logically separate issues:

(1) the chain of causation between disease and exposure at work (i.e. to be an occupational disease, material causation);
(2) the infringement of safety or precautionary measures required by law or on the basis of the general duty of care by the defendant (fault); this shifts the burden of proof and the defendant must demonstrate that it adopted all the safety measures required to avoid damage;
(3) the chain of causation between the omission and the disease: the damage would not have occurred (or would have worsened) if the defendant had complied with his own duties (legal causation);[105]

[104] As a matter of fact, criminal cases are common while private law employment cases are very few. Among the labour decisions, cf. Cass., 9 May 1998, no. 4721, *Giust. Civ.* 1999, I, 539; Cass., 23 May 2003, no. 8204, *Arch. Civ.*, 2003, 1281; Cass., 12 May 2004, no. 9057, *Rep. Foro it.*, 2004, 'Infortuni sul lavoro [3560]', n. 107; Cass., 14 January 2005, no. 644, *Giur. It.*, 2005, 1391, commentary by N. Coggiola, 'Il risarcimento dei danni da esposizione ad amianto: dall'utilizzo del concetto dell'aumento del rischio all'inversione dell'onere della prova sul nesso di causalità'; Cass., 6 November 2006, no. 23642, 'Massimario' *Foro it.*, 2006, 1920; Cass., 7 November 2006, no. 23719, *Resp. Civ. Prev.*, 2007, 648, commentary by N. Coggiola, 'Esposizione ad amianto e mancato risarcimento dell'ansia di contrarre una patologia'. As regards lower courts, see Trib. Barcellona Pozzo di Gotto, 15 April 2004, no. 501, 11 November 2004, no. 1557, 11 November 2004, no. 1558, *Giur. It.* 2005, 1168, commentary by N. Coggiola, 'Il giudice e la statistica: attività lavorative, esposizione all'amianto ed asbestosi (ovvero quando il numero di morti e malati fa la prova nel processo'; Trib. Trieste, 25 March 2004, *Giur. It.*, 2005, 497, commentary by N. Coggiola, 'L'esposizione alle polveri di amianto ed il nesso di causalità di fronte al giudice civile'.

[105] P. Trimarchi, *Causalità e danno* (Milan: Giuffrè, 1967); F. Realmonte, *Il problema del rapporto di causalità nel risarcimento del danno* (Milan: Giuffrè, 1967); P. Forchielli, *Il rapporto di causalità nell'illecito civile* (Padova: Cedam, 1960); D. Poletti, 'Le regole di (de)limitazione del danno' in N. Lipari and P. Rescigno (eds.), *Trattato di diritto civile,*

(4) the damage suffered by the victim (damage).

As regards compensation, by an action in contract the claimant is entitled to seek compensation only for damage which was foreseeable when the obligation arose (article 1225 c.c.). Both actions give compensation for non-pecuniary losses, as the recognition of a crime is no longer a requirement in case of infringement of a right or value protected at a constitutional level.[106]

While proof of the occupational nature of the disease is sufficient to obtain indemnity from the National Institute of Insurance for Occupational Accidents and Disease (INAIL), civil liability of the employer requires a full case to be made.[107] Each one of the prerequisites will be considered, examining the approach of case law and relevant legislation. As there are a substantial number of regulations related to risks connected to exposure to asbestos, reference will only be made to provisions which are connected to civil liability, e.g. to the rules that deal with the link between asbestos-related deaths and the working environment; those that define the relationship between civil liability and social

vol. IV, tomo III, *Attuazione e tutela dei diritti, La responsabilità e il danno* (Milan: Giuffrè, 2009), p. 291; R. Pucella, *La causalità incerta* (Torino: Giappichelli, 2007); R. Pucella and G. De Santis, *Il nesso di causalità. Profili giuridici e scientifici* (Padova: Cedam, 2007); A. Belvedere, 'Causalità giuridica' *Riv. dir. civ.* 2006, 7. On material causation, see in particular M. Capecchi, 'Il nesso di causalità materiale e il concorso di cause' in G. Visintini (ed.), *Risarcimento del danno contrattuale ed extracontrattuale* (Milan: Giuffrè, 1999), p. 301 and more extensively, M. Capecchi, *Il nesso di causalità da elemento della fattispecie 'fatto illecito' a criterio di limitazione del risarcimento del danno* (Padova: Cedam, 2005). On causation in industrial accidents, see F. Pontrandolfi , 'Infortuni sul lavoro e malattie professionali tra occasionalita e causalità spunti ricostruttivi del danno da lavoro' in F. Facello (ed.), *Il sistema di tutela degli infortuni e delle malattie professionali* (Milan: Giuffrè, 2005), p. 163 .

106 SU, 11 November 2008, no. 26972, 26973, 26974, 26975, *Danno e resp.* 2009, 19. Cf. the commentaries by F. D. Busnelli, 'Le Sezioni Unite e il danno non patrimoniale' *Riv. dir. civ.* 2009, 97 and E. Navarretta, 'Il valore della persona nei diritti inviolabili e la sostanza dei danni non patrimoniali' *Foro it.* 2009, I, 143. Cass., 31 March 2003, no. 8827 and no. 8828, in *Foro it.* 2003, I, 2272, note E. Navarretta, 'Danni non patrimoniali: il dogma infranto e il nuovo diritto vivente' *Danno e resp.* 2003, 816, commentary by F.D. Busnelli, 'Chiaroscuri d'estate. La Corte di Cassazione e il danno alla persona'; G. Ponzanelli, 'Ricomposizione dell'universo non patrimoniale: le scelte della Corte di Cassazione'; A. Procida Mirabelli Di Lauro, 'L'art. 2059 va in paradiso'; Corte cost., 11 July 2003, no. 233, *Rass. dir. civ.* 2003, 775, commentary by P. Perlingieri, 'L'art. 2059 c.c. uno e bino: una interpretazione che non convince' and in *Giur. It.* 2003, c. 1777, note P. Cendon-P. Ziviz, 'Vincitori e vinti (... dopo la sentenza 233/2003 della Corte costituzionale)'.

107 For this approach and this outline, cf. G. Burragato and A. Morone, 'Amianto e rapporto di lavoro. Profili attinenti alla responsabilità civile del datore di lavoro' in Various authors, *Rischio amianto* (Milan: Giuffrè, 1997), p. 82.

security (compulsory insurance); and those dealing with preventive measures whose infringement could give rise to fault and liability on the part of the employer.[108] Although these are inter-connected, clarity of exposition leads us to deal with them separately and in a different sequence.

(c) Material causation

In case of the occupational nature of the illness, the chain of material causation between work and the disease has to be proved. Article 3, first paragraph, of Presidential Decree of 30 June 1965, no. 1124 (Occupational Accidents and Diseases Act, the so-called 'TU INAIL') limited the recognition of the occupational origin and the social indemnity to the diseases listed in a table and developed in the practice and as a consequence of work. A decision of the Constitutional Court deleted this limit, extending protection to all illnesses for which occupational origin is proven.[109] As a consequence, the system was turned from a 'pure' one (with occupational origin as prima facie evidence only for the diseases contained in the list) to a 'mixed' one (the worker is permitted to prove the chain of causation for diseases not contained in the list).[110]

The same situation is maintained by the update made by Decreto legislativo (D. Lgs.) of 23 February 2000, no. 38 (provisions on insurance against industrial accidents and industrial diseases).[111] It must be stressed that article 10 entrusts a scientific committee with the task of developing and updating the table of industrial diseases. The list must also contain illnesses that are probably or possibly caused by exposure at work. Performing this task, the Ministerial Decree of 27 April 2004 separates the diseases into three lists, according to the level of probability of occupational origin. Mesothelioma, asbestosis and lung cancer are inserted in List I, i.e. it is highly probable that they developed as a consequence of work.

In order to obtain a social insurance indemnity, the causal connexion between the illness and industrial activity is presumed for the illnesses included in the statutory list. In case of a disease not included in the list,

[108] More references in R. Pisano, 'L'evoluzione storica della normativa italiana e comunitaria in tema di amianto' in Di Amato, above note 103, at p. 51.

[109] Corte cost., 18 February 1988, no. 179, *Or. Giur. Lav.* 1988, 591.

[110] E. Cataldi, *Malattie professionali (assicurazione obbligatoria controle)* (Rome: Istituto dell' Enciclopedia 1990), vol. XIX, p. 2 *et seq.*

[111] This Regulation is also well known abroad for containing the first legal definition of '*danno biologico*'.

the right of the worker to obtain indemnity depends on his proving a causal link with his employment. Although these criteria are established for social insurance and such presumptions are not considered analogous to civil liability claims (the worker must prove the occupational origin also for the disease contained in the list), this distinction tends to evaporate in practice, as judges use it to apply uniform criteria; that is to say, they refer to social insurance criteria to assess the occupational nature of the disease in civil liability claims.[112]

As a consequence of the approach of case law to the chain of causation between work and disease (to refer to social insurance rules) when the illness is contained in the table of occupational diseases or is considered highly probably connected to work, the court stretches the rules and presumes the causal connection. The employer or other defendant then has to prove either that a different intervening factor has caused or has contributed to causing the disease[113] or that the work could not have caused the disease, either in theory or on the facts of the case.[114]

In the second place, the employer may claim that the disease does not result from employment. When the illness is not contained in the established list or there is only a small probability or possibility of the connection with work, the chain of causation is not presumed, and the claimant has to prove its occupational origin. In order to assess such evidence, case law refers to criminal law theories and, more precisely, to the *conditio sine qua non* or 'but for' test, according to which if a certain fact is proved (working activity with exposure to asbestos), it is possible to state that it caused a certain effect (the illness) if such a link is confirmed by scientific evidence. In this field, courts make use of statistical probabilities which cannot give any certainty of actual causation.[115] In particular, they start from experience data (the level of concentration of

[112] Burragato and Morone, above note 107, at p. 84.

[113] For the incidence of tobacco addiction in the assessment of factual causation, see Cass., 12 May 2004, no. 9057, above note 104; Cass. Pen., 9 May 2003, *Foro it.* 2004, 69, commentary by R. Guariniello, 'Tumori professionali da legno e amianto'; Cass. Pen., 11 July 2002, no. 953, *Foro it.* 2003, II, 324, commentary by R. Guariniello, 'Tumori professionali da amianto e responsabilità penale'; Cass. Pen., 2 July 1999, no. 12333, *Foro it.* 2000, c. 260, commentary by R. Guariniello, 'Malattie professionali, tumori da amianto e asbestosi'; *Dir. prat. lav.* 2000, 706; Cass. Pen., 11 May 1998, *Foro it.* 1999, II, 236, commentary by R. Guariniello, 'I tumori professionali nella giurisprudenza'; Pret. Bergamo, 3 April 1997, *Foro it.* 1998, 484, commentary by L. Tramontano, 'Ancora in tema di morte per intossicazione da amianto: brevi rilievi problematici sulla "ri-descrizione dell'evento" nella verifica del nesso causale'.

[114] See the decisions quoted in *ibid.* p. 86.

[115] *Ibid.* 87.

asbestos in the work environment) in order to evaluate the level of risk of developing the disease at work, and compare this with the risk created by the surrounding environment.[116] As a consequence of this approach, the occupational origin is always recognised if, according to the kind of work, the exposure of the employee to asbestos is prolonged and substantial.[117] However, the occupational origin is not often recognised if the exposure of the employee to asbestos is not higher than elsewhere.[118] The identification of the chain of material causation becomes complicated outside the area of industrial diseases. An interesting decision of the Court of Cassation concerned the case of a wife of a worker who had been exposed to asbestos. She died as a consequence of mesothelioma, contracted by washing her husband's contaminated clothes. The doubts about the chain of causation and, most of all, about holding the defendant responsible for it arose from the fact that the victim had not been directly exposed to asbestos at work.[119] The reasoning of the court was that, even if the wife's illness did not result with absolute certainty, there was a high level of probability, so it was reasonable to presume that the illness was caused by contact with her husband's clothes. As regards the fault of the employer, the court stressed the poor conditions in which the husband had been working as no measures were adopted in order to try to avoid or limit the emissions of dust, nor were there minimum precautions. This assessment makes it unnecessary to look into foreseeability of the event, and the level of care which was concretely expected from the employer in consideration of the level reached by scientific knowledge and techniques in the field at that time.[120]

(d) Relations between civil liability and social insurance

If legislation has taken into consideration the dangerous effects of exposure to asbestos since the beginning of the twentieth century, the social protection related to compulsory insurance was brought in later and was, for a long time, incomplete.

[116] Cf. Pret. Pordenone, 7 July 1992, *Foro it.* 1992, II, 720; Pret. Torino, 5 May 1995, *Diritto e Lavoro-Rivista critica del Diritto di Lavoro* 1996, 1059.

[117] Pret. Torino, 5 May 1995, above note 116.

[118] Pret. Pordenone, 7 July 1992, above note 116. This conclusion is reached by Burragato and Morone, above note 107, at p. 90.

[119] See above note 103.

[120] Cass. Pen., 6 February 2001, above note 101.

Although chronic pulmonary disease, caused by asbestos, has been known since the beginning of the twentieth century, asbestosis was only accepted as an industrial disease, together with silicosis, by the Law of 12 April 1943, no. 455. The same statute extended compulsory insurance to both diseases if they were 'developed in the practice and as a consequence of work'. This provision was enacted in order to prevent claims for compensation for damage caused by silicosis and asbestosis, since the legislator assumed that such injury would be difficult to investigate given the adoption of precautionary measures by employers.[121] In particular, article 12 of the 1943 Law stated that, in addition to workers' indemnity from insurance, 'civil liability of the employer arises only when silicosis and asbestosis have been caused or increased as a consequence of the infringement of the preventive and security measures' to be introduced by special regulations under article 11 of the same statute. The Law of 1943 had retroactive application to the diseases revealed before it came into force (article 14), and it 'swept away', as was its aim, all the claims brought before its entry into force.[122] In order to reach this goal, the approach followed by the Court of Cassation was to deny the civil liability of the employer for claims brought after 1943, even before the norms of prevention and safety had been enacted.[123]

Subsequently, the provisions of the 1943 Law and subsequent amendments were consolidated in the Presidential Decree of 30 June 1965, no. 1124, which codified the Occupational Accidents and Diseases Act. It contained the Consolidated Act on provisions for compulsory insurance against industrial accidents and industrial diseases, which is still in force today, albeit with further amendments.

As regards the civil liability of the employer, the first three paragraphs of article 10 state that compulsory insurance exonerates the employer for industrial accidents. According to the same rule, civil liability arises in case of indictable offences and also when the criminal proceedings identify the liability of people entrusted by the employer to direct or oversee the work, and where the employer is to be considered liable for the employee's actions or omissions according to the Civil Code. On the one hand, every

[121] Cf. G. Alibrandi, 'La responsabilità civile del datore di lavoro in tema di silicosi ed asbestosi' *Riv. Dir. Lav.* 1950, 100.

[122] On this aim, cf. G. Gentile, 'Relazione' in E.C. Vigliani, *Atti del convegno sulla silicosi, Turin 22-23 February 1941* (Rome: Edizione dell' Ente nazionale di propaganda per la prevenzione degli infortuni, 1941), p. 168.

[123] Cass. 30 May 1949, no. 1385, *Riv. Dir. Lav.* 97, commentary by G. Alibrandi, 'La responsabilità civile del datore di lavoro in tema di silicosi ed asbestosi'.

industrial accident caused by the fault of the employer or his collaborators constitutes an indictable offence. On the other hand, the extent of liability has been increased by the decisions of the Constitutional Court for acts of employers different from the ones listed;[124] and, in the case of time-barred offences, the civil judge is allowed to progress the investigation of the facts constituting the crime autonomously,[125] and the same applies to other expired criminal proceedings.[126]

Although asbestosis has been considered an industrial disease since 1943, mesothelioma and lung cancer were only very recently covered by insurance. At first, according to the list of diseases approved by the Decreto Ministeriale (D.M.) of 18 April 1973, an indemnity for mesothelioma was allowed if it was associated with asbestosis. Finally, entry 56 of the D.P.R. of 13 April 1994, no. 336, which updated the table of diseases, expressly considers mesothelioma and lung cancer as industrial diseases.

(e) Preventive measures and fault of the employer

As regards the protection of workers, the general and wide-ranging duty of occupational safety in article 2087 c.c. ('in the development of his activity, the employer must take all the measures which, according to the kind of work, the experience and the technique, are necessary in order to protect the physical and moral integrity of employees') was specified by preventive and security measures when D. Lgs. 277/91, implementing EC Directive 477/1983, came into force.[127] Section III (articles 22–37) of D. Lgs. 277/91 regarding asbestos introduced provisions about demolition and removal of asbestos, and required employers to adopt suitable measures for the protection and decontamination of the work environment and to give information to employees about the risks connected to pathogenic agents.

[124] Corte Cost., 9 March 1967, no. 22, *Giur. Cost.* 1967, 164. [125] *Ibid.*

[126] Corte Cost., 29 April 1981, no. 102, *Giur. Cost.* 1981, 864; Corte Cost., 24 April 1986, *Giur. Cost.* 1986, 680. For more references, cf. Burragato and Morone, above note 107, at p. 77; A. Rossi, 'La responsabilità civile del datore di lavoro per gli infortuni sul lavoro e le malattie professionali: il sistema di parziale esonero' in F. Facello (ed.), *Il sistema di tutela degli infortuni e delle malattie professionali* (Milan: Giuffrè, 2005), p. 237.

[127] On these provisions, cf. A. Garlatti, 'Colpa e progresso tecnologico: il contenuto dell'obbligazione di sicurezza del datore di lavoro nell'ipotesi di tecnopatie conseguenti a esposizione ad amianto' *Diritto e Lavoro-Rivista Critica del Diritto del Lavoro* 2004, p. 756; A. Di Amato, 'Profili generali della responsabilità penale in materia di malattie professionali da amianto', in A. Di Amato (ed.), *La responsabilità penale da amianto* (Milan: Giuffrè, 2003), p. 75.

In particular, the implementation of occupational exposure limit values adopted by the European Union[128] was significant, as well as the obligation to adopt measures where they were exceeded despite general precautions.[129] Before 1991, case law had judged the behaviour of employers on the basis of the general rule in article 2087 c.c. and some rules on this question contained in D.P.R. 19 March 1956, no. 303 regarding the prevention of risks created by dust inhalation and the removal of dangerous materials.[130] The breadth of such provisions enabled courts to apply them to activities and risks not expressly listed.[131] After the special legislation came into force, the general rule on the duty of occupational safety maintained its importance. As more efficient instruments of protection are discovered, the general obligation in article 2087 c.c. compels employers to update their measures of protection, even though they are not yet imposed by law.[132]

The proof and assessment of fault in this field are not easy, for several reasons. First of all, due to the long period of latency, it is necessary to refer to the legal framework at the time of exposure. Thus, before D. Lgs. 277/91 came into force, apart from the indications contained in D.P.R. 19 March 1956, no. 303 for dust inhalation, the safety measures required of the employer were to be drawn from the general rule in article 2087 c.c. ('according to the type of work, the experience and the technique').[133]

When the measures are prescribed by legislation, it is quite easy to demonstrate whether they have been complied with or breached. In case of complaint of the violation of protective instruments not imposed by any law (or due 'according to the type of work, the experience and the technique', apart from the ones imposed by legislation), the burden of proof of their existence falls on the claimant.[134] On the other hand,

[128] Before the D. Lgs. 277/91 came into force, standards were dictated by collective labour agreements: cf. Burragato and Morone, above note 105, at pp. 105 and 117.

[129] Section III was repealed by D. Lgs. 257/2006 which inserts Title VI-bis 'Protection of Workers from the Risks Connected to Exposure to Asbestos' in the context of D. Lgs. 19 September 1994, no. 626, which implements the European Directives on the improvement of the safety and health of workers.

[130] Article 4 on the adoption of precautionary measures, duty of control, information and education; arts. 18, 19 and 21.

[131] The only exception is Cass. Pen., 11 April 2005, no. 7362, *Lav. Giuri.* 2006, 415. It excluded the application of D.P.R. 19 March 1956, no. 303, as asbestos is a fibre, not dust.

[132] In 1998, the Court of Cassation stated that to be exempted from liability, the employer has to respect not only the specific provisions introduced by legislation, but also the ones inferable by art. 2087 c.c.: Cass., 9 May 1998, no. 4721, *Or. Giur. Lav.* 1998, I, 651; *Giust. Civ.* 1999, I, 539.

[133] Burragato and Morone, above note 105, at pp. 96, 118.

[134] For more references on this issue, cf. Di Amato, above note 127, at p. 173.

judges may unconsciously apply the standard of the knowledge and skills typical at the time of the judgment, not at the time of the event.

The approach of case law to article 2087 c.c. follows the principle of the 'maximum level of safety possible according to technology'. According to this principle, employers have to take into consideration the most up-to-date knowledge and adopt the best technical instruments at the time, no matter how onerous that may be.[135]

As regards D. Lgs. 277/91, case law dealt with the problem of the role played by the occupational exposure limit values; in particular, whether adoption of the safety measures when these values were exceeded was sufficient to exonerate the defendant from liability or whether, on the contrary, the employer could be held liable if the claimant succeeded in demonstrating the availability of other measures suitable for reducing the risks. As a consequence, the demonstration of compliance with published standards and of measures to be applied where these were exceeded led not to an exemption from liability, but only a rebuttable presumption.[136] The claimant could win if he showed the existence of further precautions, inferable from experience and from the techniques at the time, suitable for reducing the level of the risk.[137]

In evaluating fault, it is moot whether the effects of the omission were foreseeable or not. The most common defence of employers is that at the time of the damage they were not sufficiently aware of the risks connected to exposure to asbestos, even at a low level,[138] or of the possibility of eradicating its negative consequences, as this required a certain level of knowledge and technique.[139] If the damage was unforeseeable and unavoidable, the defendant could be considered as being free of fault.

In reply to this objection, the Court of Cassation asserted that the foreseeability of harm in cases of diseases connected to asbestos related 'not

[135] R. Guariniello, 'Tre anni di applicazione del d.leg. 15 agosto 1991 n. 277 sui rischi lavorativi da piombo, amianto, rumore' *Foro it.*, 1994, II, 547.

[136] App. Milano, 29 June 2001, *Diritto e Lavoro-Rivista Critica del Diritto del Lavoro* 2001, 781. On this issue cf. R. Guariniello, 'Il principio della massima sicurezza tecnologicamente fattibile' *Foro it.* 1997, 339. For a different opinion, see Pret. Milano, 20 December 1999, *Diritto e Lavoro-Rivista Critica del Diritto del Lavoro* 2000, p. 485, commentaries by A. Culotta, 'Sul rapporto di causalità nei processi per malattie da lavoro' and S. Cantoni-P. Ricci, 'Scienza e diritto in un caso di tumore professionale'.

[137] Cass., sez. IV, 2 July 1999, *Foro it.* 2000, II, 259, commentary by R. Guariniello, 'Malattie professionali, tumori da amianto, asbestosi'.

[138] Cf. Pret. Torino, 9 February 1995, *Riv. It. Dir. Proc. Pen.* 1997, 1473, commentary by C. Piergallini, 'Attività produttive e imputazione per colpa: prove tecniche di diritto penale del rischio' and in *Rivistà Trimestriale del Diritto Penale Economico* 1997, 218, commentary by D. Micheletti, 'Tumori da amianto e responsabilità penale'.

[139] Cf. App. Milano, 9 January 2003, *Or. Giur. Lav.* 2003, 767.

only to the specific negative effects of a conduct, but also to all the harmful consequences which can result from a conduct well known for its danger to health'.[140] The same principle can be found in a statement of the Court of Appeal of Turin on 29 November 1995:

> In the field of liability for fault for the death of a worker it is not required that the person obliged to adopt the preventive measures has foreseen the death or the disease developed by the worker. It is both necessary and sufficient that he should foresee that by adopting the measures he could have avoided a serious damage to health or to life.[141]

(f) *Legal causation*

Once the fault of the employer has been proved, it must be demonstrated that failure to adopt protective measures caused the disease, i.e. that if (suitable) measures had been adopted, such a risk would have been avoided.

The approach of case law and scholars to the problem of legal causation in liability for omission applies the 'but for' theory, just as it does in assessing factual causation. The event is considered as a consequence of a certain failure to act, since, without the omission, it has been shown scientifically (either with a probability near to certainty or with serious and probable likelihood) that the event would not have occurred.[142] But the problem with asbestos is that there are no scientific certainties. So, for example, in case of manslaughter caused by pulmonary cancer or mesothelioma, the link of causation between the fault of the employer and the disease exists not only when it is proved that the adoption of protective measures could have prevented or reduced the damage, but also when it is proven that it could delay its discovery.[143]

(g) *Asbestos litigation to date*

Until the 1990s, claims for compensation were very few. However, case law was typically concerned with problems of interpretation and application of the social security measures to cases of asbestosis.[144]

[140] Cf., e.g., Cass. 14 January 2003, no. 988, quoted by R. Riverso, 'L'amianto negato ed impunito per i lavoratori morti per mesotelioma: ovvero quando le fibre non sono polveri' *Lav. Giuri.* 2006, 421.

[141] Unpublished decision, quoted by R. Guariniello in the commentary 'Dai tumori professionali ai tumori extraprofessionali da amianto', above note 101.

[142] Burragato and Morone, above note 107, at p. 128.

[143] Cass. Pen., 11 July 2002, *Foro it.* 2003, II, 324, commentary by R. Guariniello, 'Tumori professionali da amianto e responsabilità penale.'

[144] On this issue, cf. M. Miscione, 'I benefici per l'amianto fra norme di sanatoria e giurisprudenza' *Danno e resp.* 2005, 472.

In order to explain this discrepancy, a first justification can be found in the limitation period. As that period starts from the manifestation of the harm and not from the exposure to the noxious agent, claims tend to be remote in time from the event because of the long period of latency. In addition, for some time, among the factors inhibiting claims was a restrictive orientation expressed by the Court of Cassation in its decision of 30 May 1949, no. 1385.[145] Article 12 of Law of 1943 exempted the employer from civil liability for industrial diseases covered by compulsory insurance. Moreover, it stated that civil liability of the employer was still possible in cases of infringement of safety measures provided by special provisions. But the link between illnesses and asbestos was identified formally more than forty years later (by D. Lgs. 15 August 1991, no. 277). In the opinion of the court in 1949, in the absence of special legislation the employer had no warning and civil liability could not be based on the omission of the general rules, i.e. of the common prudence rules. As a consequence, the employer was completely exempted from civil liability.

Such an orientation has subsequently been changed by the Court of Cassation. It has restored the correct relationship between general and special rules, stating that the lack of regulations is not to be considered as an obstacle to the application of the general rules on tort liability until a special regulation derogates from them. Claims for compensation have become very common since the 1990s, together with an awareness of the harmful effects, real and potential, of exposure to asbestos. Due to the long period of latency, litigation is probably set to continue, despite the prohibition on the use of asbestos in 1992, and there is some suspicion that substitute materials have the potential to be equally harmful.

As regards claimants, most of the actions are brought by the heirs of the victims. With regard to defendants, as illnesses are always connected to the working environment, most of them are the employers of the victims. No claims have been brought against the manufacturers of the products containing asbestos or the owners of the premises insulated with materials containing asbestos.

In order to explain this, it could be useful to consider the results of research conducted by the National Association of Insurance Companies (ANIA), published in 2003, from questionnaires sent to their members dealing with the tort liability sector. According to this study, nearly half of the companies questioned had received claims for damages, and since 1998 there has been a substantial increase in these claims as a result of

[145] Cass. 30 May 1949, no. 1385, *Riv. Dir. Lav.* 1950, 97, commentary by G. Alibrandi, 'La responsabilità civile del datore di lavoro in tema di silicosi ed asbestosi'.

exposure to asbestos in the work environment and environmental pollution in general. The same study also showed that 'private insurance and the social protection bodies are the persons onto which the costs of ill health due to exposure to asbestos are primarily transferred'. This is also due to the fact that since the beginning of 2003, insurance and reinsurance markets do not include among their products coverage for this kind of risk.[146] This seems to mean that most illnesses do not reach the courts, as they are compensated by private or public insurance.

(h) Conclusion

This section has emphasised that the most recent civil decisions on compensation are turning upside down some established rules – mainly on fault and the link of causation – laid down in criminal law. Until now, the defences of employers on these issues have always been rejected in the criminal field. Furthermore, employers have always been ordered to pay compensation. In the civil field, the same defences are being accepted by the courts. For example, in one case the Court of Cassation confirmed the rejection of the claims where no fault had been found, notwithstanding the omission of the measures specified by law, since asbestos is not dust, but a fibre. In addition, the court had found an absence of the link of causation where the professional opinion of the probability of causation was not 'qualified'.[147] This is quite remarkable, since it should be easier to seek compensation in the civil law than in criminal law. (Criminal liability is limited to individuals, whilst civil liability extends to legal persons. In addition, the burden of proof is higher in criminal cases, and the rules for imputing mens rea are different.) This is relevant to the obtaining of 'additional damages', i.e. those not covered by the standard indemnification guaranteed by social insurance. As a matter of law, in order to obtain the 'differential damages', it should be sufficient that the judge verifies the criminal liability of the employer.[148]

[146] See www.ania.it/rc_generale/attivita/03_03%20introduzione%20AMIANTO.pdf.

[147] Cass., 11 April 2005, no. 7362, *Lav. giuri.* 2006, 415, quoted by Riverso, above note 140, at 415.

[148] Riverso, above note 140.

6

Technological Change and the Development of Liability for Fault in Spain

MIQUEL MARTÍN-CASALS AND JORDI RIBOT

1. Introduction

The three topics proposed for study have very different scope in the Spanish legal experience. The escape of sparks from railway engines causing damage to third parties in the mid- to late nineteenth century brought about substantial litigation. A number of cases reached the Tribunal Supremo, from the very beginning of the operation of the railways up until the twentieth century. However, accidents involving steam boiler explosions did not amount to a comparable stream of litigation. There are no cases decided by the Civil Chamber of the Tribunal Supremo in the period under examination. Landmark decisions, equivalent to the well-known *Teffaine*[1] case in France, and any specific judicial guidelines are lacking. As regards asbestos-related injuries and illnesses in the late twentieth century, litigation began to arise in Spain at the end of the 1990s. However, if compared with other jurisdictions, its impact on enterprises and on the tort law system as a whole has, up until now, been minimal.

2. Escape of sparks from steam engines

Prior to the development of the railways on the peninsula, railway lines were built in Cuba, then still belonging to the Kingdom of Spain. The first Spanish railway was that covering the Havana to Güines line, which went

[1] On the influence of this judgment on later developments of French tort law, see F. Leduc (ed.), *La responsabilité du fait des choses. Réflexions autour d'un centenaire* (Paris: Economica, 1997) and above pp. 112 ff.

into operation in 1837.[2] The first Spanish railway line was built, only with great difficulty and after several failed attempts, in 1848. This line linked Barcelona and Mataró, in Catalonia, and was an initiative of the indiano Miguel Biada, whose efforts made it possible to raise the money needed for the project and to gain political support from the government. English funding and technology were also essential.

The following years saw a relatively slow development of the industry due to financial setbacks and political conflicts. By 1855, only five railway lines had been built. From then on, however, a great development of railways took place and by 1875 most of the main lines had already been built. By the end of the nineteenth century, the radial network from Madrid to the main economic and industrial areas was complete and needed to undergo only minor or secondary improvements. Accordingly, the number of persons and goods transported increased dramatically.

From the very beginning, private initiatives were decisive in the building of the railway network. Although the relevant legislation viewed railway lines as 'public services', almost all of them were set up by means of concessions and exploited by private enterprises. This allowed the granting of some privileges during the process of putting the lines into operation in exchange for very intensive control and supervision by the responsible state agencies. Recurrent financial crises and the final collapse of the Spanish economy due to the Civil War (1936–1939) eventually meant that the state had to rescue all concessions and nationalise the railways in 1941. From then on – until the very recent new wave of privatisations – the state owned all railway lines and delegated their operation to an autonomous entity, also created in 1941, under the name Red nacional de ferrocarriles españoles (RENFE).[3]

As in other parts of Europe, steam locomotives dominated the landscape until more efficient and (then) cheaper technologies (such as, first, diesel engines and, later, electrification) came in and eventually completely displaced them. The last steam locomotive made in Spain came out of the factory in 1964, and the last service provided by one of these engines occurred in 1975. Technology used on the Spanish railways initially came from the United Kingdom, France and Germany. From the

[2] We shall make further reference to Cuban railways, since several important cases dealt with by the Tribunal Supremo related to fires in sugar cane plantations (*ingenios*) allegedly caused by steam locomotives.

[3] After 1941, Spanish railway lines were not directly operated by the state, but by RENFE, a public corporation, which was subject to private law rules concerning the transportation of persons and goods.

1860s onwards, however, Spanish industry developed and it was ready to provide most of the equipment needed to operate the railways.

(a) *Liability issues*

Claimants in actions brought after fires attributed to sparks from railway engines were normally the owners of the land or forests destroyed by them. There are cases, however, where the claimant was a person holding only a contractual or legal right to use and enjoy the land. Although defendants argued that lessees, tenants or other persons holding rights other than property rights could not recover damages, this argument did not succeed.[4]

The conflict between railway companies and land-owners was always tackled within the framework of liability law. Claimants sought compensation for direct property damage and never asked for a temporary or permanent cessation of railway activities. In fact, the Tribunal Supremo considered no case in which a claimant sought an injunction to try to prevent future fires due to the operation of a railway line. Such a hypothesis would have seemed unfeasible once it became clear that railway lines were encouraged by the government, operated under a concession of the state and managed under the state's supervision.

Moreover, claimants framed their actions as tort actions and based them upon the general liability rules of Partidas or, from 1889 onwards, upon the Civil Code (CC). No action shows the claimant asking for compensation due to interference with this property right, regardless of how reckless or negligent the conduct of the defendant railway company was. Accordingly, all reported cases were decided on the basis of fault liability rules.

Finally, despite the close supervision of private railway services by public bodies under the control of the Ministerio de Fomento, ordinary courts always dealt with actions arising from fires attributed to steam locomotives. Even in cases where a line was being operated on behalf of the state, relevant rules stipulated that any action to claim damages should be brought before ordinary courts and decided according to general tort liability rules.[5]

[4] Conversely, the defendant railway company, on occasion, contended that the forest's legal owner could not claim damages if he had leased it to a third party and that it was thus for the lessee to bring the tort action. The courts rejected such a defence (STS 13 October 1927, Gaceta de Madrid, Sentencias, 1928, no. 62).

[5] See art. 14 II ley of 14 November 1855 (as restated in art. 14 ley of 23 November 1877): 'if the railway is operated on behalf of the State, it shall be subject to the same liability as any other individual'.

(b) *Case law of the Tribunal Supremo*

The Tribunal Supremo never accepted any kind of immunity of railway companies from liability for the damage they caused by their activities. On the other hand, it never admitted, either, that specific legislation on railways, enacted in 1855, had introduced any kind of special strict liability regime.

The first argument was invoked in the first case reported on this issue: the judgment of the Tribunal Supremo (STS) 4 December 1858.[6] In this case, the locomotive Orión, belonging to the Cárdenas to Júcaro line in Cuba, was involved in the burning of hundreds of acres of sugar cane plantations (so-called *ingenios*) after a fire started on 26 March 1854, when the weather was extremely hot and windy. The owners of the *ingenios* sued the railway company. The defendant contended that the cause of the fire was unknown and that no fault was attributable to it. The trial court dismissed the claim, but on appeal the court of Havana awarded the claimants a substantial amount in damages. The railway company appealed before the Tribunal Supremo, arguing that, regardless of the fact that the cause of the fire was unknown, the company enjoyed a privilege that would exempt it from paying for the damage caused by its operation. The Tribunal Supremo recalled that the court of Havana had held that it had been shown that the defendant railway company was at fault by allowing its employees to throw the ashes out while the locomotive was in motion, in order to gain speed. As regards the alleged privilege, the Tribunal Supremo considered that a privilege 'would be contrary to any principle of justice when [the defendant] is at fault, as in this case'. In subsequent judgments, the Tribunal Supremo also increased the pressure on the railway companies by holding that they could be sued before the courts of the place where the fire started, instead of where they had their central headquarters.[7]

Conversely, the courts consistently ruled against the idea that the specific legislation on railways had set out a specific type of liability.[8]

[6] Colección Legislativa, Jurisprudencia Civil 1858, no. 31.

[7] Most of the railway companies had their legal domicile in Madrid, miles away from the place where the fires had occurred. Had the Tribunal Supremo accepted that courts in Madrid had jurisdiction, the victims of fires would most likely have given up their claims. Among others, see STS 3 April 1867, Gaceta de Madrid, 6 April 1867, vol. XV, p. 335.

[8] Article 14 ley of 14 November 1855 (restated in art. 14 ley of 23 November 1877) provided that '[t]he licensees or leaseholders of the railway lines shall be liable vis-à-vis the State and any other individual for the damage caused by the managers, the directors or any other employee in the operation of the railway and the signals'.

Indeed, to hold the railway company accountable, article 14 of ley of 14 November 1855 required proof of an unlawful act (*falta*) on the part of its employees. The innovation here was that the railway company was directly liable because of its employees' unlawful acts. Until then, according to the general rules laid down in the Penal Code (CP), the employer could be held liable only subsidiarily.[9]

A decision of 1864,[10] concerning a fatal accident at a level crossing, confirmed this approach. The train crushed the victim after he had forced the woman in charge of the barrier to open it and let him drive across the line in his cart. His widow sought compensation for non-pecuniary losses, a claim denied by the trial court. Before the Tribunal Supremo, she complained that the trial court had infringed article 14 of ley of 14 November 1855, 'together with the legal principles that anyone who causes harm shall compensate for it and that anyone who operates an industrial or speculative enterprise that causes harm to a third party must compensate him or her, *regardless of the fact that they did not act negligently*' (emphasis added). The Tribunal Supremo dismissed the appeal and replied that liability based on article 14 of ley of 14 November 1855 was dependent upon a fault being ascertained and attributable to the employees operating the railway lines.

(i) Nineteenth-century judgments

In the period between 1850 and 1900, several decisions of the Tribunal Supremo made it very clear that railway companies could be held liable for damage caused by the escape of sparks only if the claimant afforded proof that the company or its employees acted negligently, either by action or by omission. In addition, the claimant had the burden of proving the fault, as this was the general rule both according to the old law and to the CC (article 1214).

The leading case for this period is possibly STS 30 May 1865,[11] whose doctrine was applied straightforwardly in two subsequent decisions in cases of fires allegedly caused by steam locomotives on the island of Cuba.

[9] This was the specific issue addressed by the Circular written on 17 January 1863 by the Fiscal of the Tribunal Supremo and disseminated among all provincial courts. The aim of this Circular was to make public prosecutors and judges aware that railway companies should bear the consequences of their employees' crimes or minor offences. Liability flowing from acts or omissions that did not qualify as crimes was disregarded. See B. Vicente Garcés, *Apéndice a la colección legislativa que forma parte del diccionario de ferro-carriles* (Madrid: Manuel G. Hernández, 1875), pp. 251–64.

[10] STS 22 October 1864, Gaceta de Madrid, Sentencias, 1864, no. 299.

[11] Colección Legislativa, Jurisprudencia Civil 1865, no. 213.

A crop field crossed by the railway line Isabel II caught fire several times. Its owner brought a tort action against the railway company operating the line, asking for compensation for the property damage incurred. The trial court dismissed the claim, arguing that, '*despite the proven fact that sparks escaping from the defendant's locomotives caused the fire*, such an accident was not due to any fault of the defendant and was an unavoidable event for which nobody could be blamed' (emphasis added). The Tribunal Supremo upheld this decision, stressing that it was for the claimant to prove that fires had resulted from faulty conduct attributable to the engine drivers. Neither the rules on tort liability provided by the Partidas nor the wording of the railways' special rule on the licensees' or leaseholders' liabilities allowed the rules on the burden of proof of fault to be changed, for both sets of rules relied upon the principle that the defendant's fault must always be proved and could not be presumed.

In sharp contrast to the approach taken in other cases,[12] two subsequent judgments in Cuban cases took the same demanding approach. In 1881,[13] a case involved a steam locomotive whose nickname was 'the Burner' ('*La Incendiaria*'). The name makes it very clear that inhabitants of the area were well aware of its risks. No cause other than the sparks or the ashes expelled from the said locomotive could have caused the damage. Nevertheless, the trial court in Havana and the Tribunal Supremo took the view that the negligent operation of the locomotive or its faulty maintenance could not be taken for granted and, accordingly, that they had to be duly proved by the claimants. As in STS 30 May 1865, the Tribunal Supremo upheld the judgment of the trial court by contending that, since the claimants could not demonstrate the proof of negligence required, it must exonerate the defendant railway company from liability for the damage arising from the fire. Eight years later, STS 11 April 1889[14] also concerned a fire in several *ingenios* of sugar cane whose owners attributed it to the sparks escaping from a steam locomotive, in this case belonging to the Havana to Marianao line. The judgment declared, however, that fault on the part of the defendant railway company could not be presumed and that the claimants should bear the burden of proof of fault attributable to the defendant's employees.

Against this background, however, one of the most influential technical books, widely used to tackle legal questions that arose in daily practice,

[12] STS 4 December 1858 (see above note 6).

[13] STS 15 March 1881, Colección Legislativa, Jurisprudencia Civil 1881, no. 136.

[14] Colección Legislativa, Jurisprudencia Civil 1889, no. 139.

seemed to depart from the outcome of reported litigation.[15] In its specific entry for crop fires caused by sparks escaping from locomotives (*incendios de mieses*), its author stressed that railway companies had to pay huge amounts in damages unless the land-owners had infringed their legal duties concerning the placement of combustible material near the railway lines. Moreover, any defence of compliance with safety requirements was of no avail.[16] Nevertheless, other sources pointed to the opposite solution, which seems much more consistent with the reported case law of the Tribunal Supremo during the period under examination. They emphasised that if the company had complied with preventive measures – especially cutting the undergrowth – as certified by the officials in charge of the line, no liability could arise for setting fire to neighbouring fields.[17] In this regard, the judgment handed down on 3 June 1901[18] serves as an illustration of the deference the courts showed to railway companies. In that case, the claimant had lost an entire pile of straw that he had placed very near the tracks in a railway station. The claimant had not asked for authorisation to place the straw there.[19] After a spark from a nearby locomotive happened to cause it to ignite, he claimed compensation. Both the trial court and the Tribunal Supremo deemed his conduct to have been a tacit assumption of the risk of fire that could start by placing such combustible material in the proximity of a place where steam locomotives manoeuvred and sparks escaped.[20]

(ii) Early twentieth-century judgments (1900–1930)

Compared with the nineteenth-century cases, those litigated in the first quarter of the twentieth century are characterised by the intensive use,

[15] B. Vicente Garcés, *Diccionario razonado legislativo y práctico de los ferro-carriles españoles bajo el aspecto legal, técnico, administrativo y comercial de los mismos* (2nd edn, Madrid: Imprenta de Campuzano Hermanos, 1875) (ed. facsímil València: Librerías París-Valencia, 1994). Written by an engineer who worked for one of the most important Spanish railway companies, the book was published in four editions in ten years.

[16] *Ibid.* p. 269.

[17] See V. Ruiz, *Lecciones de caminos de hierro* (Madrid: Establecimiento tipográfico de Ricardo Álvarez, 1895), p. 607.

[18] Colección Legislativa, Jurisprudencia Civil 1901, no. 147.

[19] No combustible material could be placed within 20 metres of the railway line, with the only exception of grain and other produce during harvest. In this case, the law stipulated *expressis verbis* that no compensation could be claimed in cases of fire ('En caso de incendio por el paso de las locomotoras, los dueños no tendrán derecho a indemnización'; article 6.2º ley of 23 November 1877).

[20] Cf. the view of the Tribunal Supremo fifty years later in STS 25 March 1954 (RJ 1954\1001). See below p. 238.

either by claimants or by defendants, of arguments that relied upon the compliance with or, conversely, the infringement of, safety regulations. Courts took compliance with these provisions or their breach as determining whether the defendant company was at fault or not. Accordingly, judgments issued in this period took for granted that a proven infringement of safety measures constituted fault on the part of the defendant. Conversely, in those cases where the defendant railway company afforded evidence showing that all measures required by regulations had been taken, the chances for the success of the claimant's action seemed to be practically nil.

Safety regulations relating to fires caused by sparks were an important part of railway legislation. These rules focused, first, on requiring companies to introduce technical improvements to the design of steam locomotives,[21] but they also required them to keep the tracks clear of undergrowth and any other combustible material. From the very beginning, the relevant legal rules stipulated that licensees and leaseholders were responsible for line maintenance[22] and, from 1862 onwards, the government issued a number of specific recommendations and mandatory rules with the aim of preventing, or at least diminishing, the risk of fires caused by sparks.[23]

Two decisions of the Tribunal Supremo illustrate how courts dealt with the issue at this stage, although these judgments took sides only by implication, upholding the view taken in each case by the trial court.

In the case decided by STS 9 January 1922,[24] the claimant was the owner of several pieces of land adjacent to the railway line. After a fire occurred in his fields, he brought a tort action against the railway company, blaming it for operating a steam locomotive without the compulsory safety measures for preventing sparks from escaping and for having omitted to

[21] Article 32 of the Regulations of 8 July 1859 already required locomotives to be equipped with devices capable of preventing any risk of fire.

[22] See Titles II and III of ley of 14 November 1855 and Chapter II of Regulations of 8 July 1859.

[23] Accordingly, all railway companies were expected to set up a service charged with the task of opening up and clearing ditches. These tasks were performed under the supervision of civil servants appointed by the Ministerio de Fomento and assigned to the different divisions in which the railway network was structured. These officials and the local authorities were in charge of supervising the fulfilment of the preventive duties imposed upon the railway companies. R. Foyé, *Los Caminos de hierro de España: recopilación ordenada de las disposiciones legales vigentes sobre ferrocarriles y tranvías en sus diferentes periodos de estudio, construcción y explotación: legislación y jurisprudencia* (Barcelona: Litotipografía de Luis Tasso, 1894), pp. 442–3.

[24] Colección Legislativa, Jurisprudencia Civil 1922, no. 8.

clear away the undergrowth. The trial court had dismissed the claim in so far as the defendant railway company afforded proof of having complied with the safety measures required by law. The Tribunal Supremo upheld the trial court's decision, arguing that it could not re-examine the factual background to the case.

By contrast, in the case that ended with STS 13 October 1927,[25] the claimant had lost a substantial part of a very productive acorn forest in a fire that had occurred seven years before. The origin of the fire was allegedly a spark that had escaped from the steam engine of a locomotive that first ignited the undergrowth and then spread to the claimant's property. Witnesses' statements and other collateral evidence supported his claim. The railway company contended that the cause of the fire could not be definitively established. It contended that anyone crossing the tracks at that point could have discarded a cigarette butt. With regard to clearing away undergrowth from the tracks, the company put forward a written statement signed by the local authorities, certifying that prior to the fire it had cleared away undergrowth from the tracks. The trial court refused the evidence tendered by the company, considering it inconclusive; the local authorities had not, in fact, verified the cleaning works. Before the Tribunal Supremo, the defendant railway company criticised the trial court for having made it liable on the basis of the unproven fact that a spark had ignited the grass on the tracks and argued that this fact could not be presumed. The Tribunal Supremo upheld the opinion of the trial court and ruled that, having ascertained an infringement of safety duties, it was reasonable to link the accident to the fact that fires caused by the escape of sparks from steam locomotives were frequently reported.

(iii) Later developments (from 1930)

Judgments handed down in this period stressed the independence of the judicial review of the conduct of the defendant, regardless of whether he had shown that he had performed the duties imposed on him by the regulations. By so doing, the Tribunal Supremo made it possible to overcome the problems of outdated or ill-conceived safety regulations. By the same token, it abandoned an attitude of deference towards evidence afforded by official statements issued upon the request of the defendants.

A judgment of 29 June 1932 concerned a very similar case to the one decided by STS 13 October 1927. The claimant also blamed the railway company for not having fulfilled its duties regarding clearing the line of

[25] Gaceta de Madrid, Sentencias, 1928, no. 62.

grass and undergrowth, thereby allowing a fire to spread when a spark escaped from an engine of one of its steam locomotives. The claimant lost almost all the crops of the adjacent fields that he had leased, together with vehicles and other equipment that he kept in a house destroyed by the same fire. The railway company based its defence strategy on two lines of reasoning: first, that the claimant provided no definitive proof that the fire had started as he claimed; secondly, that the company had complied with the duties imposed by the regulations, as shown by several written statements in which officials of the Ministerio de Fomento and the local authorities declared that before the fire the railway company had cleared the tracks of undergrowth. Moreover, the company also provided the court with an official certificate stating that the steam locomotive had all safety devices in good order. Both lines of reasoning failed. The trial court and the Tribunal Supremo held the company liable, notwithstanding the fact that it had managed to prove prompt compliance with all safety requirements. Against the view of the defendant, who contended that no fault could be established in a tort action once the relevant regulatory duties had been fulfilled,[26] the Tribunal Supremo replied that:

> railway companies cannot escape liability by proving that they have complied literally with all the conditions established by regulations ... since when reality shows that the measures taken to foresee and prevent damage which was foreseeable and preventable brought no result, this clearly demonstrates the imperfection and the negligence that accompanied the adoption of these measures; and this gave rise to damage whose reparation is absolutely unavoidable on the grounds of justice.

What does this dictum mean? One must concede that what the court probably meant to say was that, in the case at issue, there was insufficient proof of compliance with regulatory duties, just as there had been in the case of 1927.

Due to a number of reasons, not least the lack of proper means to perform their role, statements made by the officials responsible for the state of the tracks deserved little credibility. Sources from this period often use derogatory terms to describe the supervisory body. One authority recalls that although few questioned its existence (since Spaniards 'have little respect for the law, but some fear of authority'), '[t]he supervisory body has to be everywhere' and it therefore 'can do more in some matters and

[26] '[I]t is evident that no fault or negligence can be ascertained when the activity is subject to regulations and the measures laid down in them have been fulfilled ... no one could be required to behave more carefully than the regulations demand.'

less in others'. Moreover, as funding was scarce and the selection of the staff arbitrary, such a body did not deserve to qualify as 'honourable … responsible and provided with effective support'. [27]

Indeed, one may doubt whether the state of the tracks was as safe as the risk posed by locomotives required. A Royal Order issued on 23 August 1872[28] had already reminded railway companies about the obligations imposed on them regarding clearing the tracks of undergrowth 'by means that are technically the most appropriate for reaching this goal'. This Order was followed by a Royal Order of 30 June 1881[29] and by a specific report on the issue prepared by the relevant governmental body, the Junta Consultiva de Caminos, Canales y Puertos. Both sought to make it clear to railway companies that the fulfilment of safety duties was mandatory. Needless to say, these reminders would not have been necessary had the companies fulfilled their duties regularly.

Against this background, it was only to be expected that courts took it for granted that the preventive measures allegedly taken by the defendant were, in fact, executed only in form. Accordingly, the Tribunal Supremo provided the lower courts with full leeway as regards the determination and assessment of the railway companies' behaviour in such incidents.

(c) Concluding remarks

In Spain, the operation of railway lines seems to have been promoted initially at the expense of land-owners and other potential victims. Mostly through indirect (procedural) means, courts denied compensation in almost all of the cases tried during the nineteenth century. It is also true, however, that decisions issued later on appear to be much more 'neutral' and, in a final stage, courts lost their faith in how public bodies were supervising the activities of railway companies. Accordingly, courts showed themselves more sceptical towards the accounts provided by the railway companies when incidents occurred. The trend was to allow an independent assessment of the railway company's behaviour, according to the court's own free examination of the evidence provided by the parties, without relying exclusively or primarily upon alleged compliance with the safety measures required by the regulations in force when the accident occurred. From that point of view, one may venture to suggest that awareness was widespread about the inaccurate performance of supervision

[27] Vicente Garcés, above note 15, at pp. 275–6.

[28] Gaceta de Madrid, 30 August 1872.

[29] Gaceta de Madrid, 20 July 1881.

tasks by the responsible public bodies and that ex post facto statements produced by local authorities also lacked any credibility.

At this stage, the Tribunal Supremo spoke about formal compliance with safety measures and allowed the injured party to provide such elements of evidence that could lead the court to conclude that, despite claims that it had complied with regulations, the railway company was, indeed, at fault.

Although this line of reasoning had previously been used in STS 13 February 1928,[30] in a case of derailment involving personal injuries and death, in the years that followed the burden of proof was to become very difficult for defendants. Courts tended to mix the willingness to examine the substance of compliance with a slightly different approach, requiring the defendant to prove that it had taken all preventive measures available to avoid the damage, regardless of whether they were required by safety regulations or not. With regard to sparks causing fires, STS 25 March 1954[31] could thus hold a railway company liable for the ignition of a pile of wood placed by its owner near the tracks. The court recalled that 'formal compliance' with the safety measures required by the regulations did not exonerate the defendant company. Indeed, 'the very fact of the fire demonstrated that these regulations were insufficient to prevent the fire from happening and that there were measures that the defendant company could and should have taken to prevent it'.

3. Steam boiler explosions (1850–1900)

Steam engines appeared quite early in Spain. At the very beginning of the nineteenth century, several textile factories located in Catalonia were already being operated by steam engines. The first steam boiler made in Spain was put into operation in 1805 and from then on a number of textile factories were equipped with steam engines.[32] The Napoleonic War and the political and social instability that followed suddenly interrupted this process. By the 1830s, most of the original equipment was severely damaged

[30] Colección Legislativa, Jurisprudencia Civil, 1928, no. 63. In this case the court mentioned the fact that the defendant company had probably complied with the indispensable requirements provided for by the Regulations, but it had not executed other preventive measures that were needed and that had been required by the public bodies supervising its activities after previous accidents that had taken place.

[31] RJ 1954\1001. Cf. STS 3 June 1901 (Colección Legislativa, Jurisprudencia Civil 1901, no. 147), referred to above.

[32] J. Nadal i Oller, 'La economía española 1892–1931' in A. Moreno (ed.), *El Banco de España. Una historia económica* (Madrid: Banco de España, 1970), p. 67.

and industrialists had to make new investments. In 1832, a new steam engine was installed in a factory located in Barcelona (that of the Bonaplata brothers) after which many more followed. In a few years, Barcelona and other neighbouring cities, such as Sabadell and Terrassa, became the centre of the technological development that kept pace with the success of textile industries established there. A number of Spanish enterprises started to produce steam engines and steam boilers, in order to provide for the needs of the growing industrial network.[33] These enterprises were located very near the centres of textile industries and supplied the manufacturers with the engines needed to equip their factories. In the space of very few years, steam engines were adopted in manifold industrial activities and allowed the siting of industries in locations other than on the banks of rivers, where they had hitherto been sited to use natural water power.

The development of steam technology, however, never reached the level that it attained in other Western countries.[34] It has been estimated that by 1900, Spain had steam engines capable of providing 1.2 million horsepower, whereas figures for the United Kingdom are 13.7 million, for Germany 8.08 million, for France 5.92 million, for Russia 3.1 million and for Italy 1.52 million. '[T]he forests of tall smoke chimneys, so characteristic of the European industrial landscape, appear like an exception in our country.'[35] Together with the above-mentioned political and social hurdles, the requirements inherent in the use of steam technology substantially hindered its development. First and foremost, the lack of coal was decisive, it being essential for the effective operation of steam engines. In 1865, coal cost 2.95 times more in Catalonia than it did in England.[36] The lack of fast and cheap connections from and to the few mining areas had a negative influence as well.

This relatively poor and belated development of steam technology might explain why the number of accidents and explosions seems to be lower than in other countries. Nevertheless, steam boiler explosions took

[33] For instance, La Maquinista Terrestre i Marítima SA, created in 1855 with the explicit intention of constructing 'terrestrial and marine, as well as locomotive steam engines for the railways'. See J. Nadal i Oller, 'Les màquines de vapor fixes de La Maquinista Terrestre i Marítima, S.A.' in J. Maluquer de Motes i Bernet (ed.), *Tècnics i tecnologia en el desenvolupament de la Catalunya contemporània* (Barcelona: Enciclopèdia Catalana, 2000), pp. 234–41. Two other local undertakings were Nuevo Vulcano (1835) and Alexander y Hnos (1849). See S. Riera i Tuebols, 'La màquina de vapor, una nova font d'energia en la Revolució Industrial' in *Elements d'història de la tècnica* (Barcelona: Associació d'Enginyers Industrials de Catalunya, 1995), p. 189.

[34] See J. Nadal i Oller, *España: 200 años de tecnología* (Madrid: INI, 1988), pp. 39–42.

[35] *Ibid.* p. 40. [36] *Ibid.* p. 41.

place occasionally, causing substantial damage, both in the number of people dead or injured and in damage to property. The contemporary daily press reported the most catastrophic incidents, which usually killed the men operating the boiler and other people working next to the site of the explosion.[37] Relevant information, although indirect, was also provided by the professionals who were best placed to see and assess the causes and the consequences of steam boiler explosions – the engineers.[38]

From a legal point of view, the surveillance and supervision of dangerous or noisy activities has always been the responsibility of local authorities.[39] Since the Middle Ages, municipalities have issued health and safety regulations. They determined where and under which conditions industries and other activities could settle in the city. From the very beginning of the nineteenth century, both rules at a national level and updated local regulations have laid down the requirements that new industries, and especially those deemed dangerous, should meet in order to start their activities. Some rules forbade certain types of industries from settling in the city centre[40] in order to prevent accidental harm to neighbours due to fires and explosions.[41] Yet the first national regulation was not enacted until the second quarter of the twentieth century.[42] Moreover, contemporary witnesses stated that supervision by local officials was scant and never managed to ensure the effective

[37] Between 1887 and 1905, for example, the local newspaper *Diario de Gerona* gave accounts of many accidents involving steam boilers in Spain with very serious consequences in terms of deaths, injuries and damage to property. On many occasions, the explosion of the boiler also caused the destruction of the factory and the loss of the main economic activity of the area. That happened, for instance, when a cork workshop caught fire after the explosion of the steam boiler (*Diario de Gerona*, 17 February 1894, p. 10) and in the case of the destruction of a factory in Algeciras (*Diario de Gerona*, 29 June 1902, p. 22).

[38] These references have been compiled by R. Garrabou, *Enginyers industrials, modernització econòmica i burgesia a Catalunya (1850–Inicis del segle XX)* (Barcelona: L'Avenç, 1982), pp. 290–1.

[39] Article 72 Municipal ley of 2 October 1877, developed by Royal Decree of 7 February 1913. See also J. Gordley, *The Development of Liability Between Neighbours* (Cambridge: Cambridge University Press, 2009), ch. 7.

[40] See the references compiled in E. Algarra Prats, *La defensa jurídico-civil frente a humos, olores, ruidos y otras agresiones a la propiedad y a la persona* (Madrid: McGraw-Hill, 1995), p. 158.

[41] See, e.g., the local regulations of Barcelona passed in 1891. Regarding explosive and powder mills, see Royal Orders of 11 January 1865, 7 October 1866 and 14 July 1844; cf. J.M. Manresa y Navarro, *Comentarios al código civil español* (7th edn, Madrid: Reus, 1972), vol. IV, p. 1001.

[42] Later on substituted by Regulation 2414/1961, of 30 November, on annoying, dangerous, unhealthy and harmful activities (BOE no. 292, 7 December 1961), which, to a great extent, is still in force whenever regional legislation is lacking.

prevention of accidents.[43] Moreover, unlike their European counterparts, Spanish owners of steam boilers were never willing to set up associations to share maintenance costs and ensure periodic safety checks, although engineers officially exhorted the industrialists to action several times, in 1882,[44] 1895, and again in 1901.[45]

Engineers were indeed very concerned about the great number of explosions that was taking place at the end of the nineteenth century – accidents that they deemed preventable.[46] These accidents attracted frequent comment by the specialised press[47] and engineers complained about the need for general safety legislation, including measures such as the establishment of a competent body with strong powers to control and punish the owners of deficient steam boilers. After one accident in Barcelona, the Asociación de Ingenieros Industriales even set up a commission whose object was 'to obtain and translate as soon as possible to our language all preventive decrees and other repressive laws that, in the learned parts of Europe, keep at bay the selfishness of some heartless industrialists'.[48]

[43] See 'Otra explosión y otras víctimas' *Revista tecnológico-industrial* 1884, 309–14 (blaming the local authorities of L'Hospitalet for failing to carry out supervisory duties and the owner of the factory for reckless disregard for obvious safety measures and the overburdening of machinery used in the factory). See also M. Lafont i Pont, in *El porvenir de la industria* 1884, 360–1 and the critical appraisal made by a leading engineer Francisco Vila y Lletjós on the inadequacy of local regulations to prevent boilers' explosions and their consequences (*La Dinastía*, n. 1536, 7.5.1886, p. 5).

[44] *Revista tecnológico-industrial* 1882, 166–7.

[45] Garrabou, above note 38, at p. 291. See also M. Tarrés Vives, *Normas técnicas y ordenamiento jurídico* (Valencia: Tirant lo Blanch, 2003), pp. 104–5.

[46] Risks related to the explosion of steam engines and steam boilers were widely known. Nonetheless, for many years the causes of these explosions remained unascertained. As in other countries, it was not until the last quarter of the nineteenth century that the causes of explosions were discovered and effective preventive measures recommended. See, e.g., J. Bolívar, 'Una explosión' *Revista tecnológico-industrial* 1881, 84–90. See also *Revista tecnológico-industrial* 1882, 165 and 'Las explosiones de calderas de vapor' *Diario de Gerona*, 7 February 1897, p. 14.

[47] Several works were made public during these years, for instance a 'Memoria' presented at the Real Academia de Ciencias Naturales y Artes of Barcelona by C. Sintas, 'Explosiones fulminantes de los generadores de vapor: estudio de las causas que las producen', 28 June 1882. See also G.J. de Guillén-García, *Explosiones de generadores de vapor* (Barcelona: Imprenta de Pedro Ortega, 1895), which was awarded the first prize in the competition organised in 1893 by the Asociación de Ingenieros Industriales of Barcelona.

[48] *Revista tecnológico-industrial* 1884, 388. Indeed, the same journal published a number of Spanish translations of the main regulations in force in Europe (Britain, France, Belgium, Italy, Austria, Switzerland and Germany) about steam engines. See, e.g., *Revista tecnológico-industrial* 1882, 317 (Italy) and 311 (Germany).

For years, engineers insisted on the need for a rigorous inspection of the boiler when installing it and that periodic control should be compulsory.[49] Yet no general legislation on how to handle steam engines was passed during the nineteenth century. Only when the development of legislation on industrial accidents occurred were the local services of industrial supervision (*servicios provinciales de inspección industrial*) established.[50] In addition, this service, whose aim was to prevent industrial accidents, was not effectively put into operation until 1924,[51] and although inspectors supervised factories and workshops, examined boilers and verified gas and fluid meters, the enactment of official technical guidelines was delayed until 1929.[52]

(a) Civil litigation on steam boiler explosions

In striking contrast to the fact that a huge number of very serious accidents occurred involving steam boilers during the last third of the nineteenth century, these accidents did not trigger civil litigation. At any rate, no case dealing with steam boiler explosions was decided by the Civil Chamber of the Tribunal Supremo in the period under examination (1850–1900).

This lack of civil litigation could be due to the fact that, in most cases, those injured were employees of the owner of the steam engine. They or their surviving dependants typically lacked the ability and the means to pursue a civil action for damages against the factory owners. More often than not, the negligent behaviour that had caused the explosion could be attributed to the operatives[53] or to other workers who were killed

[49] See *Revista tecnológico-industrial* 1882, 112. More often than not, the owners of the boilers frequently opposed the inspections. See *Revista tecnológico-industrial* 1887, 355–64 (supervision of ship steam engines). In *El Porvenir de la Industria* 1884, 275–9 and 283–5, a regulation concerning the installation and operation of steam engines was published, which was passed on 10 February 1871 by Barcelona's public authorities but never officially published, for unknown reasons ('cuya aplicación no se hizo pública por razones que ignoramos', *ibid.* 275).

[50] L.E. de la Villa Gil, *La formación histórica del Derecho español de Trabajo* (Comares: Granada, 2003), pp. 152–3. On the Act on Accidents in the Workplace 1900, whose arts. 6, 7 and 8 provide for the constitution of a committee charged with the task of studying and proposing the technical means able to prevent industrial accidents, see below.

[51] By means of Royal Order issued on 25 January 1924: see D. Canals i Ametller, *El ejercicio por particulares de funciones de autoridad (Control, inspección y certificación)* (Granada: Comares, 2003), p. 27.

[52] By means of Royal Decree issued on 21 November 1929, approving the regulations on the examination and testing of receptacles containing pressurised fluids.

[53] STS 10 December 1915, Colección Legislativa, Jurisprudencia Civil, vol. LII, no. 134.

or severely injured as a result of the explosion, thereby depriving any claim brought on their behalf or by their dependants of any chance of success.

The determination of the employer's civil liability in criminal proceedings was never a possibility. Indeed, in the context of the late nineteenth century's unbridled capitalism, owners of factories as well as managers responsible for the proper maintenance of steam boilers enjoyed an effective immunity for harm done during the operation of their factories.[54]

Contemporary sources regretted this lack of effective means to make these persons accountable for the damage resulting from the bad conditions in which the steam boilers were kept. To those who objected to the idea of the law establishing the compulsory supervision of any boiler, the state of affairs in Spain was compared with the means available in Britain:

> [We] are friends of the autonomy of industrialists, but in this case it is necessary and unavoidable that the government intervene in the affair and let them know about its paternal interest. Some will say that in England the government does not worry about the conditions in which the generators work; but we can answer them that, there, the moral and material responsibility for this type of accidents does exist, whereas not in this country. Moreover, one must recall that last year, 1881, the Houses voted a law giving rules to make more effective the enquiries that follow every explosion to ensure that the liable person is held accountable.[55]

For decades, industrial accidents did not give rise to civil litigation,[56] but to social unrest and, eventually, to the development of unions and workers' revolutionary movements. From the legal point of view, the response to the workers' claims is to be found in the passing of the ley

[54] Commenting on the impunity of the persons liable for one of the explosions, see *Revista tecnológico-industrial* 1887, 357.

[55] *Revista tecnológico-industrial* 1882, 163. On the proceedings opened as a result of steam boiler explosions and the Boiler Explosions Act 1881 referred to in this quotation, see above p. 62.

[56] See the preliminary note written by R. García Ormaechea, *Jurisprudencia del Tribunal Supremo y de la Comisión Superior de Previsión sobre accidentes de trabajo 1902–1934* (Madrid: Minuesa de los Rios, 1935), p. 8, who stated that from 1838 to 1900 only one decision of the Tribunal Supremo concerned damage suffered by an employee in the workplace (STS 14 December 1894). Nevertheless, we may point to STS 2 March 1897, 18 March 1898 and 30 April 1898 as decisions that have prevented injured workmen or their families from receiving compensation because they could not prove the employer's negligence. In these three cases, the accidents involved the derailment of trains.

on Accidents in the Workplace in 1900.[57] The Preamble to the legislation emphasised that:

> no aspect of more vital interest is offered for the situation of the worker in the modern industry than the one related to the accidents to which he is constantly exposed when using the powerful aid rendered by the uninterrupted advance of mechanics, physics and chemistry. It was not possible to close one's eyes to the frequent spectacle of human beings wounded, mutilated or killed by the force of the machines or to the expansive power of powerful and even more dangerous substances, without the hope, at least, that injuries would be cured and that the resulting invalidity would not amount to hunger and helplessness during their sad and enforced idleness.

That ley tried to provide protection to those harmed in the industry by setting up a no-fault liability scheme, together with compulsory liability insurance. Among the beneficiaries of this new legislation there were, of course, workers injured by steam boiler explosions (or the dependants of those who had died) (article 5). Unsurprisingly, the first 'industries and activities' mentioned as being subject to the strict liability regime under article 2 were 'factories, workshops and any other industrial equipment where a power source other than human energy is being used' (article 3).[58]

(b) *Steam boilers and other dangerous things in the Civil Code 1889*

Cases involving damage to the property of neighbouring land-owners or personal injuries to people unrelated to the operation of the steam boiler were very rare. Indeed, only a handful of judgments of the Tribunal Supremo can be linked to this topic. These decisions, however, are important when assessed against the background of the legislation, especially in so far as the CC expressly mentions steam technology in one provision (article 590 CC) and refers implicitly to it in another (article 1908, paragraph 1 CC).

Article 590 CC concerns legal distances required when installing 'wells, sewers, aqueducts, furnaces, chimneys, stables, wares of corrosive material, artefacts driven by steam, or any other industry that is dangerous or noxious or whose products are'. The same provision refers to 'the

[57] Ley de 30 de enero de 1900, de accidentes de trabajo en los establecimientos industriales, *Gaceta de Madrid*, 31 January 1900.

[58] '[Fá]bricas y talleres y los establecimientos industriales *donde se hace uso de una fuerza cualquiera distinta de la del hombre*' (emphasis added).

regulations and the uses of the land' in order to determine the distances and the requirements that had to be met before siting these industries. The regulations in force at the time of the passing of the CC defined as dangerous activities, for instance, factories producing roofing tiles, bricks, plaster, varnishes, tin, soap; bread furnaces, confectionery, lime and similar items; large depots for coal, tar, straw, powder or phosphorus; workshops of chandlers and locksmiths; and distilleries of brandy and similar spirits.[59] 'In default of ordinances, precautions considered necessary shall be taken after expert inspection, for the purpose of preventing all injuries to the neighbouring tenements or buildings' (article 590 *in fine* CC).

According to article 1908 CC, owners would be liable for the damage caused in the following cases:

(1) by the explosion of machines that have not been kept with due care and by the inflammation of explosive substances that have not been stored in a safe and suitable place;
(2) by excessive smoke that is injurious to persons or property;
(3) by the collapse of trees placed in thoroughfares, where this is not caused by force majeure;
(4) by emanations from sewers or deposits of toxic matter, built without the precautions required by the place in which they are sited.

Among Spanish legal scholarship, it is commonly held that article 590 CC, together with articles 389 and 391 CC, are focused on the conditions and limitations imposed by law on the use of land, whereas article 1908 CC refers to liability issues that arise from the realisation of certain risks inherent in the things owned by the defendant.[60]

The references to steam engines and to dangerous substances in articles 590 and 1908.1 CC are symptomatic of the issues of the time. One of the main features of the Spanish Civil Code is that it often echoes a society in which farming is the essential livelihood of those persons to whom the Code is addressed.[61] When the Code entered into force, the country was still anchored in an economy that depended less upon the price of coal

[59] As listed in F. Baudín, 'Establecimientos peligrosos' in *Enciclopedia Jurídica Española* (Barcelona: Seix, 1912), vol. XIV, p. 938.

[60] Inter alia, see A. Hernández Gil, *Obras completas*, tome IV, *Derechos reales. Derecho de sucesiones* (Madrid: Espasa-Calpe, 1989), p. 144. See also Algarra Prats, above note 40, p. 160.

[61] As L. Díez-Picazo put it, 'it was a world in which, even though the first industrial revolution had already taken place, society had still not felt its effects or had not felt them in depth' (*Derecho de daños* (Madrid: Civitas, 1999), p. 84).

than on the weather. Articles 590 and 1908, paragraph 1 CC, therefore, can be seen as signs that legislators were beginning to pay attention to new sources of risks, which were linked to the incipient industrial development that Spain was then undergoing.[62]

From this point of view it should not be a surprise that (although relying upon certain Roman precedents)[63] article 1908 CC appears to be a novelty, brought in at the last stage of the complicated process that led to the enactment of the CC.[64] The current wording of article 1908 did not appear in the draft of the CC of 1851. Its origins can be traced back to the preliminary draft written in 1882–1888.[65]

No specific explanation of the reasons that led to its introduction in the CC can be derived from the preliminary works. In the draft of the CC of 1851, articles 1907 and 1909 were one article on the owner's liability for damage caused by the collapse of a building.[66] After the public consultation that followed the presentation of the Bill, the drafting commission agreed to extend the owner's liability to 'other similar works, and to the collapse of trees'.[67] This suggestion was consistent with the duties that the Code imposed on the owners of buildings or trees that were at risk of being ruined or of collapsing.[68] It is very likely that the drafting commission then searched for a way to link the owner's duties laid down in article 590 CC to the specific instances of liability that were to be introduced in article 1908. To achieve this, the most influential source appears to be article 1133 of the Argentine Civil Code of 1870,[69] which compiled as items of tort liability several instances mentioned by the Roman sources

[62] See Algarra Prats, above note 40, pp. 153–4.

[63] *Ibid.* p. 156 concedes that one of the subsections of art. 1908 CC could be linked to D.8, 5, 8, 5–7.

[64] Article 590 CC was also amended at this final stage to include, at the end, the phrase 'for the purpose of preventing any damage to neighbouring tenements or buildings'. The meaning of the provision remained unchanged, but the legislators thereby stressed the preventive character of the rule, which obviously links art. 590 with art. 1908 CC. See Algarra Prats, above note 40, p. 161.

[65] Cf. M. Peña Bernaldo de Quirós (ed.), *El anteproyecto del código civil español (1882–1888)* (Madrid: Reus, 1965), pp. 728–9.

[66] See art. 1903 of the Draft Civil Code of 1851.

[67] See J.F. Lasso, *Gaite, Crónica de la codificación española*, tome 4, *Codificación civil: génesis e historia del Código* (Madrid: Ministerio de Justicia, Comisión General de Codificación, 1970), vol. I, p. 283.

[68] See arts. 389–90 CC.

[69] M. de Bofarull, *El Código civil español según la edición oficial: anotado y concordado con la anterior legislación y jurisprudencia española y los códigos extranjeros* (Madrid: Imprenta de Antonio Pérez Dubrull, 1888), p. 593, note 3.

on nuisance law.[70] In any case, the origins of paragraph 1 of article 1908 remain unknown.[71] We may only assume that the drafting commission took as a source of inspiration the list of faulty conducts, that were defined as minor offences in the Penal Code 1870, then in force.[72]

(c) *Liability for damage caused by certain dangerous things*

A crucial question is whether the existence of a specific provision on liability flowing from damage caused by certain dangerous activities or things offers an alternative rule to the general fault liability principle laid down in article 1902 CC.

Following the French precedent of article 1386 of the Code Civil, article 1907 CC was devoted to the owner's liability in the case of the collapse of a building. According to García Goyena, this rule was not to be deemed an exception to the general rule on fault liability. It only sought to afford a tort claim to the victim in spite of the fact that he was fully aware of the risk of collapse and said nothing about it.[73] This point of view is consistent with the fact that article 1907 CC literally requires the collapse to have been caused 'by a lack of necessary repair' to hold the owner liable.

The first commentators on the new codified rules, however, considered that article 1907 CC amounted to a reversal of the burden of proof of fault and that this regime was to be applied (as the provision literally declares)[74] to all cases referred to in article 1908 CC. Moreover, there was not a *numerus clausus* of such cases and the regime laid down by article

[70] For example, cases of fallen trees or excessive smoke. The provision also mentioned emanations from sewers, where they had been constructed without the necessary precautions.

[71] See, recently, E. Algarra Prats, 'La responsabilidad civil por la explosión de máquinas y la inflamación de sustancias explosivas. Algunas reflexiones sobre el artículo 1908.1º del Código Civil y su aplicación por los Tribunales' in E. Llamas Pombo (Coord.), *Estudios de Derecho de Obligaciones. Homenaje al profesor Mariano Alonso Pérez* (Madrid: La Ley, 2006), vol. I, p. 27 at p. 34.

[72] See art. 601.1 CP 1870, which punished the infringement of safety measures as regards the establishment or operation of steam engines, boilers, furnaces, or any other similar devices with a danger of fire or explosion. Article 601.4 referred to inflammable substances and chemical components. See S. Viada y Vilaseca, *Código penal reformado de 1870 con las variaciones introducidas en el mismo por la Ley de 17 de julio de 1876 concordado y comentado para su mejor inteligencia y fácil aplicación* (4th edn, Madrid: Fernando Fe, 1890), pp. 747–51.

[73] See F. García Goyena, *Concordancias, motivos y comentarios del Código civil español* (Madrid: Imprenta de la Sociedad Tipográfico-Editorial, 1852; reprinted Barcelona: Base, 1973), p. 257.

[74] Article 1908 CC starts with the following sentence: 'Likewise, the owners shall be liable ...'.

1908 could, therefore, be applied to other settings, provided that the same justification applied.[75]

Although the interpretation of those provisions triggered the usual controversy among legal writers, it must be stressed that during the first half of the twentieth century the ideas held by the first commentators eventually prevailed.[76] By the same token, some treaties and academic books began using the category 'liability for damage caused by inanimate things' to refer systematically to the rules of articles 1907–1910 CC. The standard view now holds that paragraphs 1 and 4 of article 1908 CC establish fault liability, with the peculiarity that 'fault is presumed because of the risk that has been created'[77] and that the burden of proof has shifted to the owner of the thing.[78] Moreover, paragraphs 2 and 3 deem him accountable for any damage caused, unless he can prove that the damage resulted from force majeure.[79]

In the 1970s and at the beginning of the 1980s, some scholars tried to give even more weight to article 1908 CC. They offered a more ambitious construction of this provision, which sought to interpret it so that whenever damage arises in the settings the article described – explosions, excessive emissions, fallen trees – this very fact would demonstrate the negligent behaviour of the owner (*res ipsa loquitur*).[80] Other scholars also stressed that, in these cases, proof of circumstances freeing a defendant from liability would, in fact, be almost impossible.[81] Against this framework, de Ángel claimed that article 1908 CC had originally been drafted

[75] See J.M. Manresa y Navarro, *Comentarios al código civil español* (2nd edn, Madrid: Imprenta de la Revista de Legislación, 1908), vol. XII, pp. 650–1. See also J. Castán Tobeñas, *Derecho civil español, común y foral* (3rd edn, Madrid: Reus, 1931), vol. II, p. 808. But see C. Valverde y Valverde, *Tratado de derecho civil español* (4th edn, Valladolid: Talleres Tipográficos Cuesta, 1935), vol. III, p. 801 (emphasising the idea of 'wrongdoing' that in his opinion lay behind the situations referred to in art. 1908 and connecting them with the general rule of fault liability laid down in art. 1902 CC).

[76] See especially A. Borrell Maciá, *Responsabilidades derivadas de culpa extracontractual civil* (Barcelona: Bosch, 1941), p. 222.

[77] L. Díez-Picazo and A. Gullón, *Sistema de derecho civil* (9th edn, Madrid: Tecnos, 2000), vol. II, p. 632 and E. Roca Trías, *Derecho de daños. Textos y materiales* (4th edn, València: Tirant lo Blanch, 2003), p. 307.

[78] M. Albaladejo, *Derecho civil* (10th edn, Barcelona: Bosch 1997), II-2, p. 521. See also C.I. Asúa, in *Manual de derecho civil* (2nd edn, Madrid: Marcial Pons, 1998), p. 503.

[79] Most recently, see F. Peña López, 'Comentario de los artículos 1907, 1908 y 1909' in R. Bercovitz (ed.), *Comentarios al Código Civil* (Aranzadi: Elcano, 2001), pp. 2132–5.

[80] J. Puig Brutau, *Fundamentos de derecho civil* (Barcelona: Bosch, 1956), pp. 697–700.

[81] C. Rogel Vide, *La Responsabilidad civil extracontractual: introducción al estudio jurisprudencial* (Madrid: Civitas, 1977), p. 184.

as a rule of liability for risk, i.e. based upon the dangerousness of the installation or of the activity:

> The only possible explanation [of article 1908], in our opinion, consists in recognising that the original spirit of article 1902 was that the victim had to prove the fault of the defendant. From that point of view, *the special rule of article 1908 para. 1 would find its rationale in a reversal of the burden of proof* ... these special rules [thus] would come to be like a precursory formulation of today's doctrine of dangerous activities enshrined in article 2050 of the Italian Civil Code.[82]

It is far from clear whether the drafters of the Code wanted to introduce a special rule with respect to the liability regime in the cases referred to in article 1908, especially those connected with steam boiler explosions. The Argentine precedent, quoted above, seems to confirm what later came to be the prevailing opinion, i.e. that these cases, although mentioning the infringement of duties of care, imply a reversal of the burden of proof against the owner of the dangerous thing. Unlike the Argentine provision, however, article 1908 CC does not start with a sentence laying down a rule such as: '[w]hen damage to somebody results from any inanimate thing, its owner shall be liable unless he affords the proof that no fault existed on his part'. One may well argue that this omission is conscious and rules out the shifting of the burden of proof to the owner, let alone the possibility of grounding a general clause of strict liability upon this provision.

(d) Case law of the Tribunal Supremo

Whatever the intention of the legislator, from the very beginning the Tribunal Supremo denied any special rule in the situations mentioned in article 1908 CC relating to the general rule of liability for fault. Furthermore, the court never held that the rule of article 1908 could play a role as a general clause of strict liability for damage caused by dangerous things or activities. Instead, the article was applied as part of the general rules of fault liability and only many years after the entry into force of the Civil Code did the Tribunal Supremo declare that paragraphs 1 and 4 of article 1908 CC laid down a reversal of the burden of proof of fault.

A clear illustration for holding such a point of view is the judgment handed down by the Tribunal Supremo on 23 June 1900.[83] The court dealt with a

[82] R. de Ángel Yagüez, *Tratado de responsabilidad civil* (Madrid: Civitas, 1993), p. 550 (emphasis added). See also Rogel Vide, above note 81, at p. 188.

[83] *Colección Legislativa, Jurisprudencia Civil*, tome 8, vol. II, p. 151.

claim for property damage resulting from the explosion of inflammable material and explosives carried by the steamship *Cabo Machichaco* while moored in Santander harbour. The explosion severely damaged a number of buildings and the first party insurer of them sought the contribution of the ship-owner on the basis of articles 1902, 1903, paragraph 4 and 1908 CC. It argued that these provisions allowed the conclusion to be drawn that 'whenever an accident occurs that harms someone as a result of the lawful act of a third party ... *that fact in itself reveals, according to the natural order of things, that there cannot be any cause other than fault on the part of the defendant*' (emphasis added). The trial court dismissed the claim because the claimant had not proved the acts or omissions of the defendant who had caused the explosion. In its appeal to the Tribunal Supremo, the insurance company insisted upon the fact that requiring the claimant to prove such circumstances deprived the above-mentioned legal provisions of any effect. The Tribunal Supremo dismissed the appeal, making no mention of article 1908 CC and answering the claimant's contention by arguing that 'any action for damages on the basis of acts or omissions deemed negligent necessarily requires proof of the defendant's fault ... and this proof is up to the claimant according to the general principle of the proof of obligations laid down in article 1214 CC' (emphasis added).

This judgment was not isolated. Many years before, a judgment of the Tribunal Supremo on 27 October 1866[84] had considered the damage caused by the explosion of a powder warehouse. The court then expressly stated that the claimants ought to prove the defendant's fault to obtain the compensation requested. The Tribunal Supremo and the trial court dismissed the claim that the defendant had been at fault by infringing the administrative rules about the placing of explosives in urban areas. On appeal, the claimants alleged that the trial court had infringed 'the principle of law that nobody can obtain benefit with risk to a third party', the principle that 'harm caused must be repaired', as well as 'the ideal that requires the public administration to adopt the measures considered advisable to prevent the risks of dangerous industries'. The Tribunal Supremo upheld the trial court's decision as regards the burden of proving the defendant's fault and rejected the allegation that infringement of administrative rules or of their 'spirit' could form the basis of an appeal before a civil court.

After the entry into force of the CC, STS 23 June 1913[85] dismissed a claim based upon the defendant's infringement of article 390 CC. The claim-

[84] Gaceta de Madrid, Sentencias, 1866, no. 305.

[85] Colección Legislativa, Jurisprudencia Civil, tome 45, vol. II, p. 862.

ants asked for compensation for the damage caused by the continuous emission of smoke from a neighbouring bread furnace. Although the case clearly came under one of the headings of article 1908 CC, both the trial court and the Tribunal Supremo based the liability of the defendant upon article 1902 CC only. Furthermore, STS 1 February 1916[86] dealt with the serious injuries suffered by a girl when a window frame fell on top of her. In the lawsuit it was discussed to what extent the house was about to collapse and whether the owner had been warned about the necessity of carrying out repair works. The Tribunal Supremo upheld the opinion of the trial court which had rejected the liability of the house-owner on the basis that no proof of his fault was offered. The accident 'could, instead, have followed from an act of God or been due to the intervention of a third party' and the Tribunal Supremo held that such consideration did not infringe article 1907 CC.

This approach seems to have prevailed, at least during the first quarter of the twentieth century. At the end of the 1920s, however, the Tribunal Supremo made some minor steps towards a different interpretation of article 1907 and the following articles of the Civil Code. In spite of what the court had said in 1916, STS 22 October 1929[87] held that '[t]he omission of the preventive measures [required by article 1907] shall be presumed by the mere fact that the wall collapsed unless [the defendant] proves otherwise'. Later, the Tribunal Supremo stated that article 1908 CC did not require proof of the negligence of the defendant.[88] The judgment concerned property damage flowing from the mining activities developed by the defendant, who contended in his appeal to the Tribunal Supremo that 'the theory of strict liability was obsolete'. The court rejected such a claim and asserted, instead, that this theory suggested '[an] aspiration that the legal order shall evolve towards the idea that everybody is liable for all damage caused, even without fault ... and although the actor had acted with all due care and prudence'. Although the case was decided on the grounds of article 1902 CC, the court stressed that the same result would have been achieved by applying article 1908 CC, since the latter holds the owner liable without fault. Quoting this judgment, a subsequent decision implicitly held that all instances of article 1908 CC trigger liability, unless the owner demonstrates the intervention of force majeure.[89]

[86] Colección Legislativa, Jurisprudencia Civil, tome 53, vol. I, p. 330.

[87] Colección Legislativa, Jurisprudencia Civil, tome 108, vol. IV, p. 974.

[88] STS 14 May 1963 (RJ 1963\2699).

[89] STS 14 March 1968 (RJ 1968\1737).

(e) Concluding remarks

In its case law of the early years of the application of the CC, the Tribunal Supremo never based its decisions other than on the general rules on fault liability applied to the cases mentioned in article 1908 CC. Instead, the Tribunal Supremo simply included it among the application of the general rules on fault. To begin with, the defendant's fault had to be proved by the claimant. Later on, in the heat of the generalisation of the reversal of the burden of proof of fault under article 1902 CC, the particular cases referred to in article 1908 (and specifically cases of steam engine explosions) needed no special mention. They had become unnecessary to support a radical pro-victim transformation of tort liability. As De Ángel has conceded:[90]

> Bearing in mind the prevailing trend of case law as regards the interpretation of article 1902, and in particular the generalised reversal of the burden of proving fault (when applying the said article 1902), the cases [referred to in article 1908] lack any special feature and have lost the historical meaning that probably justified them.

Finally, it is worth mentioning that the last stage of the drafting of the CC brought in another almost imperceptible but important change in the text of the Code of 1851. Whereas the latter still followed the French model and made a general reference to liability 'for the things that one uses or has under one's care' (article 1901, paragraph 1),[91] in 1882–1888 the drafting commission thought that this reference was unnecessary, since this liability was subsequently regulated by articles 1905 (animals), 1906 (hunting game), 1907 (buildings) and 1908 (other instances). Needless to say, the elimination of that phrase prevented the development of a general rule on liability for damage caused by inanimate things, as happened in France with the last sentence of the first alinéa of article 1384.[92]

[90] De Ángel Yagüez, above note 82, at p. 551.

[91] '[La] obligación expresada en el artículo precedente no se limita a la reparación de los perjuicios ocasionados por un hecho propio, sino que se extiende a la de los causados por el hecho de las personas que uno tiene bajo su dependencia, *o por las cosas de que uno se sirve o tiene a su cuidado*' ('[T]he obligation expressed in the preceding article does not apply to damage caused by one's own act only, but also to compensation for damage caused by persons dependent upon you, *or by the things which one uses or has under one's care*) (emphasis added).

[92] See France, above p. 114. See also J. Garnier, 'De l'arrêt Teffaine aux arrêts Jand'heur' in F. Leduc (ed.), *La responsabilité du fait des choses. Réflexions autour d'un centenaire* (Paris: Economica, 1997), pp. 25–34.

4. Asbestos-related injuries and illnesses in the late twentieth century

As in other industrialised countries, from the 1940s until the late 1980s asbestos was widely used in Spain in a large number of industrial processes and to manufacture many products in current use. Imports of asbestos reached their peak in 1977 (around 140,000 tons). By 1999, when awareness of the health risks posed by exposure to asbestos had increased at an international level, the figure was still about 35,000 tons.

Complying with European Directives, Spain forbade the distribution and processing of any type of asbestos in 2001.[93] Indeed, at the beginning of the 1980s its use was banned during the manufacture of food and foodstuffs[94] and in any textile product for personal or domestic use, or in any product aimed at children.[95] Since 11 October 2006, all asbestos-related working activities have been banned, with the sole exception of those related to the decontamination of buildings and equipment.[96]

Concerning the number of victims of asbestos-related illnesses, available data on deaths from mesothelioma show that the number of persons who died from such illnesses was 2,265 between 1978 and 1992, whereas between 1989 and 1998 the number was 1,647.[97] Some sources claim that more than 60,000 workers would have been exposed to asbestos dust before 1985 and that by 2002 more than 1,000 would have died as a result of lung cancer, mesothelioma or asbestosis. In a Bill discussed a few years ago, its proponents recalled data foreseeing 50,000 deaths over the next thirty years.[98]

[93] See Order of 7 December 2001 (BOE no. 299, 14 December 2001), implementing European Directive 77/1999/EC (OJ L207, 6 August 1999), which led to the absolute prohibition of the distribution and use of any variety of asbestos for whatever purpose. Prior regulations had been laid down to implement European Directives 83/478/EEC (OJ L263, 24 September 1983), 85/610/EEC (OJ L375, 31 December 1985) and 91/659/EEC (OJ L363, 31 December 1991): see Order of 30 December 1993 (BOE no. 4, 5 January 1994).

[94] By Royal Decree 1351/1983 of 27 April 1983 (BOE no. 126, 27 May 1983).

[95] See Royal Decree 106/1985 of 23 January 1985 (BOE no. 27, 31 January 985) and Royal Decree 2330/1985 of 6 November 1985 (BOE no. 300, 16 December 1985).

[96] In these workplaces the maximum level of concentration of asbestos dust is 0.1 fibres per cm^3. See Royal Decree 396/2006 of 31 March 2006 (BOE no. 86, 11 April 2006), implementing European Directive 2003/18/EC (OJ L97, 15 April 2003).

[97] See A. Azagra Malo and M. Gili Saldaña, *Guía InDret de jurisprudencia sobre responsabilidad civil por daños del amianto* (Working Paper no. 277) available at www.indret.com, and more references to medical literature therein.

[98] *Proposición no de ley relativa a la adopción de medidas que ayuden a paliar la situación de los trabajadores afectados por la exposición al amianto* (BOCG, Congreso, VII Legislatura, Serie D, no. 435 at 21).

Until very recently, however, the problem of asbestos-related diseases and risks seems to have been confined to the debate on the coverage of occupational diseases and tackled accordingly.[99] Until the mid-1990s, litigated cases mainly involved questions about eligibility for social security incapacity allowances or surviving dependants' benefits. In this respect, asbestos-related diseases qualified as occupational diseases for the purposes of social security allowances and health provision as early as 1947, when asbestosis was included on the official list by the Decree of 10 January 1947.[100] In 1978, lung cancer and pleural and peritoneal mesothelioma were also included.[101]

As regards regulations on safety standards, Spanish legislation basically followed the path taken at international level in order, first, to diminish and, later, to prevent all health risks related to asbestos dust inhalation or exposure to asbestos. Since 1940, employers have had to provide special protection to employees working with substances that may produce dust and install devices in workplaces which ensure that the air is kept free of fibres. Yet the levels of exposure allowed in workplaces were not determined until 1961. The maximum level of concentration then was 175 fibres per cm^3. Between 1984 and 1993, this level was eventually reduced to 0.60 fibres per cm^3.[102] For a long time, nevertheless, the enforcement of these rules was poor. It should not be ignored that during Franco's regime (1939–1975) free unions were forbidden and any kind of critical comments about the lack of safety measures in the workplace or about

[99] The prospect of some sort of alternative compensation system only recently seems to be taking shape. By 2006, the Spanish Parliament required the government to 'study' the possibility of setting up a 'mechanism' that could 'satisfy the justified claims of the victims' of asbestosis, including workers' relatives and third parties exposed to asbestos dust from neighbouring factories (BOCG Congreso, Serie D, no. 360, 29 March 2006, pp. 25–6). At any rate, the motion falls short of what was initially asked for by the MP: 'the establishment of a public system of compensation for victims' (BOCG Congreso, Serie D, no. 337, 20 February 2006, pp. 10–11).

[100] BOE no. 21, 21 January 1947.

[101] See Royal Decree 1995/1978 of 12 May 1978. As regards other illnesses linked to asbestos exposure, the application of legal provisions on occupational diseases has often been very restrictive vis-à-vis victims and their dependants. More recently, see STS 13 November 2006 (*La Ley 2006, 441*) on cancer of the larynx.

[102] Since 1983, Spanish authorities have implemented European Directives aimed at preventing asbestos exposure in the workplace. See, especially, the Order issued on 31 October 1984, which approved the Regulations for workplaces involving asbestos-related risks (BOE no. 267, 7 November 1984) and, later, the Order of 26 July 1993 (BOE no. 186, 5 August 1993). In the case of crysolite, any exposure was banned in 1987. Concerning other varieties of asbestos, the level of exposure was left at 0.30 fibres per cm^3 in 1993. See Azagra Malo and Gili Saldaña, above note 97, Table 2 at p. 13.

working conditions, either in the press or by means of demonstrations, received severe punishment. Some cases litigated after the dictator's death revealed that the supervisory powers of public bodies were often not exercised as forcefully as expected.[103]

(a) *Asbestos litigation*

At the time of writing, asbestos-related litigation in Spain has not had the legal, social and political impact comparable to that in other jurisdictions. Neither legal scholarship (including courts and academic writers) nor public opinion at large has taken the risk of asbestos and its consequences for health as an issue of great legal concern and discussion.[104] Litigation arising from asbestos-related diseases has been confined to cases involving former employees (or their dependants or relatives after their death) of industries manufacturing products containing asbestos (such as fibreboard) or those that use it intensively during their manufacturing process (such as car or shipbuilding industries). These cases usually involved claimants aged between 50 and 70, who had worked for the defendant during the 1960s and 1970s and who first retired and then died because of asbestos-related diseases.

Since 1996, the number of claims against employers (and, on occasion, their liability insurers), seeking additional or supplementary compensation on the basis of contractual or tort liability, has increased dramatically. A recent account of reported cases between 1996 and 2004 found no fewer than sixty judgments on appeal on this issue.[105] Most of these claims have been dealt with by labour courts under the specific framework provided by social security law and the law on industrial relations.[106]

[103] STS 2 October 1982, RJ 1982\5728.

[104] But see now A. Azagra Malo, *La tragedia del amianto y el Derecho español* (Barcelona: Atelier, 2007).

[105] See Azagra Malo and Gili Saldaña, above note 97, at p. 21. Updated data can be found in A. Azagra, *Compensación de las victimas del amianto: litigación o aseguramiento* (2007), p. 4 (available at www.indret.com).

[106] Endless discussion has taken place over whether the competence to try actions brought by the injured worker or his family against the employer lay exclusively with the labour courts or, on the contrary, whether ordinary courts could still try these cases on the basis of the general rules on tort liability law. The prevailing opinion, held by most scholars and by the Social Section of the Tribunal Supremo in many recent decisions, is that labour courts are the only courts allowed by law to decide the question of damage caused by an employer's fault, when the issue is raised in the context of a labour relationship. The Civil Section of the Tribunal Supremo, however, continues to hold that whenever damage is the result of conduct that may give grounds for a tort claim, the action can

On the other hand, no reported decisions have been found where the claimant was the victim of so-called 'domestic' exposure or 'environmental' exposure, the latter concerning exposure while living in close proximity to mines or asbestos industries, or working or living on contaminated premises. Nevertheless, some groups of victims seem to be currently organising themselves to seek compensation for permanent injuries allegedly suffered as a result of their being exposed to asbestos dust from a factory located in the area where they lived.[107]

As to the role of criminal courts, their intervention in asbestos-related litigation has remained insignificant. However, in 1998 the managers of one of the factories of the main manufacturer of asbestos-related products (URALITA), together with the physicians responsible for the workers' periodic health examinations, were convicted for eight accidental deaths and twenty-nine serious personal injuries, allegedly due to asbestos dust exposure caused by their negligent behaviour consisting in not providing workers with safety devices and ignoring the maximum levels of concentration of fibres allowed by the law.[108] They were eventually found not guilty. In spite of the fact that the trial court held that the injuries and deaths resulted from the victims' exposure to asbestos, the court freed the defendants from criminal liability, drawing on their compliance with health and safety regulations that were in force when the factory was operational.[109]

(b) *Liability issues*

(i) Fault as a ground for liability

One of the main general features of Spanish case law on liability is that courts apply the reversal of the burden of proof of fault to all kinds of activities, regardless of their being especially dangerous or risky.[110]

always be brought before civil or ordinary courts. None of these debates, however, is exclusive to the asbestos-related diseases litigation.

107 See www.ictisp.com/-msanchez/afectats/afectats.htm. This is the website of an association founded in Cerdanyola and Ripollet (Barcelona), where URALITA has had one of its main factories since 1902. A civil action against URALITA has been brought on behalf of 48 former neighbours of the factory (see *El País*, 25 November 2007). On the availability of social security allowances, see above note 99.

108 See *El País*, 3 June 2006 (www.elpais.es).

109 Judgment of Penal Court Alicante, no. 3, 4 August 2006 (ARP 2006\525). According to the daily press, this decision is final, since the recourse brought by the victims and by the Ministerio Fiscal was rejected by the relevant court (*El País*, 23 January 2007 (www.elpais.es)).

110 See, in this respect, the conclusions stated in M. Martín-Casals and J. Solé Feliu, 'Fault under Spanish Law' in P. Widmer (ed.), *Unification of Tort Law: Fault* (The Hague/London/New York: Kluwer, 2005), pp. 227–64 at p. 229. Recent case law, however, shows

Accidents in the workplace and occupational diseases are two of the main areas where the so-called 'objectivisation' of tort liability has taken place, either by the technique of increasing the level of required care or by reversing the burden of proof of fault on the basis of the risk created for the defendant's benefit or of the need to internalise the cost of entrepreneurial activities.[111] In spite of this general trend, case law on the responsibility for damage flowing from exposure to asbestos dust reveals that most of the judgments have rejected any recourse not only to strict liability, but also to a mere reversal of the burden of proof of fault.

Following a leading judgment handed down by the Social Chamber of the Tribunal Supremo on 30 September 1997[112] that directly dealt with damage caused to workmen by exposure to asbestos, the prevailing view of courts holds that:

> in the area of work accidents and occupational diseases, which are under the protection of strict liability, the duplication of this liability by way of fault liability, either in contract or in tort ... instead of being a social improvement becomes an element of instability and inequality. For this reason, *in this area fault liability is confined to its classical and traditional sense, without extension, since this is already foreseen and established in a safer and fairer way within the Social Security system.*[113]

a more restrictive approach in the appreciation of defendant's fault. See J. Ribot/A. Ruda, 'Spain', in H. Koziol and B. Steininger (eds.), *European Tort Law 2007* (Vienna/New York: Springer, 2008), pp. 555–6.

[111] For instance, see STS 29 April 2004 (RJ 2004\2092) and STS 22 April 2003 (RJ 2003\3545). In legal writing, see the overview of N. Álvarez Lata, in L.F. Reglero Campos (Coord.), *Lecciones de Responsabilidad Civil* (Cizur Menor: Aranzadi, 2002), p. 463 and, in more detail, B. Gutiérrez-Solar Calvo, *Culpa y riesgo en la responsabilidad civil por accidentes de trabajo* (Madrid: Thomson Cívitas, 2004).

[112] STS Social 30 September 1997 (RJ 1997\6853). The case that gave rise to this judgment is an archetypical asbestos claim case. The victim had been working in the same clothing industry for twenty-one years and during this time she had worked in different departments of the factory, exposed to blue asbestos or crocidolite. She obtained declaration of permanent disability on the grounds of occupational disease, but she died shortly afterwards as a result of mesothelioma which she had developed together with asbestosis. Although the company had duly complied with all health and safety measures in the workplace, the Court of Appeal decided against the company on appeal, asserting that the defendant company 'had maintained, in its own interest, an activity in spite of the fact that it knew that there was a very reasonable risk that it would cause lung injuries resulting from the substance with which the victim worked' (STSJ País Vasco 16 November 1996 (Aranzadi Social (AS) 1996\3767)). The Social Section of the Tribunal Supremo quashed this judgment and considered instead that the creation of a risk approach makes sense only when 'one refers to damage to third parties who are alien to the social framework that benefits from this progress and development, i.e. when social risks are valued with regard to persons who are predominantly considered on an individual basis, as is the case in civil law'.

[113] Emphasis added. See, among many others, STSJ País Vasco 25 October 2005 (AS 2005\2851) and 24 May 2005 (AS 2005\1529), and STSJ Catalunya 27 September 2001 (AS 2001\4156).

Judgments ruling against the defendant on the grounds of a reversal of the burden of proof are thus unusual[114] and very few courts dare to ground the defendant's liability upon the idea of 'entrepreneurial risk'.[115]

(ii) Role of health and safety regulations

Compliance with health and safety regulations concerning working conditions is frequently at stake in ascertaining whether the employer is to be considered negligent or not. A number of judgments find for the defendant company on the grounds that safety regulations did not require adopting any measures other than those effectively taken. As long as neither the legislature nor the public authorities were aware of the carcinogenic properties of exposure to asbestos dust, the courts have held that not preventing the level of exposure legally permitted at that time cannot give rise to fault of the employer.[116]

A judgment of the Superior Court of Catalonia of 11 April 2004[117] is a good illustration of this approach. According to this decision, 'there cannot be required from the company a level of care that was so high as to foresee working risks so scientifically unknown in our country that even the rules then in force did not require a higher standard of diligent conduct'. Along the same lines, the judgment of the Court of Appeal of Seville of 18 November 1999[118] dismissed the claim on the basis that:

> the claimant does not indicate which other preventive measures could have been taken according to the state of scientific knowledge at the time or which alternative, complementary or more effective measures could have been taken other than those that had been taken and which would have prevented the ensuing harm.

Furthermore, other decisions argue that the competent public bodies never found the defendant guilty of any infringement of safety regulations and that other claims seeking the imposition upon him of punitive measures for alleged omission of safety measures have always been rejected.[119]

[114] Such as STSJ Murcia 30 September 2002 (JUR 2002\256296) and Aragón 4 December 2000 (JUR 2001\54158). But see STSJ Catalunya 10 November 2005 (JUR 2006\77904).

[115] One exception is SAP Barcelona 18 September 2002 (JUR 2003\7800). More recently, see SAP La Coruña 10 March 2006 (JUR 2006\132886).

[116] STSJ País Vasco 26 October 2005 (JUR 2005\20863).

[117] AS 2005\1339. But see the rulings of the same court quoted below note 129.

[118] AC 1999\8671.

[119] STSJ Catalunya 29 October 2002 (AS 2002\3987).

Conversely, some judgments hold employers liable by means of expanding their preventive duties beyond what was expressly established by the safety regulations in force. They point out that the requirement to keep the air in the workplace at a certain level of purity, in force since the 1940s, and the classification of asbestosis as an occupational illness since 1947 indicate that companies were obliged to take steps in order to prevent the sort of damage for which compensation is being claimed,[120] even when these illnesses were not fully known when exposure to asbestos took place.[121] In the problematic cases of mesothelioma, the argument links with the idea that 'if adequate measures to prevent asbestosis had been taken, mesothelioma would also have been prevented'.[122]

Other decisions recognise that 'although at that moment the knowledge on this matter and the prevention of harm were not as advanced as they currently are, an obligation of safety resulted from the labour contract and the company did not completely fulfil it'.[123] This is related to what is called the 'dynamic character' of the employer's obligation of safety.[124] The employer had the duty to remove, or reduce as much as possible, the risk that a substance posed to workers in as much as he knew its health risk, without waiting for a specific rule to be issued.[125] Moreover, the employer must even 'anticipate these risks or at least act in accordance with the technical advances that pointed to these risks'.[126] Occasionally, this rule is justified by what the defendant company has done in other work centres, where it has already introduced improvements or where

[120] STSJ Madrid 20 April 2004 (AS 2004\2478).

[121] STSJ Aragón 25 February 2004 (JUR 2005\104917).

[122] STSJ Murcia 2 November 2005 (AS 2005\3551). See previously, from the same court, 7 May 2001 (AS 2001\1868).

[123] STSJ Aragón 13 October 1999 (AS 1999\3379); commented on by M.T. Igartua Miró, 'Obligación de seguridad y responsabilidad "contractual" del empresario', *Aranzadi Social* 3, 1999, 2973.

[124] STSJ Cantabria 15 July 2005 (JUR 2005\183600) and Murcia 30 October 2006 (JUR 2007\15230).

[125] Dissenting opinion to STSJ Catalunya 11 April 2005 (AS 2005\1339). In this case the responsible public authorities considered that it had been proved that the worker performed his activity 'in an asbestos-dusty environment, entailing the risk of occupational disease, which could have been removed or reduced by adopting certain technical means, which the company did, indeed, adopt later on by carrying out an important reorganisation in the performance of its economic activity'.

[126] STSJ País Vasco 25 October 2005 (AS 2005\2851), although the dissenting opinion of one of the judges emphasised that there was no reference in the decision of first instance to the safety measures that should have been adopted for over twenty years.

the company has long been counting asbestos fibres or applying other preventive measures.[127]

(iii) Causation

Causation sometimes appears as the main hurdle to compensation, notably when it is doubtful whether the disease suffered by the victim can be attributed to asbestos exposure at work or not. Decisions finding for the defendant usually resort to the premise that the alleged infringement of safety measures is not sufficient to hold the defendant liable, since 'the levels of exposure authorised by the legal norms then in force would have caused the harm all the same'.[128] A similar line of reasoning is applied as regards the breach of ancillary duties (such as periodic medical examinations or measurements of the number of asbestos fibres per cubic centimetre).[129] Since medical examinations could not have detected the risk to which workers were exposed, the harm would have occurred all the same, even if all medical examinations required by legislation had been conducted.[130]

Judgments finding against the defendant apply, instead, quite flexible yardsticks when establishing the causal link between the conduct of the company and the damage suffered by the worker. Some draw on the so-called 'judicial presumption', which considers that a certain fact has been proved from other facts which have been duly proved in the proceedings, such as the fact that the victim had worked for the defendant company, manipulating or handling objects containing asbestos (for instance, tubes, tissues, the asbestos-contaminated parts of a ship, etc.).[131] Other judgments apply a causation test similar to that used to award social security benefits for occupational diseases.[132] Accordingly, the defendant company was held liable, notwithstanding the fact that the worker had also worked in a car repair shop for four months, where brake shoes

[127] SAP Valencia 27 April 2005 (AC 2005\972).

[128] STSJ Catalunya 29 October 2002 (AS 2002\3987).

[129] STSJ Galicia 6 February 2004 (AS 2004\1442) and STSJ País Vasco 24 May 2005 (AS 2005\1529).

[130] Amongst many, see STSJ Madrid 17 January 2002 (AS 2002\1583), País Vasco 29 April 2003 (AS 2003\2137) and Catalunya 29 October 2002 (AS 2002\3987).

[131] STSJ Cantabria 15 July 2005 (JUR 2005\183600). By contrast, see STSJ Madrid 30 May 2002 (JUR 2002\259216), where the court refused to follow this approach.

[132] On survivors' allowances paid by social security, see STSJ Catalunya 15 November 2002 (AS 2002\4240), and STSJ Castilla-La Mancha 25 September 2000 (AS 2000\4392). By contrast, for a case in which the award of the allowance was rejected, see STSJ País Vasco 19 October 1999 (AS 1999\3674).

containing asbestos were replaced, or that he lived in a town where ships were dismantled.[133] Moreover, uncertain causation did not prevent the courts from holding the employer liable in a judgment issued on 7 May 2001 by the Tribunal Superior of Murcia.[134] The court held liable only one of the three employers for whom the victim of mesothelioma had been working. It must be stressed that the two other former employers were not even sued.[135]

(iv) Other defences

Finally, a number of claims have been dismissed on the basis that the one-year limitation period laid down either by the specific labour law provision (article 59 of the Estatuto de los Trabajadores) or by the general rules on tort law (article 1968, paragraph 2 CC) has elapsed.[136] In these judgments the prescription period is considered to run from the moment of the final diagnosis of the disease[137] or (in the case of dependants) from the date of the death of their relative.[138] The prevailing (and even more pro-victim) view considers, instead, that the starting date is the time at which the consequences of the illness are definitively established by means of the official declaration of total or permanent disability.[139]

(c) Concluding remarks

The Spanish experience regarding asbestos-related injuries or diseases reveals that the case-by-case analysis conducted (mainly) by (labour) courts provides only a scant, fragmentary and underdeveloped analysis of the main issues at stake.

It is to be stressed that the regulatory framework regarding a minimally (but erroneously) safe work environment has very much

[133] STSJ Asturias 12 March 2004 (JUR 2005\127106). Against this approach, see STSJ Comunidad Valenciana 21 April 2005 (AS 2005\1553).

[134] STSJ Murcia 7 May 2001 (AS 2001\1868).

[135] See a brief comment on this decision by A. Ruda González, Spanish case note to *Fairchild* v. *Glenhaven Funeral Services Ltd and others* (2004) 2 ERPL 245 at 250.

[136] Concerning the application of art. 59 of the Estatuto de los Trabajadores rather than art. 1964 CC, which provides for a general fifteen-year limitation period for contractual liability claims, see STSJ Madrid 30 May 2002 (JUR 2002\259216) (quoting settled case law of the Tribunal Supremo).

[137] See STSJ País Vasco 24 July 2001 (AS 2001\4434), STSJ Madrid 5 November 2002 (JUR 2003\95549) and 6 February 2003 (JUR 2003\186372).

[138] See STSJ Murcia 11 May 1998 (JUR 1998\99298).

[139] For instance, see SAP Barcelona 18 September 2002 (JUR 2003\7800) and STSJ Murcia 2 November 2005 (AS 2005\3551).

influenced the final outcome of many claims. Courts have shown themselves very reluctant to cross the boundaries of fault liability, although many of them have increased the duties of care of employers in a way that has led them to impose liability upon the industry. On the other hand, a sort of 'state-of-the-art defence' is frequently invoked, even if implicitly, by defendants as well as by courts, freeing defendants from liability. As is well known, this contention takes for granted that damage was not foreseeable, due to limitations in the scientific and technical knowledge available. What is striking in some Spanish judgments, however, is that health and safety regulations often provide the standard terms of that knowledge, instead of being determined by the actual scientific knowledge available at the time of exposure. Strangely enough, all parties seem to concede that the minimum standard of precaution is the battlefield, no one daring to ask whether these provisions could be updated or up to what point the enterprise could and should have taken extra measures in addition to those required as a minimum by the relevant rules.[140]

Finally, asbestos litigation in Spain provides a good illustration of how the legal system deals with harms that impinge upon a considerable number of victims. First and foremost, it is social security that provides economic support and health assistance to victims, without substantial litigation to recover expenses from some potentially liable person. Secondly, it is within the regulatory framework provided by health and safety regulations that courts have dealt with the issue of whether the employer was at fault or not and therefore if he should pay damages to former workers or to their dependants. Thirdly, amounts paid as a result of actions brought against employers have never reached figures comparable to those awarded in other parts of the world. Indeed, a rough estimate of data available from published judgments reveals that the average for a death is €60,000 and that, as a maximum, some judgments have awarded €200,000.

Moreover, the fact that, up until now, the litigation has remained within the limits of the workplace invites speculation about the success of

[140] An alternative approach is nevertheless making its way through the courts and gaining ground in recent rulings. STSJ Catalunya 10 November 2005 (JUR 2006\77904), for instance, has held URALITA liable, drawing on the availability of 'knowledge among the scientific community on a worldwide scale' about the danger of exposure to asbestos. The lack of specific health and safety regulations concerning either the level of exposure to asbestos in the workplace or any other specific preventive measure would be irrelevant. In the same sense, see STSJ Catalunya 25 January 2006 (AS 2006\655).

any claim being brought by those living in areas neighbouring the asbestos industries.[141] The restrictive approach taken by labour courts during almost twenty years of litigation about related social security allowances and the employer's civil liability could be an obstacle to any recovery.

[141] At the end of 2007, it was mentioned in the media that some legal firms were promoting the constitution of a national association to promote legal rights of asbestos victims (Plataforma Estatal de Defensa Juridica de Afectados por el Amianto). See *El País*, 25 November 2007 (www.elpais.com).

APPENDIX: MAIN CODE PROVISIONS CITED

The texts provided in this Appendix provide the main texts of national codes dealing with Delict that were applied in relation to changes in technology.

1. French Civil Code (1804)

Delict

1382 Any human act whatever which causes damage to another obliges him by whose fault it occurred to make reparation.

1383 Every person is liable for the damage which he causes not only by his own act but also by his negligence or imprudence.

1384(1) He is liable not only for the damage which he caused by his own act, but also for that which is caused by the act of persons for whom he is responsible, or by things which he has in his keeping.

(2) [1] However, he who possesses by whatever right all or part of a building or of personal property in which a fire occurs is liable vis-à-vis third persons for damage caused by such fire only if it is proved that it should be attributed to his fault or to the fault of persons for whom he is responsible.

(3)[2] This provision does not apply to relations between owners and tenants which remain regulated by Articles 1733 and 1734 of the Civil Code.

(4)[3] The father and mother, to the extent that they exercise the right of custody, are jointly liable for damage caused by their minor children living with them.

(5) Masters and principals [are liable] for damage caused by their domestics and employees in the functions for which they have been employed.

(6) Teachers and artisans [are liable] for damage caused by their pupils and apprentices during the time when they are under their surveillance.

[1] Text resulting from the Law of 7 November 1922.

[2] Text resulting from the Law of 7 November 1922.

[3] Text resulting from the Law no. 70-459 of 4 January 1970.

(7)[4] The above liability arises unless the father and mother and the artisans prove that they could not prevent the act which gave rise to such liability.

(8) As to teachers, the faults, imprudence or negligence alleged against them as having caused the damaging act must be proven according to ordinary law by the plaintiff in the suit.

1385 The owner of an animal or he who avails himself of it while it is being put to his use is liable for the damage which the animal causes, whether the animal was in his keeping or whether it had strayed or escaped.

1386 The owner of a building is liable for damage caused by its ruin, when this occurred through want of maintenance or by a construction defect.

2. German Civil Code (1900)

Delict

823(1) A person who wilfully or negligently injures the life, body, health, freedom, property, or other right of another contrary to law is bound to compensate him for any damage arising therefrom.

(2) The same obligation attaches to a person who infringes a statutory provision intended for the protection of others. If according to the purview of the statute infringement is possible even without fault, the duty to make compensation arises only if some fault can be imputed to the wrongdoer.

...

826 A person who wilfully causes damage to another in a manner *contra bonos mores* is bound to compensate the other for the damage.

...

828(1) A person who has not completed his seventh year of age is not responsible for any damage which he does to another.

(2) A person who has completed his seventh but not his tenth year is not responsible for the harm which he inflicts on another in an accident with a motor vehicle, a railway or a hover rail. This does not apply if he has caused the injury deliberately.

(3) A person who has not yet competed his eighteenth year is, in so far as his responsibility is not excluded under paragraphs 1 or 2, not responsible for the harm which he inflicts on another if, on the commission of the action causing the harm, he does not have the necessary intelligence to realise his responsibility.

...

831 (1) A person who employs another to do any work is bound to compensate for damage which the other unlawfully causes to a third party

[4] Text resulting from the Law of 5 April 1937.

in the performance of his work. The duty to compensate does not arise if the employer has exercised ordinary care in the choice of the employee, and, where he has to supply appliances or implements or to superintend the work, has also exercised ordinary care as regards such supply or superintendence, or if the damage would have arisen, notwithstanding the exercise of such care.

(2) The same responsibility attaches to a person who takes over the charge of any of the affairs specified in paragraph 1, sentence 2, by contract with the employer.

...

839 (1) If an official wilfully or negligently commits a breach of official duty incumbent upon him towards a third party, he shall compensate the third party for any damage arising therefrom. If only negligence is imputable to the official, he may be held liable only if the injured party is unable to obtain compensation elsewhere.

(2) If an official violates his official duty by a decision in a legal issue, he is only responsible for the harm arising from this if the violation of duty consists of a criminal act. This provision has no application to a refusal or delay which is contrary to duty in exercise of the office.

(3) The duty to compensate does not arise if the victim has intentionally or negligently refrained from averting the harm by the use of a legal remedy.

...

847 [*repealed in 2002, see now 5.253(2)*] In the case of injury to body or health, or in the case of deprivation of liberty, the injured party may also demand an equitable compensation in money for damage which is not a pecuniary loss.

...

Property

903(1) In so far as the law or the rights of third parties do not conflict with this, the owner of a thing can deal with the thing at his discretion and can exclude others from any interference with it ...

...

906(1) The owner of land cannot forbid the introduction of gases, steam, smells, smoke, soot, heat, noise, vibrations and similar things coming from the land of others in so far as these do not interfere with his use of his land or interfere only to a negligible extent. A negligible interference typically occurs where the limits or targets established by statute or regulation are not exceeded by the interferences laid down and assessed by these provisions. The same applies to standards in general administrative provisions that have been promulgated under article 48 of the Federal Environmental Impact Protection Act and represent the state of the art.

(2) The same applies to the extent that a material interference is caused by a use of another plot of land that is usual in that location and cannot be

prevented by measures that are economically reasonable for users of this kind. Where an owner is obliged under these provisions to tolerate an interference, he may require the user of the other land to pay compensation, if the interference impairs a use or revenue of the former's land that is usual in that location beyond a degree that the owner can be expected to tolerate ...

907(1) The owner of a plot of land may require that no facilities are created or maintained on a neighbour's plot of land where it can be predicted with certainty that their existence or use will have an impermissible effect on the owner's land. Where an installation satisfies the provisions of the law of the *Land* concerning the distance from the boundary or other protective measures, then the removal of an installation may only be required if the impermissible interference actually occurs.

(2) Trees and bushes are not facilities within the meaning of these provisions.

...

1004(1) Where ownership is interfered with through the removal or deprivation of property, the owner can request that the interfering person remove the interference. Where further interferences are anticipated, then the owner can demand a prohibitory injunction.

(2) Such claims are excluded where the owner is bound to tolerate the interference.

3. Italian Civil Code (1943)

Delict

2043 Any intentional or negligent act, which causes unlawful harm to another obliges the person who committed the act to provide compensation.

...

2048 The father and mother, or the guardian, are liable for the harm caused by the unlawful act of the minor, unemancipated children, or of persons within their custody, who reside with them. The same applies to a parent by affiliation.

Teachers and others who teach an art, trade or profession are liable for the harm caused by the unlawful act of the pupils or apprentices under their supervision.

The persons mentioned in the preceding paragraphs are only relieved of liability if they prove that they were unable to prevent the act.

2049 Masters and employers are liable for the harm caused by an unlawful act of their servants or employees in the exercise of the functions to which they are assigned.

2050 Whoever causes harm to another in the performance of activities hazardous by their nature, or by reason of the means employed, is liable in damages, unless he proves that he has taken all appropriate measures to avoid the harm.

2051 Everyone is liable for harm caused by things in his keeping, unless he proves that the harm resulted from a fortuitous event.

2052 The owner of an animal, or a person using it for the period of his use, is liable for harm caused by the animal, regardless of whether the animal was in his custody or had strayed or escaped, unless he proves that the damage resulted from a fortuitous event.

2053 The owner of a building or structure is liable for the harm caused by its collapse, unless he proves that such harm was not caused by defective maintenance or defect in construction.

2054 The custodian of a vehicle that is not guided by rails is liable for the harm caused to person or property by the operation of the vehicle, unless he proves that he did all that was possible to avoid the harm.

In the case of the collision of vehicles, it is presumed, in the absence of proof to the contrary, that each operator contributed equally to the cause of the harm suffered by each vehicle.

The owner of the vehicle, or in his stead the usufructuary or purchaser with reservation of ownership is liable *in solidum* with the custodian of the vehicle, unless he proves that the vehicle was being operated against his will.

In any case, the persons mentioned in the preceding paragraphs are liable for the harm resulting from defects in the manufacture or maintenance of the vehicle.

...

2056 The damages owed shall be determined in accordance with the provisions of articles 1223, 1226 and 1227.

The damages arising from loss of earnings shall be assessed by the judge according to the circumstances of the case.

...

2059 Non-patrimonial damages may be awarded only in cases provided by law.

4. Spanish Civil Code (1889)

Property

7.2 [5] The law does not protect the abuse of rights or the antisocial use of them. All action or omission that, by the intention of its author, by its object or by the circumstances in which it is done, would exceed the normal limits of a right, with nuisance to another individual, will give rise to compensation or

[5] As inserted by Decree of 31 May 1974.

to the necessary administrative or judicial measures to prevent the abuse from continuing.

. . .

590 No one may build sewers, aqueducts, ovens, forges, chimneys, stables, stores for corrosive materials, steam-powered machinery or installations which, of themselves or by their outputs, are dangerous or noxious, near to a communal well or a neighbour's well without following the distances prescribed by regulations and local custom and without taking any protective measures necessary for the work, as also laid down by regulations. Where there is no regulation, the builder shall take all precautions which are deemed necessary, according to the established opinion of experts, to prevent any damage to neighbouring tenements or buildings.

Delict

1902 The person who by action or omission causes damage to another by fault or negligence is obliged to repair the damage caused.

1903 The obligation imposed in the preceding article applies not only for a person's own acts or omissions, but also for those persons for whom a person is responsible.

The father and the mother are responsible for the harm caused by their minor children who are living with them.

Tutors are liable for the harm caused by minors or incapacitated persons under their authority and living with them.

Employers or directors of an establishment or firm are liable for the harm caused by their employees in the performance of tasks that are entrusted to them or of their functions.

The state is liable under this head for the acts of its special agents, but not when the harm has been caused by a civil servant entrusted with practical administration, for which case the preceding paragraph applies.

Masters and directors of arts and trades are liable in respect of the harm caused by their pupils or apprentices, as long as they remain under their control.

The liability under this article ceases when the persons mentioned prove that they have shown the diligence of the good father of the family to avoid the harm.

. . .

1905 The possessor of an animal or a person who makes use of one is liable for the harm that it causes, even where it had strayed or escaped. This liability only ceases when he proves that the harm resulted from a fortuitous event or from the fault of the person who suffered the harm.

1906 The owner of a building is liable for the harm caused to neighbouring property when he has not taken the necessary steps to prevent it increasing . . .

1907 The owner of a building is liable for the damage resulting from its total or partial ruin, if this was caused by the failure to make necessary repairs.

1908 The owner of objects is also liable for harm caused:

(1) by the explosion of machines that had not been kept with due care and by the burning of explosive substances that were not stored in a safe and suitable place;

(2) by excessive smoke, that is injurious to persons or property;

(3) by the falling of trees located on thoroughfares, when this is not caused by force majeure;

(4) by emanations from sewers or deposits of toxic matter, built without the precautions appropriate to the place in which they were.

INDEX

Specific cases and statutes will be found under entries at the specific countries to which they refer (e.g., French case law, German statute law).

For EU product safety concerns, contact us at Calle de José Abascal, 56–1°,
28003 Madrid, Spain or eugpsr@cambridge.org.

www.ingramcontent.com/pod-product-compliance
Lightning Source LLC
LaVergne TN
LVHW052255200926
848481LV00002B/376
* 9 7 8 1 1 0 7 4 7 5 8 0 9 *